Vegetation and

the Atmosphere

Volume 2
Case Studies

Vegetation and the Atmosphere

Volume 2
Case
Studies

EDITED BY

J. L. MONTEITH

Department of Physiology and
Environmental Studies,
University of Nottingham,
School of Agriculture,
Sutton Bonington,
Loughborough, England

1976

ACADEMIC PRESS
London · New York · San Francisco

A Subsidiary of Harcourt Brace Jovanovich, Publishers

ACADEMIC PRESS INC. (LONDON) LTD
24–28 Oval Road,
London NW1

U.S. Edition published by
ACADEMIC PRESS INC.
111 Fifth Avenue,
New York, New York 10003

Library of Congress Catalog Card Number: 75-19664

ISBN: 0-12-505102-6

PRINTED IN GREAT BRITAIN BY
J. W. ARROWSMITH LTD, BRISTOL BS3 2NT

Contributors to Volume 2

L. HARTWELL ALLEN   *USDA, Department of Agronomy, University of Florida, Gainesville, Florida, 32611, USA*

J. E. BEGG   *Division of Plant Industry, P.O. Box 1060, Canberra, ACT 2601, Australia*

K. W. BROWN   *Department of Soil and Crop Sciences, Texas A and M University, College Station, Texas 77843, USA*

T. V. CALLAGHAN   *Nature Conservancy, Merlewood Research Station, Grange-Over-Sands, Lancs LA11 6JU, England*

O. T. DENMEAD   *Division of Environmental Mechanics, CSIRO, P.O. Box 821, Canberra, ACT 2601, Australia*

M. FUCHS   *Institute of Soils and Water Agricultural Research Organization, The Volcani Center, POB 6, Bet Dagan, Israel*

G. B. JAMES   *Department of Botany, University of Aberdeen, St Machar Drive, Aberdeen AB9 2UD, Scotland*

P. G. JARVIS   *Department of Forestry and Natural Resources, University of Edinburgh, King's Buildings, Edinburgh EH9 3JU, Scotland*

J. D. KALMA   *Division of Land Use Research, CSIRO, P.O. Box 1666, Canberra ACT 2601, Australia*

J. J. LANDSBERG   *Long Ashton Research Station, Long Ashton, Bristol BS18 9AF, England*

E. R. LEMON   *USDA Microclimate Investigations, Bradfield Hall, Cornell University, Ithaca, New York 14850, USA*

M. LEWIS   *York University, Downsview, Ontario M3J 1P3, Canada*

E. LINACRE   *School of Earth Sciences, Macquarie University, Sydney, Australia*

JU. L. RAUNER   *Institute of Geography, Academy of Sciences, of the USSR, Staromonetny, 29, Moscow, USSR*

R. E. REDMANN   *Department of Plant Ecology, University of Saskatchewan, Saskatoon, Saskatachewan, Canada*

E. A. RIPLEY   *Department of Plant Ecology, University of Saskatchewan, Saskatoon, Saskatachewan, Canada*

C. W. ROSE   *Griffith University, Nathan, Queensland 4111, Australia*

B. SAUGIER   *Centre d'Etudes Phytosociologiques et Ecologiques, B.P.1018, Route de Mende, Montpelier, France*

G. STANHILL   *Institute of Soils and Water, Agricultural Research Organization, The Volcani Center, POB 6, Bet Dagan, Israel*

B. W. R. TORSSELL   *Division of Land Use Research, CSIRO, P.O. Box 109, Canberra, ACT 2601, Australia*

Z. UCHIJIMA   *National Institute of Agricultural Sciences, Nishigahara, Kita-Ku, Tokyo, Japan*

Preface

Vegetation and the Atmosphere is a review, in two volumes, of those concepts and measurements in micrometeorology which have a direct bearing on the problems of plant ecology. Some of these problems have been revealed by developments in agricultural science, some have emerged from hydrology, and some from the conservation and management of land resources. In the last 20 years, progress in collecting and analysing micrometeorological records has been very rapid. It is opportune to take stock of current knowledge and to ask whether ecological science is getting the full benefit from all the information now available about physical processes and mechanisms in plant communities.

The first volume dealt with basic principles of radiation, heat, mass and particle transfer in vegetation, with water in the soil-plant-atmosphere continuum, with modelling and with instrumentation. Volume 2 contains a number of case studies for different types of vegetation illustrating how the principles discussed in Volume 1 are put into practice.

The first five chapters deal with a number of agricultural crops whose micrometeorology has been intensively studied: wheat and other temperate cereals, maize, rice, sugar beet, potatoes, sunflower, and cotton. Chapter 6 is less concerned with measurements of the physical environment than with their practical application in a dynamic model of growth for Townsville Stylo growing in a mixed pasture. Chapters 6 to 9 describe work on different types of forest: coniferous stands, deciduous woodland, and tropical rainforest. In Chapter 10, the modification of micro-climates is discussed with reference to citrus orchard. The last three chapters show how micrometeorological concepts can be applied to whole ecosystems represented by swamps, prairie grassland, and tundra. Stimulus and support for the two latter studies came from the International Biological Programme.

When the treatment of each chapter is compared, obvious differences emerge in the amount of detailed information which was available to each contributor. In general, progress has been most rapid when a multidisciplinary group has been able to concentrate on one particular stand of vegetation for a period of at least five years. This is the rough time-constant for a full-scale experiment in which all the relevant physical and biological measurements are properly integrated.

In the Preface to Volume 1, I suggested that plant ecology has reached the point where it is often possible to estimate, without serious error, the physical state of a plant organ given the relevant environmental factors. Several of the chapters in this volume illustrate this point. I also suggested that within the general subject of physical ecology, the biology needs to catch up with the physics. In particular, I believe we must think more deeply about how the development and differentiation of a specific plant organ depends on its physical state.

One of the distinguishing features of *Vegetation and the Atmosphere* is the uniformity of symbols used throughout both volumes. All the main symbols for the second volume are listed on pp. ix–xii and the rationale for their choice is discussed in the Preface to Volume 1. Two sets of units follow most of the symbols in the list: the basic SI unit, and a unit of convenient size used in the text.

I am most grateful to the authors of both volumes for their cooperation and patience during a long gestation period; to my secretary, Miss Edna Lord for help with editorial matters; and to the staff of the Academic Press for the courteous and efficient way in which they handled all our negotiations.

J.L.M.   *December 1975*
Sutton Bonington

Main Symbols

a Foliage area density ($m^2\,m^{-3}$).

A Area (m^2).

$\mathbf{A}$ Flux density of CO_2 ($kg\,m^{-2}\,s^{-1}$; $g\,m^{-2}\,h^{-1}$ or $mg\,m^{-2}\,s^{-1}$).

c_p Specific heat of air at constant pressure ($J\,kg^{-1}\,K^{-1}$; $J\,g^{-1}\,K^{-1}$).

C Transfer coefficient or trapping efficiency (dimensionless):

        subscripts H heat;
                       i impaction;
                       M momentum;
                       P particles;
                       V water vapour.

                       These subscripts alone refer to transfer between an isolated object and the surrounding air. The additional subscript a implies transfer between a canopy and the air above it.

$\mathbf{C}$ Flux density of sensible heat in air (positive upwards) ($W\,m^{-2}$).

d Zero plane displacement (m; cm).

D Diffusion coefficient ($m^2\,s^{-1}$; $cm^2\,s^{-1}$):

        subscripts C CO_2;
                       H heat;
                       V water vapour.

e Vapour pressure (Pa; mbar).

$e_w(T)$ Saturation vapour pressure at temperature T (Pa; mbar).

$\mathbf{E}$ Flux density of water vapour in air (positive upwards) ($kg\,m^{-2}\,s^{-1}$; $g\,m^{-2}\,s^{-1}$ or h^{-1}).

$\mathbf{F}_C$ Flux of CO_2 in air per unit ground area (positive upwards) ($kg\,m^{-2}\,s^{-1}$; $g\,m^{-2}\,h^{-1}$ or $mg\,m^{-2}\,s^{-1}$).

g Conductance ($m\,s^{-1}$; $cm\,s^{-1}$).

           For subscripts see C above.

$\mathbf{G}$ Flux density of heat in soil (positive downwards) ($W\,m^{-2}$).

h Crop height (m).

H   Total heat flux into air per unit ground area $= \mathbf{C} + \lambda\mathbf{E}$ (W m^{-2}).

J   Rate of heat storage in canopy per unit ground area (W m^{-2}):
     subscripts H sensible;
                      V latent;
                      veg by change in temperature of vegetation.

k von Karman's constant $(= 0\cdot41)$.

K Turbulent transfer coefficient; eddy diffusivity $(\text{m}^2 \text{ s}^{-1})$:
     subscripts C CO_2;
                      H heat;
                      M momentum;
                      V water vapour.

$\mathscr{K}$ Extinction coefficient in canopy.

l Length (m).

L Leaf area index; area of foliage per unit ground beneath $(\text{m}^2 \text{ m}^{-2})$.

L   Long-wave radiation flux per unit ground area (W m^{-2}):
     subscripts d downward;
                      e emitted by leaves;
                      n net;
                      u upward.

m Mass of vegetation per unit canopy volume (kg m^{-3}).

p Atmospheric pressure (Pa; mbar); shelter factor:
     subscripts c canopy;
                      M momentum;
                      S shoot;
                      V water vapour.

P Gross precipitation (m; mm).

P   Gross photosynthesis rate per unit ground area $(\text{kg m}^{-2}\text{ s}^{-1}$; $\text{g m}^{-2}\text{ h}^{-1}$ or $\text{mg m}^{-2}\text{ s}^{-1})$.

q Specific humidity weight of water per unit weight of moist air $(\text{kg kg}^{-1}$; $\text{g kg}^{-1})$.

$q_w(T)$ Saturation specific humidity at temperature T $(\text{kg kg}^{-1}$; $\text{g kg}^{-1})$.

r Diffusion resistance $(\text{s m}^{-1}$; $\text{s cm}^{-1})$.
     subscripts c canopy;
                      e cuticle;
                      H heat;
                      i isothermal (or climatological);
                      m mesophyll;
                      M momentum;

q	pore;
s	stomatal;
ST	bulk stomatal;
V	water vapour.
	These subscripts alone refer to exchange between an isolated object and the surrounding air. The additional subscript a implies exchange between a canopy and the air above it.
R	Rate of respiration by canopy per unit ground area $(\text{kg m}^{-2}\,\text{s}^{-1}; \text{g m}^{-2}\,\text{h}^{-1}$ or $\text{mg m}^{-2}\,\text{s}^{-1})$.
$\mathbf{R}_n$	Net radiation flux per unit ground area (W m^{-2}).
S	Total area of foliage (i.e. leaves, stems, etc) per unit ground area $(\text{m}^2\,\text{m}^{-2})$.
S	Flux density of solar radiation (W m^{-2});
subscripts b	direct beam, horizontal surface:
c	complementary;
d	diffuse;
n	net;
p	direct beam, normal incidence;
t	total.
T	Temperature $(\text{K}; {}^\circ\text{C})$; throughfall of rain $(\text{m}; \text{mm})$.
T'	Wet-bulb temperature $(\text{K}; {}^\circ\text{C})$.
u	Horizontal wind speed (m s^{-1}).
u'	Fluctuation in horizontal wind speed $(= u - \bar{u})\,(\text{m s}^{-1})$.
u_*	Friction velocity $(=\sqrt{\tau/\rho})\,(\text{m s}^{-1})$.
w'	Fluctuation of vertical velocity (m s^{-1}).
z	Vertical distance (m).
z_0	Roughness length $(\text{m}; \text{cm})$
β	The Bowen ratio $\mathbf{C}/\lambda\mathbf{E}$.
γ	The thermodynamic value of the psychrometer constant $(\text{Pa K}^{-1}; \text{mbar}\,{}^\circ\text{C}^{-1})$.
Γ	Dry adiabatic lapse rate (K m^{-1}).
δe	Vapour pressure deficit of air $(= e_w(T) - e)\,(\text{Pa}; \text{mbar})$.
Δ	Rate of change of saturation vapour pressure with temperature $(\text{Pa K}^{-1}; \text{mbar C}^{-1})$.
ε	Emissivity; ratio of densities of water vapour and dry air $(= 0.622)$.
λ	Latent heat of vaporization of water $(\text{J kg}^{-1}; \text{J g}^{-1})$.
μ	Heat absorbed by formation of carbohydrate per unit weight of CO_2 assimilated $(\text{J kg}^{-1}; \text{kJ g}^{-1})$.
ν	Kinematic viscosity of air $(\text{m}^2\,\text{s}^{-1})$.

ρ Density of moist air ($kg\,m^{-3}$; $g\,m^{-3}$); reflectivity.

σ Stefan–Boltzmann constant ($W\,m^{-2}\,K^{-4}$); standard deviation.

τ Transmissivity (dimensionless); relaxation time (s).

τ Momentum flux per unit ground area (Pa or $N\,m^{-2}$).

χ Weight of water vapour in unit volume of moist air ($kg\,m^{-3}$; $g\,m^{-3}$).

φ Weight of CO_2 in unit volume of air ($kg\,m^{-3}$; $g\,m^{-3}$).

ψ Water potential ($J\,m^{-3}$ or $J\,g^{-1}$).

subscripts   leaf;
             soil;
             plant.

Acknowledgements

In Chapter 2, Fig. 14 is reproduced from E. Inoue *et al.* (1968), *J. agric. Met. (Tokyo)* **23**, 165–176 with permission from the senior author and publishers.

In Chapter 3, Fig. 1A is reproduced from K. W. Brown and N. J. Rosenberg (1971/2), *Agric. Met.* **9**, 241–263 with permission from the authors and Elsevier Scientific Publishing Company.

Figure 1B is reproduced from I. F. Long and H. L. Penman (1963) *in* "The Growth of the Potato" with permission from the authors and Butterworths, London.

Figure 3 is reproduced from K. W. Brown and N. J. Rosenberg (1971a), *Agron. J.* **63**, 207–213.

Figure 6 is reproduced from W. Shepherd (1972), *Water Resources Res.* **8**, (4), 1092–95, with permission of the author and the American Geophysical Union.

Figure 7 is reproduced from F. J. Burrows (1969), *Agric. Met.* **6**, 211–226, with permission from the author and Elsevier Scientific Publishing Company.

Figure 10 is reproduced from J. L. Monteith and G. Szeicz (1960), *Quart. J. R. met. Soc.* **86**, 205–214, with permission of the author and F. Darton and Company Ltd., Watford, Herts.

In Chapter 5, Fig. 1 is reproduced from M. Schupp *et al.* (1969). SWC Research Report 412, with permission from the authors and publishers.

Figure 2 is reproduced from H. W. Gausman *et al.* (1971a), SWC Research Report 423, with permission from the authors and publishers.

In Chapter 6, Figs. 1 and 2 are reproduced from C. W. Rose *et al.* (1972a), *Agric. Met.* **9**, 385–403, with permission from the authors and Elsevier Scientific Publishing Company.

Figures 3, 4a and 6 are reproduced from C. W. Rose *et al.* (1972b), *Agric. Met.* **10**, 161–183, with permission of the authors and Elsevier Scientific Publishing Company.

Figure 7 is reproduced from R. L. McCowan (1973), *Agric. Met.* **11**,

53–63, with permission from the authors and Elsevier Scientific Publishing Company.

In Chapter 7 material has been reproduced from the following sources with permission from the authors and publishers:

Figure 1 from J. L. Monteith and G. Szceicz (1961), *Quart. J. R. met. Soc.*, **87**, 159–170;

Figures 3 and 5 from J. M. Norman and P. G. Jarvis (1974), *J. appl. Ecol.* **11**, 395–396;

Figure 6a from J. T. Woolley (1971), *Pl. Physiol.* **47**, 656–662;

Figure 9 from J. B. Stewart and A. S. Thom (1973), *Quart. J. R. met. Soc.* **99**, 154–170;

Figure 14 from Y. Yamaoka (1958), *Trans. Am. Geophys. Union*, **39** (2), 266–272;

Figure 15 from P. E. Waggoner and N. C. Turner (1971), *Bull. Conn. Agric. Exp. Station*, No. 726.

Figure 17 from M. M. Ludlow and P. G. Jarvis (1971), *J. appl. Ecol.* **8**, 925–953;

Figure 22 from J. A. Helms (1970), *Photosynthetica* **4**, 243–253;

Additional material reproduced from S. Linder (1971), *Physiol. Plant* **25**, 58–63; S. Freyman (1968), *Can. J. Pl. Sci.* **48**, 326–328; W. G. Egan (1970), *For. Sci.* **16**, 79–94.

In Chapter 9 material has been reproduced from the following sources, with permission from authors and publishers:

L. H. Allen, Jr *et al.* (1972), *Ecology*, **53**, 102–111 (Duke University Press);

L. H. Allen, Jr *et al.* (1974), *Photosynthetica* **8** (3), 184–207 (Institute of Experimental Botany, Czechoslovak Academy of Sciences);

E. Lemon *et al.* (1970), *Bioscience* **20**, 1054–1059 (American Institute of Biological Science).

Contents

4. Sunflower

B. SAUGIER

5. Cotton

G. STANHILL

6. Townsville Stylo (*Stylosanthes humilis* H.B.K.)

C. W. ROSE, J. E. BEGG and B. W. R. TORSSELL

7. Coniferous Forest

P. G. JARVIS, G. B. JAMES and J. J. LANDSBERG

8. Deciduous Forest

JU. L. RAUNER

9. Carbon Dioxide Exchange and Turbulence in a Costa Rican Tropical Rain Forest

L. H. ALLEN, JR. and E. R. LEMON

10. Citrus Orchards

J. D. KALMA and M. FUCHS

11. Swamps

E. LINACRE

12. Grassland

E. A. RIPLEY and R. E. REDMANN

13. Tundra

M. C. LEWIS and T. V. CALLAGHAN

Contents to Volume 1

1. Temperate Cereals

O. T. DENMEAD

Division of Environmental Mechanics, CSIRO, Canberra, Australia

I. INTRODUCTION

The case studies reported in this chapter have been chosen to fulfil four broad aims: (i) to explore important aspects of whole plant physiology for temperate cereals in the field, and the influence of the physical environment on them; (ii) to illustrate the potentialities of physical methodology for studying plant growth processes; (iii) to provide field data which will allow comparison and extrapolation between controlled and outdoor environments; and (iv) to provide mechanistic or empirical descriptions of the climatology of temperate cereals for use in model analyses or estimating procedures. A particular concern has been to examine physical processes and their related physiological causes or consequences *within* the canopy.

Because of space limitations, the chapter is necessarily incomplete in its selection of topics, and even in those aspects which have been examined, I have drawn heavily on my own research, often to the exclusion of the work of others. This has been done for reasons of access to and interpretation of the original data, but it is regrettable that much good work has had to be omitted.

II. RADIATION

Although the influence of any one factor in the field environment cannot be considered in isolation, radiation receipt is central to the development of the micro-climate of plant stands and their physiological response to weather. It has an obvious, direct energetic effect on convection, evaporation and photosynthesis; it has a further indirect effect on these processes through its control of stomatal aperture; and together with other factors such as the extent of vertical mixing, wind speed and soil-water supply, it is responsible for the vertical distribution of temperature and humidity in the stand and the attainment of tissue and soil temperatures.

The principles of radiative transfer as they relate to plant stands are discussed in Vol. 1, Ch. 2; here it is sufficient to list some relevant parameters of the radiation balance of temperate cereals. More attention is given to case studies of radiation transmission and absorption in the stand and to examining the radiation dependence of a number of physiological processes.

A. Climatology

1. *Radiation Balance*

Table I lists representative measurements of the day-time radiation balance of temperate cereals, which are typical of many agricultural crops: reflection coefficients for short-wave radiation are close to 0.2 and the net radiation amounts to half to three-quarters of the daily radiation load. To the reflection coefficients in Table I can be added measurements by Monteith (1959) of 0·20 to 0·27 for wheat and by Fritschen (1967) of 0·23, 0·23 and 0·21 for barley, oats and wheat respectively.

The table also contains an entry from King and Evans (1967) for the radiation balance of a wheat crop in a growth cabinet. Facilities of this type differ in their characteristics but the radiation components measured by King and Evans are probably typical of many such installations. Apart from differences between cabinet and field in the spectral composition of the short-wave radiation, the main difference is the substitution of a hot

Table I
Components of the radiation balance of temperate cereals
summed over the daylight hours ($S_t = 100$ units)

Crop	S_r	L_u-L_d	R_n	Comments	References
Oats	24	18	58		Impens and Lemeur (1969)
Wheat	18	9	73	Mature droughted crop; green $L = 0.7$	M. C. Anderson (unpublished, 1971)
Wheat	22	24	54	Irrigated crop; $L = 4.7$	Paltridge et al. (1972)
Wheat	19	16	64	High nitrogen fertilizer	Stanhill et al. (1972)
Wheat	20	-18	98	Phytotron cabinet	King and Evans (1967)

ceiling for a cold sky with the consequence that incoming long-wave radiation exceeds the outgoing flux and net radiation is a very much higher proportion of the 'solar' radiation load.

Figures 1a and 1b (from Denmead, unpublished 1968) show the diurnal variation in some of the radiation components of a wheat crop. The change in reflection coefficient with solar elevation is typical of many measurements over crops, as is the slightly greater net long-wave loss in the afternoon which results from surface heating.

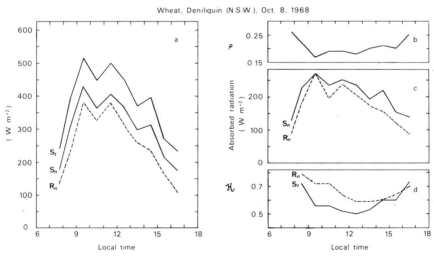

Wheat, Deniliquin (N.S.W.), Oct. 8, 1968

Fig. 1. (a) Total solar radiation, net short-wave radiation, S_n, and net radiation above a wheat crop. (b) Short-wave radiation reflection coefficient. (c) Short-wave and net radiation absorbed by foliage between top of crop (60 cm) and 15 cm. (d) Extinction coefficients for short-wave and net radiation. From Denmead (unpublished, 1968).

2. *Estimating Net Radiation*

In climatological studies, the need often arises to estimate net radiation from solar radiation, the latter measurement being more generally available. Table II presents the results of several studies in which net and solar radiation

Table II
Parameters of equations for estimating net radiation from
solar radiation ($R_n = a + bS_t$; a in W m^{-2})

Crop	a	b	Reference
Barley	-84	0·655	Fritschen (1967)
Oats	-117	0·750	Fritschen (1967)
Oats	-63	0·683	Impens and Lemeur (1969)
Wheat	-81	0·697	O. T. Denmead (unpublished, 1964)
Wheat	-122	0·808	Fritschen (1967)
Wheat	-124	0·797	O. T. Denmead (unpublished, 1968)
Wheat	-89	0·682	Paltridge *et al.* (1972)

were measured simultaneously over temperate cereal crops. Linear regressions of the form:

$$R_n = a + bS_t$$

have been calculated, either by the original investigators or from their published data. Despite the apparent disparity in the parameters of the estimating equations, and in the proportionality between R_n and S_t evident in Table I, all equations give estimates of net radiation which differ between themselves by at most 20%. This level of accuracy is often acceptable; for instance, in estimating daily evaporation rates.

3. *Radiation in the Canopy*

a. Canopy structure. The radiation climate within a plant stand is markedly affected by stand geometry (see Volume 1, Chapter 2). Many temperate cereals have an erectophile leaf habit, at least in stands not yet at full maturity. Examples of the canopy structure of immature and mature wheat crops are shown in Fig. 2a from measurements with inclined point quadrats by M. C. Anderson (unpublished, 1968). Young leaves are steeply inclined to the horizontal with average foliage angles close to 80°. Leaf inclination gradually decreases with depth in the canopy as a result of ageing, and old leaves near the ground have average foliage angles of about 40°.

Figure 2b depicts some results of an analysis of the frequency distribution of leaf angles in the immature crop. It shows the leaf density function defined

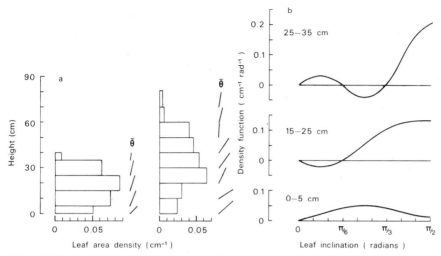

Fig. 2. (a) Vertical distribution of leaf area density, and mean leaf inclinations in immature (left) and mature wheat crops. (b) Leaf area density function for various foliage layers in immature wheat crop. From M. C. Anderson (unpublished, 1968).

by Philip (1965), which is the contribution to leaf area density of leaves inclined at angles between θ and $\theta + d\theta$. The negative values are a consequence of the smoothing involved in fitting Fourier series to the raw data. They are, of course, physically unrealistic but do not affect the main results significantly. The analysis indicates clearly how the top leaves are confined to a narrow range of steep inclinations, and how the range widens at lower positions in the canopy. It is a structure which leads to an even distribution of incident radiation on the leaves of the stand, and avoids excessively large radiation loads. Some long-term studies of the significance of such a structure for total grain yield and dry matter production have been made, e.g. Tanner (1969) and Angus *et al.* (1972), but there is a need for comprehensive studies of its influence on many other facets of crop physiology, for instance, the diurnal course of stomatal opening, plant-water relations, transpiration, and photosynthesis.

   b. Radiation transmission. For convenience, the discussion here is in terms of extinction coefficients. It is recognized, however, that the extinction coefficient model formalized by Beer's law is only an approximation even for short-wave radiation, and is still more empirical for net radiation. The reader is referred to Volume 1, Ch. 2, and to Anderson (1969) and Monteith (1969) for formal descriptions of radiation transfer in plant stands.

The extinction coefficient $\mathscr{K}$ is defined by the equation

$$\Phi(L) = \Phi(0)\exp(-\mathscr{K}L)$$

where Φ is a general radiation flux density on a horizontal surface below a leaf area index of L in the stand, and $\Phi(0)$ is the flux density on a horizontal surface above the stand.

(i) Short-wave radiation. Measurements of the transmission of short-wave radiation in temperate cereals have been reported by Monteith (1965a, 1969) and Angus et al. (1972) for barley, and by Anderson (1969) and Paltridge et al. (1972) for wheat. Figure 1, from unpublished work of Denmead, gives another example for wheat. In this case, measurements inside the canopy were made with strip radiometers. The essential features of these studies are:

(1) an extinction coefficient model fits the observations reasonabiy well for the central part of the day. In the measurements of Paltridge et al. (1972) it applied only during a period within ±2 hours of midday, but in Monteith's study and in the wheat of Fig. 1, it applied within ±4 hours of midday;

(2) the fractional transmission changed regularly through the day in the wheat of Fig. 1 (see Fig. 1d), but remained essentially constant in Monteith's barley;

(3) there is no general agreement on the value of $\mathscr{K}$. In barley, Angus et al. (1972) report a value of 0·2 (at a solar altitude of 45°) for a short, erect variety and 0·6 for a variety with long, drooping leaves. Monteith (1969) gives a constant value of 0·69 for solar altitudes between 30° and 60°. In wheat, the midday observations of Paltridge et al. (1972) give an extinction coefficient of 0·3 at a solar altitude of 60°. For the wheat of Fig. 1, $\mathscr{K}$ varied from 0·75 at a solar altitude of 35° to 0·5 at 60°;

(4) the extinction coefficients for the wheat of Fig. 1 are close to those predicted by geometrical probability (as given, for instance, by Anderson, 1966) except early and late in the day. Those of Paltridge et al. and Angus et al. are very much less than this theory predicts.

(ii) Net radiation. Usually, temperatures of foliage at different positions in the canopy are similar, so that the net exchanges of long-wave radiation between foliage elements are small. As Fig. 1c shows, most of the radiation absorption is in the short-wave. Thus it is not surprising that an extinction coefficient model often describes the transmission of net radiation with reasonable accuracy. In the wheat crop of Fig. 1, the value of $\mathscr{K}$ for net radiation is slightly higher than that for short-wave (Fig. 1d), but it exhibits much the same diurnal variation. In another study of net radiation in wheat, Denmead (unpublished, 1968) found that the appropriate value of $\mathscr{K}$ was predicted well enough for most of the daylight hours by the theoretical value of $\mathcal{K}$ for direct radiation, which varied by a factor of 2 between 0800 hours and midday.

Monteith (1969) reports measurements of the transmission of net radiation in barley which can also be described by an extinction coefficient model, but as in his short-wave measurements, $\mathcal{K}$ appears to be constant over most of the day.

B. Radiation and Leaf Conductance

Many environmental factors affect stomatal opening but as Slayter (1967) points out, the main controls in the field appear to be exerted by light intensity and water stress. The role of plant–water relations in stomatal opening is discussed later in the chapter. This section examines the response of stomatal opening to irradiance in the leaves of plants not under water stress.

Figure 3, which summarizes results of Denmean and Millar (1976b), shows the relation between the average conductance of wheat leaves in the field and their net radiation absorption. Conductance is the reciprocal of the more familiar leaf diffusive resistance. The measurements come from two different crops and were made at various times of day on all green leaves in

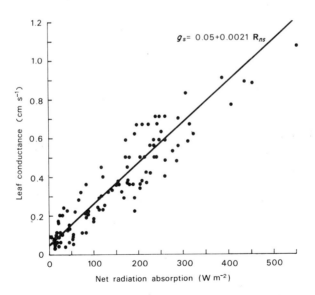

Fig. 3. Conductance of wheat leaves, g_s, as function of net radiation absorbed by leaf, $\mathbf{R}_{ns}$. From Denmead and Millar (1976b).

the canopy. Net radiation includes some thermal radiation but as Fig. 1c shows, most radiation absorption is in the short wavelengths. A linear relationship has been fitted to the data although there is no *a priori* reason to expect any particular form. The 95 % prediction confidence interval is approximately 0.15 cm s^{-1} either side of the regression line.

A notable feature is that there is no indication of the on–off response in stomatal opening which has been reported in some other species; leaf conductance appears to increase linearly with increasing radiation up to the highest irradiances which were measured. These corresponded to global radiation flux densities on a horizontal surface above the crop of around 1000 Wm^{-2}.

The range of conductances in Fig. 3 is representative of the most common values reported for temperate cereals, although there are some notable exceptions—see, for example, Milthorpe and Penman (1967), Downes (1970) and Paltridge *et al.* (1972). The author has not found another set of observations as comprehensive as those of Fig. 3 with which to test the generality of the relationship given there, but some confirmation comes from measurements of the diffusive resistance of barley leaves in the field by Monteith *et al.* (1965). These show equivalent conductances in bright light of the same magnitude as the wheat measurements in Fig. 3. Their highest conductance was 0.9 cm s^{-1} with a standard deviation of 0.1 cm s^{-1}. Conductance also increased with irradiance but a direct comparison with the wheat data is not possible because their radiation measurements were of global radiation on a horizontal surface above the crop.

III. TRANSFER PROCESSES

The formal background to this section is the description of turbulent transfer given in Vol. 1, Ch. 3. Other relevant descriptions of the physical mechanisms involved and current methods of study are contained in recent reviews by Lemon (1969) and Denmead and McIlroy (1971). The section examines some general aspects of the transfers of heat, mass and momentum in the air layers above and within the canopies of cereal crops and discusses in more detail the effects of turbulence on transfers between leaf surfaces and the air about them. Later sections give specific examples of the use of micrometeorological techniques in the study of physiological processes in the field.

A. The Similarity Assumption

Of the variety of micrometeorological techniques available for measuring evaporation and CO_2 exchange, two have been used widely in agricultural meteorology: the aerodynamic method which uses momentum as a tracer of gas movement, and the energy balance method which uses total heat as a tracer (see Vol. 1, Ch. 3). Central to both methods is the assumption of similarity between the atmospheric transport mechanisms for the gas and the tracer, specifically that the transfer coefficients for the latter maintain a constant ratio to each other (usually unity). Some case studies with temperate cereals, in which this assumption has been tested, are presented below.

Denmead and McIlroy (1970) examined equality between K_H (for heat) and K_V (for water vapour) by comparing energy balance estimates of evaporation from wheat (assuming $K_H = K_V$) with the water loss from a weighing lysimeter in the same field. The measurements encompassed a very wide range of atmospheric conditions and soil–water deficits. The good agreement between the two estimates (within the limits of instrumental error) confirmed that $K_H \simeq K_V$.

In the same study, the energy balance method was used to calculate the net uptake of CO_2 by the wheat on the further assumption that $K_H = K_V = K_C$. The results are discussed in more detail in the section dealing with CO_2 exchange. Over a period of 33 days, the average net uptake of CO_2 per day was 25.1 g m^{-2}. Measurements of the change in dry weight of the tops over the same period gave an average growth rate which corresponded to the daily uptake of 19.2 g m^{-2} of CO_2. Allowing for the growth of roots, and considering the probable errors in both sets of measurements, the agreement between them is quite reasonable. It could hardly be regarded as proof of equality between K_H, K_V and K_C, but it does suggest that there is no great disparity between them.

The applicability of the aerodynamic and energy balance methods have been examined by E. F. Bradley and myself in a recent trial in wheat. In applying the aerodynamic method, we followed Deacon and Swinbank's (1958) approach of using a drag coefficient, C_{aM}, estimated close to the surface in order to minimize stability effects:

$$C_{aM} = \left(\frac{u_*}{u}\right)^2 = k^2 \Big/ \left[\ln\left(\frac{z-d}{z_0}\right)\right]^2,$$

where all symbols have their usual significance (see Vol. 1, p. 66). C_{aM} was calculated at 50 cm above the top of the wheat from 43 neutral wind profiles and then used to predict the momentum flux in both stable and unstable conditions. Figure 4a compares u_* calculated in this way with u_* measured directly by an eddy correlation technique (Haugen et al., 1971). The aerodynamic method was obviously very successful in predicting momentum flux, regardless of atmospheric stability.

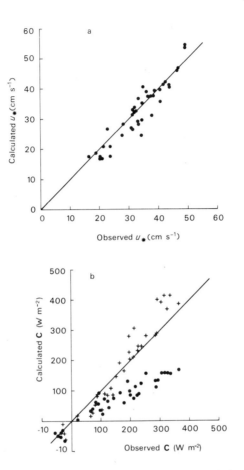

Fig. 4. (a) Comparison of friction velocity, u_*, above wheat crop calculated by aerodynamic method with u_*, calculated by eddy correlation (observed u_*). (b) Comparison of convective heat flux, C, above wheat crop calculated by aerodynamic method (dots) and an energy balance method (crosses) with C calculated by eddy correlation (observed C). From E. F. Bradley and O. T. Denmead (unpublished, 1971).

The drag coefficient was also used to calculate the convective heat flux, C. Assuming K_M (for momentum) $= K_H$, it follows that

$$C = -\rho c_p C_{aM} u^2 \, \Delta T / \Delta u,$$

where $\rho =$ air density, $c_p =$ specific heat of air, and ΔT and Δu are differences between two heights of temperature T and wind speed u, in this case at 24 and 198 cm above the top of the crop. Simultaneous measurements of other meteorological parameters enabled us to calculate the heat flux in two other, independent ways: (a) from the energy balance, assuming $K_H = K_V$ and

$$\mathbf{C} = \beta(\mathbf{R}_n - \mathbf{G})/(1 + \beta)$$

with $\beta = c_p \Delta T/\lambda \Delta q$ where Δq is the appropriate difference of specific humidity; and (b) by eddy correlation, with

$$\mathbf{C} = \rho c_p \overline{w'T'}$$

where the prime denotes a deviation of vertical velocity w from its mean value. As Fig. 4b shows, the aerodynamic method predicted the heat flux reasonably well in stable and near neutral conditions, but failed in unstable conditions. The energy balance method gave acceptable accuracy in all conditions.

Assuming that the eddy correlation measurements of the heat flux and the momentum flux, τ, were correct, and that the flux density of water vapour was $\mathbf{E} = (\mathbf{R}_n - \mathbf{G} - \mathbf{C})/\lambda$, the ratios of the fluxes to the associated gradients in *unstable* conditions were as follows:

$$\mathbf{C}/(-\rho c_p \, \mathrm{d}T/\mathrm{d}z) : \mathbf{E}/(-\rho \, \mathrm{d}q/\mathrm{d}z) : \tau/(\rho \, \mathrm{d}u/\mathrm{d}z) = K_H : K_V : K_M = 1 : 1 \cdot 1 : 0 \cdot 4.$$

In near neutral and stable conditions, the transfer coefficients were nearly equal.

The general character of these results is, of course, not peculiar to temperate cereals. Discrepancies between K_M and the other transfer coefficients have been demonstrated in many experiments over various agricultural surfaces—see, for instance, Swinbank and Dyer (1967), Wright and Brown (1967) and Businger *et al.* (1971) for recent examples. This case study has been presented because few such detailed comparisons appear in the literature of agricultural meteorology and the aerodynamic method is often used to measure mass transfer without due regard to the limits of its applicability. Those cases in which it has been found to give acceptable answers appear to be ones in which atmospheric conditions have been close to neutral, as, for instance, in Penman and Long (1960). In many areas in which temperate cereals are grown, highly unstable conditions occur often. Then, the aerodynamic method will underestimate the true fluxes.

B. Transfer Within the Canopy

The formal description of turbulent transfer in the air spaces within the canopy still lacks an adequate theory. Here, some experimental studies of the shape of the wind profile and the variation with height of transfer coefficients in temperate cereals are given.

1. *The Wind Profile*

Figure 5a shows profiles of horizontal wind speed within the canopies of wheat crops measured by various workers. Rather surprisingly, the profiles have quite similar shapes despite large difference in crop height, foliage density and crop structure (see figure caption). In the three crops, the wind

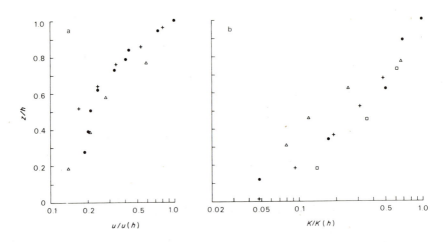

Fig. 5. Normalized profiles of (a) wind speeds $u/u(h)$ and (b) transfer coefficients $K/K(h)$ in wheat crops. For wind speed: dots from Denmead and Bradley (1973) with $h = 89$ cm, $L = 2\cdot2$; crosses from Baines (1972) with $h = 86$ cm, $L = 9$; triangles from E. Inoue, E. R. Lemon and O. T. Denmead (unpublished, 1962) with $h = 130$ cm, $L = 2$. For transfer coefficients: dots from Denmead (unpublished, 1971) with $h = 89$ cm, $L = 2\cdot2$; crosses from Denmead (unpublished, 1968) with $h = 81$ cm, $L = 3\cdot2$; triangles from Penman and Long (1960) with $h = 60$ cm, $L = 2$ and high wind speeds; squares from Penman and Long for low winds.

profile in the upper half of the canopy can be described empirically by the equation:

$$u(z) = u(h) \exp\left[-n(1 - z/h)\right],$$

where n is an attenuation coefficient (see also p. 44). The value of n varies from $2\cdot6$ to $4\cdot0$. Below $h/2$, wind speeds decrease at a slower rate than the equation predicts.

Cionco (reported by Lemon, 1969) gives a similar range for n in wheat and indicates a tendency for it to increase with wind speed. Long *et al.* (1964) also measured wind profiles of this general shape in the top of a barley canopy but found that wind speeds increased in the bottom canopy because of the sparse vegetation there.

2. *Transfer Coefficients*

Figure 5b summarizes the results of three studies in wheat canopies, in which the height dependence of the transfer coefficient K, was investigated. In the measurements by Denmead, K was calculated by an energy balance method for different times on a number of days. Penman and Long's values were estimated by constructing a profile of the flux density of water vapour from calculations of the evaporation rates of the soil and the leaves, and dividing the flux densities by the corresponding humidity gradients. In all cases, the variation of K with height can be described by an empirical relationship similar to that describing the wind profile, namely,

$$K(z) = K(h)\exp\left[-m(1 - z/h)\right],$$

where the attenuation coefficient, m, varies from 2·2 to 3·3.

C. Transfer at Leaf Surfaces

So far, this section has been concerned with transfer processes in the air layers above and within plant stands. Operating in parallel, are complementary transfers of heat, mass and momentum at foliage surfaces. The exchanges of heat and water vapour between wheat leaves and the air surrounding them have been studied in the field by Denmead and Bradley (1973). They defined a leaf transfer coefficient, g_l, by the relationship

$$F = g_l(\Theta_s - \Theta_a)$$

in which F is the flux density of heat or mass averaged over the two leaf surfaces, Θ_s is the appropriate concentration at the leaf surface and Θ_a is the concentration in the ambient air. It will be recognized that g_l is the reciprocal of the commonly used leaf boundary layer resistance, i.e. a conductance.

For combined transfer of heat and water vapour, the relationship above can be manipulated to give an expression relating the average leaf temperature, T_s, to the average flux density of net radiation on the leaf, $\mathbf{R}_{ns}$, and the ambient temperature and humidity:

$$T_s - T_a = \frac{\mathbf{R}_{ns}(1 + x)/\rho - \lambda\,\delta q}{c_p(1 + x) + \lambda\Delta}$$

where $x = g_l/g_s$ (g_s being, as previously, the sum of the conductances of the two leaf surfaces); Δ is the slope of the curve relating saturation specific humidity to temperature, i.e. $dq_w/dT = q_w(T_a) - q_a$; and the other symbols are conventional.

These equations were used to calculate values of h_a in a number of experiments. In one series, leaves were coated with black paint which prevented evaporation and confined all the heat loss to convection. Then

$$g_l = \mathbf{R}_n/\rho c_p(T_s - T_a).$$

Other experiments were conducted with unpainted leaves which were able to exchange heat with the air by both convection and evaporation. In these cases, g_l was obtained by solving the second of the equations above, firstly from measurements of the appropriate parameters for individual leaves, and secondly from average measurements for leaf layers. The results are shown in Figure 6 where g_l is plotted as a function of wind speed. The

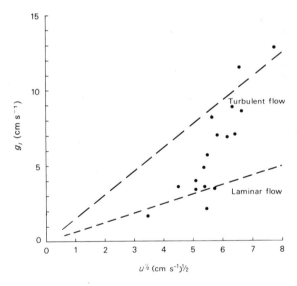

Fig. 6. Leaf-air transfer coefficients, g_l, for wheat leaves as function of wind speed over leaf, u, from Denmead and Bradley (1973). Lines are predicted relationships for laminar and turbulent flow from Parlange *et al.* (1971).

dashed lines in the figure correspond to the relationships between h_a and wind speed suggested by Parlange *et al.* (1971) for forced convection from flat plates in laminar and turbulent flow, namely,

$$g_l = (2 \times 0.332/Pr^{\frac{2}{3}})(uv/l)^{\frac{1}{2}}$$

for laminar flow, and

$$g_l = (5 \times 0.332/Pr^{\frac{2}{3}})(uv/l)^{\frac{1}{2}}$$

for fully turbulent flow. In these equations, Pr is the Prandtl number, u is wind speed, v is the kinematic viscosity of air and l is the average width of the 'leaf' (taken as 1 cm for wheat leaves).

These are difficult measurements to make in the field and considerable scatter is expected in the calculated values of g_l. Nevertheless, they form a reasonably consistent set, at low wind speeds conforming to the predictions for laminar flow and at high wind speeds approaching the expectation for turbulent flow. Most of the measured transfer coefficients are intermediate between the theoretical predictions, due perhaps to transition from laminar to turbulent conditions, but due also to other effects such as varying orientations to the wind, the influences of neighbouring leaves and the effects of streamlining.

An interesting ecological comparison can be made between these measurements and measurements of g_l for the large leaves of snap bean ($l \simeq 7.5$ cm) obtained in the field in a similar way by Kanemasu *et al.* (1969), and subsequently revised by Pearman *et al.* (1972). The leaf transfer coefficients for snap bean span the same range of values as for wheat but were measured at much higher wind speeds, 34 to 224 cm s^{-1} compared with only 12 to 60 cm s^{-1} for wheat. In this sense, transfers of heat and mass from wheat leaves might be considered more 'efficient'.

The transfer coefficients measured in the field by Denmead and Bradley (1973) and by Kanemasu *et al.* (1969) range from about 2 cm s^{-1} to 14 cm s^{-1}. Many growth chambers are designed for low, steady air flows which result in much smaller transfer coefficients. Gaastra (1969), for instance, quotes some typical leaf boundary-layer resistances which correspond to transfer coefficients between 0.7 and 2 cm s^{-1}. Examination of the equations above will show that the magnitude of g_l influences not only the temperatures of leaves but also their evaporation rates. Since many growth processes are affected by temperature and water stress, it is relevant to examine the energy balance of leaves in the two environments. Table III gives some indication of possible differences. It compares calculated temperatures and evaporation rates of leaves in typical field and growth chamber conditions. Net radiation, air temperature, humidity, and leaf conductance are assumed to be the same in both cases. The only difference is the magnitude of g_l (taken as 7 cm s^{-1} for the field and 1 cm s^{-1} for the chamber). The table shows that this difference can produce appreciable effects, particularly on leaf temperature, suggesting that extrapolation of the results of growth studies between laboratory and field should be approached with caution.

Table III
Effects of leaf transfer coefficient on leaf temperature and evaporation
rate for typical field and growth chamber environments

$$(\mathbf{R}_n = 300 \text{ W m}^{-2}; T_a = 25°\text{C}; \delta q = 10 \text{ mg g}^{-1}; g_s = 1 \text{ cm s}^{-1})$$

	g_l cm s^{-1}	$T_s - T_a$ °C	**E** g m^{-2} h^{-1}
Field	7	0·3	400
Chamber	1	5·2	350

IV. EVAPORATION

The physical and climatological aspects of evaporation are discussed in Volume 1 and many published case studies of evaporation from temperate cereals are available, e.g. Rider (1954), Penman and Long (1960), Long *et al.* (1964), Fritschen (1966), Denmead (1969), Denmead and McIlroy (1970), and Paltridge *et al.* (1972). More detailed studies of transpiration from the various leaf layers of temperate cereal crops have been described by Penman and Long (1960), Monteith (1965b) and Denmead (1968, 1970). Penman and Long (1960) and Denmead (1968) have also estimated the evaporation from the soil surface as a separate component of the water loss from wheat crops, and a formal analysis of the relative magnitudes of soil evaporation and transpiration with experimental verification from wheat has recently been given by Denmead (1973).

No special attention need be given to these aspects here. Instead, the section will examine water transport within plants in the transpiration stream and the effect on transpiration of plant water status. This is an area of research in which there is a large amount of scattered information but few attempts to link the physiological and environmental factors in a realistic plant model.

A. Water Transport in the Plant

Physical–mathematical models of water movement in the soil–plant–atmosphere continuum have been extremely useful for developing a qualitative understanding of water transport in plants and for examining the influences of both the aerial and soil environments on transpiration. They have necessarily been formulated in terms of a fairly simple plant model in which water moves from the soil, through a simplified 'root' and 'the stem' to 'the leaf' from which all the transpiration occurs. Real plants have branching vascular systems and many leaves along the stem, which transpire at different rates and whose water relations interact. Although this complexity is generally recognized, many field investigations are still predicated on one-leaf plant models. The adaptation of micrometeorological techniques for studies inside plant canopies and the development of improved methods for measuring the water potentials of plant tissues can now provide the detailed observational framework necessary to extend one-leaf models to real plant stands.

Denmead and Millar (1976a) have applied these techniques in a recent study of water transport in wheat plants. They used an energy balance method to measure the flux density of water vapour in the canopy. With the aid of supplementary measurements of crop structure and anatomical models of the vascular system of the stem, they were then able to calculate the fluxes of liquid water through the roots, the different stem sections, the leaves, and the ears of individual tillers. Simultaneous measurements of water potentials in the root zone and leaves gave the potential drops across the different segments of the transpiration pathway and allowed calculation of the resistances to flow through them. Figure 7 portrays the various steps in the analysis for one particular set of observations.

The study indicated large resistances to water transport in the roots, stems and leaves. The resistance to water transport across the root system of an individual tiller was $(1.00 \pm 0.37) \times 10^4$ bar s cm^{-3}, and the resistance per unit length of stems and leaves was $(1.59 \pm 0.19) \times 10^3$ bar s cm^{-4}. Similar root resistances were measured in laboratory experiments of E. F. Cox (reported by Cowan and Milthorpe, 1968). The resistances of whole root systems of young wheat plants were between 1.2×10^4 and 5×10^4 bar s cm^{-3}. Passioura (1972) has also calculated comparable resistances in the roots of cereal plants by assuming Poiseuille flow along the xylem elements.

These large plant resistances may often be more important in determining leaf–water potentials than the transmission resistance of the soil. In Denmead and Millar's observations for instance, even a moderate transpiration rate of 10^{-5} cm s^{-1} (0.36 kg m^{-2} h^{-1}) resulted in a potential drop of 3 to 4 bars across the root system and a further drop of 8 or 9 bars to the top leaf. A feature of this study was that the water potential of a particular leaf depended more on the total transpiration load on the whole plant and its vertical distribution than on the transpiration from that leaf. In this connection, transpiration from the ear was particularly important: on occasions it was half the total transpiration. The development of large water potential gradients in the stem in order to transport this water contributed substantially to low water potentials throughout the plant.

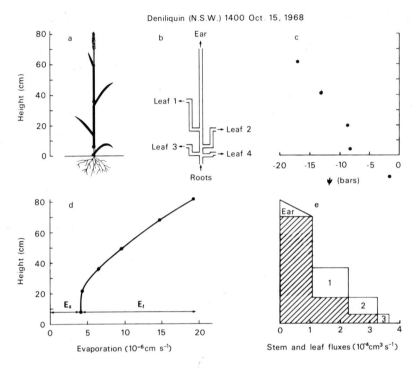

Fig. 7. (a) A wheat tiller to scale. (b) Model of the water conducting system of the stem on same scale. The xylem vessels carrying the flow to a particular leaf are separated from those carrying the main flow at the node *below* the one of leaf attachment. Diagram shows water pathway from the roots, through the stem, to each of the leaves and the ear. Leaves are numbered from the top. (c) Water potentials, ψ, in root zone and in each of the leaves. (d) Flux density of water vapour in the canopy. (e) Calculated stem and leaf water fluxes based on (b) and (d). Numbers denote fluxes to each leaf. From Denmead and Millar (1976a).

B. Leaf-Water Relations

The components of water potential in the leaves of wheat plants and the effects of leaf–water status on stomatal opening were studied by Millar and Denmead (1976). They found substantial differences in the relationship between turgor potential and water potential for leaves at different positions on the stem. These appeared to be due to changes in osmotic potential which was lowest (more negative) in the top leaf and increased progressively (i.e. became closer to zero) from top to bottom of the stem (Fig. 8). Since the water potential is the algebraic sum of the turgor and osmotic potentials, the consequences are that lower water potentials can be developed in the

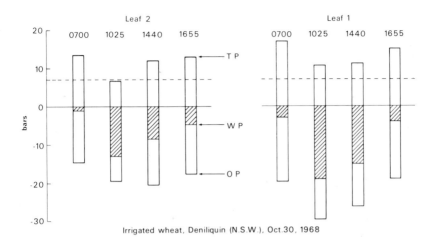

Fig. 8. Turgor potentials, TP, osmotic potentials, OP, and water potentials, WP, in top two leaves of wheat crop at different times of day. Leaf 1 at the top. From Millar and Denmead, (1976).

top leaves and at any water potential, there is a smaller loss of turgor, than in lower leaves. Figure 8 illustrates these differences for the top two leaves of a wheat crop.

Stomatal opening in all leaves was found to be strongly dependent on irradiance (following the relationship of Fig. 3) until turgor potential was reduced to about 8 bars. At lower turgor potentials, stomata began to close irrespective of irradiance. Because of the differences between leaves in the relation of turgor to water potential, the critical water potentials at which stomatal closure was induced were -19, -13 and -7 bars for leaves in the first (top), second and third positions on the stem. It was also apparent that stomatal closure was induced first in bottom leaves.

C. Transpiration and Stomatal Control

Denmead and Millar (1976b) also studied the effectiveness of stomatal closure in regulating transpiration in wheat. They found somewhat different modes of control depending on how water stress was induced.

When soil–water potential was high and stress resulted from an excessive atmospheric evaporation demand, the adjustment in stomatal opening was just sufficient to keep leaf–water potentials at their critical levels. This is the type of feedback control envisaged by Cowan (1965). In consequence, there exists a maximum rate of transpiration that can be sustained by the plant—

the rate equivalent to the flux of liquid water when the water potentials in the leaves are at their critical levels and soil water is freely available. In this case, the maximum rate was about $4 \times 10^{-4}\,cm^3\,s^{-1}$ for a tiller, or 0.6 $kg\,m^{-2}\,h^{-1}$ when transpiration is expressed on a ground area basis.

This rate is likely to depend on variety, crop density and root development, but it is interesting that almost all the published measurements of evaporation from wheat crops appear to be consistent with it. The fastest rates of water loss (soil evaporation plus transpiration) recorded for wheat appear to be $8.4\,kg\,m^{-2}\,day^{-1}$ (Fritschen, 1966) and $1\,kg\,m^{-2}\,h^{-1}$ (Paltridge et al., 1972).

When water stress results from inadequate soil-water supply, the feedback control appears to be ineffective. Denmead and Millar studied the course of transpiration when the average soil-water potential in the root zone was -14 bars. Within a few hours of daybreak, leaf–water potentials fell below

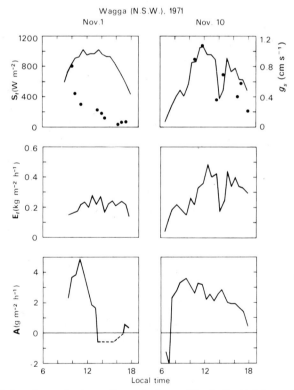

Fig. 9. Solar radiation above wheat crop S_t, conductance of flag leaf g_s (dots), crop transpiration E, and CO_2 exchange A when average soil-water potential in root zone was -14 bars (Nov. 1) and -0.1 bar (Nov. 10). From Denmead and Millar (1976b).

their critical levels (from -14 bars at 0700 to -26 bars at 1030), and leaf conductances decreased from values typical of unstressed leaves to values characteristic of leaves with closed stomata. The course of transpiration and the changes in leaf conductance are shown in Fig. 9, where they are contrasted with the situation nine days later when soil water was plentiful and leaf–water potentials did not exceed critical levels. The lowest leaf–water potential was -19 bars in the flag leaf at midday. Leaf conductance then depended on irradiance rather than leaf–water status and transpiration responded throughout the day to the fluctuating radiation load.

D. Water Relations and Plant Growth

These changes in water relations from leaf to leaf and the response of the plant to water stress seem to be rational adaptations to the field environment. In the wheat plant, the flag leaf has a better light climate for photosynthesis than lower leaves and the physiological evidence is that it is the main external supplier of assimilate for grain production (Wardlaw, 1965). On the other hand, transport of water up the stem along gradients of decreasing water potential requires the lowest potentials to be developed in the top leaves. If all leaves had similar water relations, the consequences would be that the flag leaf suffered water stress first. In fact, not only can low water potentials be developed in the flag leaf without the loss of turgor and stomatal closure that would occur in lower leaves, but during stress, stomatal closure occurs in the lower leaves first. This behaviour has the effect of lessening the potential gradient through the plant (since the total transpiration load is reduced) and water stress in the flag leaf is thereby forestalled.

V. CARBON DIOXIDE EXCHANGE

An important development in agricultural meteorology in recent years has been the use of micrometeorological techniques for the study of carbon dioxide exchange in the field. Undoubtedly, measurement errors can be large, particularly in the canopy where, for reasons discussed in Vol. 1, Ch. 3 and by Lemon (1969) and Denmead and McIlroy (1971), errors may be some tens of per cent. Nevertheless, the methods are potentially very useful for studying the photosynthesis and respiration of crops without disturbance over short time periods; for detailed examination of the activity of the various parts of the plant; and for pinpointing the influences of the field environment on plant growth. Here, the emphasis will be on these physiological and ecological aspects rather than on the methodology itself.

The technique should not be regarded as a substitute for the relatively simple and inexpensive methods of serial harvesting used in conventional growth analysis. Rather, its most profitable use is to complement the conventional methods by supplying details of the spatial and temporal variation of plant growth.

A. Sources of CO_2

The carbon dioxide absorbed by plants is supplied by turbulent diffusion from the atmosphere and by evolution from the soil. The soil contribution can be large. For instance, Monteith *et al.* (1964) found that soil respiration under a barley crop released an average of $10.6 \, g \, m^{-2} \, day^{-1}$ of CO_2 and supplied about 22% of the carbon assimilated by the crop throughout the growing season. Denmead (1969, and Fig. 10a, this chapter) has measured soil respiration rates of up to $16.4 \, g \, m^{-2} \, day^{-1}$ of CO_2 under wheat, with peak contributions of as much as 43% of the daytime carbon assimilation. Of course, the plant is insensitive to the source of its CO_2, but the accurate measurement of soil respiration is an important but weak point of current methodology for measuring CO_2 exchange in the field—see Monteith (1962) and Denmead and McIlroy (1971) for further discussion. In the present context, there is more interest in the significance of soil respiration to the carbon balance of a crop.

Some of the CO_2 evolved from the soil undoubtedly comes from the breakdown of soil organic matter by microorganisms. For instance, the rates of soil respiration just quoted correspond to daily conversions of less than 0.1% of the native soil carbon in the root zone and less than 3% of the standing dry matter of a mature crop. But some of the CO_2 appears to come from root activity as well. By comparing respiration rates on adjacent areas of bare and cropped soil, Monteith *et al.* (1964) estimated that root respiration amounted to 2 to $3 \, g \, m^{-2} \, day^{-1}$. Figure 10a shows daily cycles of soil respiration under wheat crops measured by Denmead (unpublished)* on successive clear and cloudy days. There is a strong suggestion that the evolution of CO_2 is linked with the current rate of assimilation in the crop although interpretation is complicated by the possible effects on respiration rate of soil temperature. Figure 10b which shows the daily contributions from atmosphere and soil to the CO_2 uptake of wheat crops (Denmead, unpublished) also suggests a proportionality between crop assimilation rates and

---

*A number of the measurements of CO_2 exchange discussed in this section are referenced as Denmead (unpublished). They were obtained by the author using techniques described by Denmead and McIlroy (1971). For measurements inside the canopy, total heat was used as a tracer of CO_2 movement. Above the canopy, either an energy balance method or an aerodynamic method based on a low-level drag coefficient was employed.

soil respiration. Neales and Davies (1966) have obtained indirect evidence
for such a linkage from growth chamber experiments with wheat seedlings.
They found that root respiration increased in response to light supplied to the
tops of the plant and that the increase commenced within an hour of the

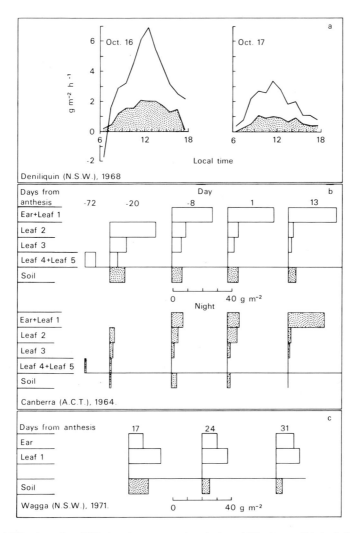

Fig. 10. (a) Total uptake of CO_2 by wheat crop and evolution of CO_2 from soil (stippled) on clear
(Oct. 16) and cloudy (Oct. 17) days. (b) CO_2 evolved from soil, and daytime assimilation and
night-time respiration of foliage layers in wheat crop. (c) CO_2 evolved from soil and assimilation
by ear and flag leaf in wheat crop. From Denmead (unpublished)—see footnote, opposite.

start of the light period. It seems likely that root respiration is an important component of the plant's carbon balance and its study should be a rewarding, if difficult, field of research.

B. Assimilation, Respiration and the Carbon Balance

1. *Crops*

Experiments in growth cabinets and leaf chambers have shown that respiration in the dark can be large relative to assimilation. King and Evans (1967), for instance, measured dark respiration rates from a dense 'crop' of wheat in a growth cabinet which amounted to 30% of the assimilation rate in the brightest light (roughly the equivalent of the visible radiation in full sunlight). Rapid respiration in the dark appears to occur in the field also. Figure 11 from Denmead (unpublished) shows micrometeorological measurements

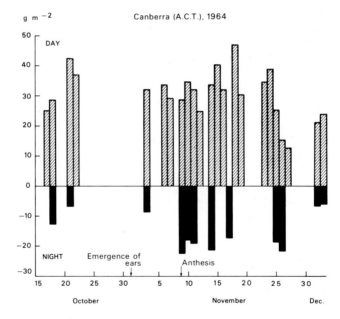

Fig. 11. Day and night exchanges of CO_2 between wheat crop and atmosphere. From Denmead (unpublished)—see footnote, p. 22.

of the day and night exchanges of CO_2 between the atmosphere and a wheat crop. The measurements span a period from 27 days before anthesis to 23 days afterwards. A point of particular interest is that dark respiration almost doubled after the emergence of the ears.

Simultaneous measurements of CO_2 exchange at the soil surface and supplementary harvests made it possible to construct an approximate balance sheet for CO_2 over the 33 days between October 24 and November 25. The averages of the measurements during this period were:

		$g\,m^{-2}\,day^{-1}$
Day:		
	Uptake from atmosphere	33·3
	Uptake from soil respiration	8·6
	Day-time assimilation	41·9
Night		
	Release to atmosphere	18·4
	Release from soil	1·6
	Dark respiration of tops	16·8
Net assimilation of tops		25·1
Increase in dry weight of tops		
(from serial harvests)		19·2
Balance assimilated by roots		5·9

Because of missing periods and errors in the measurements, particularly at night, the analysis is only approximate, but it serves to illustrate the relative magnitudes of the various processes of assimilation in a field crop. In particular, we note that dark respiration was 40% of the daytime CO_2 uptake.

2. *Plant Parts*

Denmead (unpublished) has employed micrometeorological methods inside the canopy in more detailed studies of assimilation and respiration by the leaves and the ears of wheat plants. As mentioned previously, errors can be large in studies of this type, and the results, although consistent, should be taken as preliminary. Figure 10b from one of these studies, shows the relative magnitudes of assimilation and respiration measured at different times of the growing season in foliage layers which encompassed either leaves only or leaves and ears. Three features are prominent:

(i) the dominance of assimilation (after ear emergence) by the flag leaf and the ear;

(ii) the increase in dark respiration after anthesis [which was probably due to grain respiration—cf. Carr and Wardlaw (1965) and Evans and Rawson (1970)];

(iii) the apparent balance between assimilation and respiration—the older leaves, while not assimilating rapidly, were not parasitic in so far as they assimilated more than they respired.

In another field study, Denmead (unpublished) was able to examine the separate CO_2 exchange by the ear. In this case, the measurements were made late in the growing season and only the ear and the flag leaf were green. Carbon dioxide exchange in older leaves was negligible. Some of the results are shown in Fig. 10c where it can be seen that the total assimilation was not very much less than that of the leafier crop in the first study, confirming the dominance of assimilation by the flag leaf and the ear. Assimilation by the ear itself accounted for a substantial part of the total CO_2 uptake.

Carbon dioxide exchange in ears and flag leaves has also been studied by Thorne (1965) who used an assimilation chamber in the field, and by Carr and Wardlaw (1965) and Evans and Rawson (1970) who used assimilation chambers in a glasshouse and a growth cabinet. Table IV compares the four sets of measurements. The micrometeorological measurements are of a comparable magnitude with those from the chambers and intermediate between them. There is not space here to investigate the comparative physiology. The main purpose of Table IV and other observations described in

Table IV

Rates of assimilation (**A**) and dark respiration (**R**) of ears and flag leaves of wheat plants (units are mg h^{-1})

Method	Ear		Flag leaf		Ear + Flag leaf		Reference
	A	**R**	**A**	**R**	**A**	**R**	
Assimilation chamber in glasshouse (cv. Sabre)	−0·7	2·9	2·0	0·3	1·3	3·2	Carr and Wardlaw (1965)
Assimilation chamber in field (cv. Jufy 1 and Atle)	0·5	0·3	3·0	0·1	3·5	0·4	Thorne (1965)
Assimilation chamber in growth cabinet (cv. Sonora)	3·0	2·0	7·9	—	10·9	—	Evans and Rawson (1970)
Energy balance in field (cv. Heron)	2·1	—	4·6	—	6·7	3·7	Denmead (unpublished, see footnote p. 22)

this section is to demonstrate the potentialities of micrometeorological techniques for supplying the detailed information needed in studies of the physiology of crop growth in the field. In passing, it is worth noting that assimilation chambers are not without their troubles. Thorne, for instance, found that the light intensity in her chamber was 75 % of that outside; and in sunny periods, temperatures in the chamber were 5 to 7°C higher than those in the ambient air.

C. Effects of Environment

1. *Radiation*

Laboratory experiments with plants and leaves of temperate cereals have demonstrated a strong dependence of photosynthetic rate on irradiance, [for instance, King and Evans (1967), Evans and Dunstone (1970)], and it is not surprising to find that when soil water is plentiful, radiation flux density is an important determinant of the assimilation rates of temperate cereals in the field. This is evident for instance in the studies of Angus *et al.* (1972) with barley and those of Denmead (1969) and Puckridge and Ratkowsky (1971) with wheat. Figure 12 from Denmead (unpublished) shows more explicitly the dependence on irradiance of the CO_2 uptake by crops and individual leaves of field wheat. The rates shown are averages for periods of one hour.

The crop measurements come from two studies: the first (Fig. 12a) in a dryland crop with a leaf area index of 1·6, the second (Fig. 12b) in an irrigated crop with a leaf area index of 3·2. Even in the sparse dryland crop, there is little evidence of light saturation. Much of the scatter arises from the fact that the measurements were spread over a period of five weeks in which time there were variable soil-water supplies and a change from vegetative to reproductive growth. The measurements in the irrigated crop were made over a much shorter period of five days when soil water was plentiful, and the dependence on irradiance is much stronger. In fact, 87 % of the variation in assimilation rate can be explained by a simple linear regression on irradiance. The regression line is shown in Fig. 12b.

An explanation of the linear response of crop photosynthesis to radiation receipt is found in the response of individual leaves, which is also close to linear (Fig. 12c). In this case, CO_2 uptake is expressed on the basis of unit leaf area and is plotted against the net radiation absorbed by the leaves, which as we have shown earlier is mostly in the short wavelengths. The highest values of absorbed radiation correspond to solar radiation flux densities on a horizontal surface above the crop of about 1000 W m^{-2}. Because the leaves are steeply inclined (cf. Fig. 2) and lower leaves are shaded, the average radiation incident on the leaves is very much less than

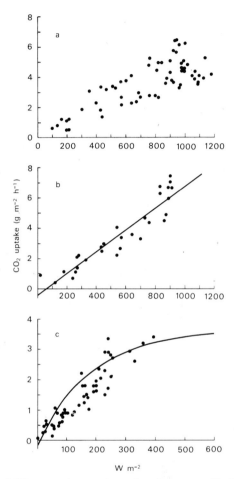

Fig. 12. (a) Uptake of CO_2 per unit ground area by wheat crop with $L = 1.6$, as function of solar radiation above crop. (b) Ditto from crop with $L = 3.2$. Line is regression of CO_2 exchange on solar irradiance. (c) Uptake of CO_2 per unit leaf area by individual leaves of wheat crop as function of net radiation absorbed by leaf. Curve is uptake by wheat leaves in assimilation chamber as function of visible radiation incident on leaf, from Evans and Dunstone (1970). All other measurements from Denmead (unpublished)—see p. 22, footnote.

on a horizontal surface, and the light saturation that is typical of laboratory measurements [and exemplified by the light response curve for wheat leaves of Evans and Dunstone (1970)—the line in Fig. 12c] does not occur.

Finally, we can note that the increase in CO_2 uptake with increasing irradiance is also in accord with the tendency for leaf conductance to increase with radiation absorption, which is shown for the same leaves in Fig. 3.

2. *Water Stress*

The sensitivity of photosynthesis in field wheat to water stress has been demonstrated by Denmead (1970), who showed that fast rates of potential evaporation could inhibit photosynthesis by inducing partial stomatal closure even when soil water was plentiful. Figure 9, from data of Denmead and Millar (1976b), provides a more striking example. On November 1, limited availability of soil water resulted in the development of leaf-water potentials as low as -26 bars and induced almost complete stomatal closure. Transpiration was greatly restricted and the uptake of CO_2 ceased altogether. On November 10, soil water was plentiful and leaf-water potentials did not fall below -19 bars. Leaf conductance was unaffected by water stress and transpiration and assimilation proceeded at fast rates throughout the day.

Other environmental factors have been shown to affect photosynthetic rates in controlled environments, e.g., temperature (Downes, 1970) and CO_2 concentration. Denmead (1970) presented evidence to show that the photosynthetic rate of wheat crops may be affected by day to day variations in ambient CO_2 concentration. Generally, however, lack of control makes it difficult to study the effects of these variables in the changing field environment, and in any case, the very strong controls exerted by radiation and water stress make it likely that these other effects will usually be second order.

3. *Other Factors*

One other aspect deserves consideration—the possible effects of wind speed on photosynthesis. It has been argued that the low leaf–air transfer coefficients which are associated with low wind speeds may lead to reduced concentrations of CO_2 at leaf surfaces and so to reduced photosynthetic rates. Lemon (1963) and Wright and Lemon (1966) produce evidence suggesting such an effect in corn and Yabuki et al. (1972) claimed a similar effect in rice.

Figure 6, which shows the relation between h_a and wind speed for wheat leaves, allows sample calculations of likely effects for that crop, and these can be extrapolated to other temperate cereals which have generally similar leaf shapes and canopy structures. For an extreme example, let us assume a wind speed over a leaf of only $10 \, \text{cm s}^{-1}$. From Fig. 6, the appropriate value of h_a is about $2 \, \text{cm s}^{-1}$. If the photosynthetic rate was equivalent to the highest values of Fig. 12c, i.e. about $10^{-7} \, \text{g cm}^{-2} \, \text{s}^{-1}$, the CO_2 concentration at the leaf surface would be reduced below the ambient concentration by approximately $30 \, \mu\text{l/l}$. As the reduction would usually be much less it seems probable that the direct effect of wind speed on the photosynthetic rate of temperate cereals in the field is trivial. In growth rooms where, according to Gaastra's (1969) figures, h_a may be as small as 0.7 cm s^{-1}, the reduction at the leaf surface might be as much as 80 μl/l, and a noticeable effect of air movement on photosynthetic rate might then be expected.

REFERENCES

Anderson, M. C. (1966). *J. appl. Ecol.* **3**, 41–54.
Anderson, M. C. (1969). *Agric. Met.* **6**, 399–405.
Angus, J. F., Jones, R. and Wilson, J. H. (1972). *Aust. J. agric. Res.* **23**, 945–957.
Baines, G. K. (1972). *Agric. Met.* **10**, 93–105.
Businger, J. A., Wyngaard, J. C., Izumi, Y. and Bradley, E. F. (1971). *J. atmos. Sci.* **28**, 181–189.
Carr, D. J. and Wardlaw, I. F. (1965). *Aust. J. biol. Sci.* **18**, 711–719.
Cowan, I. R. (1965). *J. appl. Ecol.* **2**, 221–239.
Cowan, I. R. and Milthorpe, F. L. (1968). *In* 'Water Deficits and Plant Growth' (T. T. Kozlowski, ed.), Vol. 1, pp. 137–193. Academic Press, New York.
Deacon, E. L. and Swinbank, W. C. (1958). *In* 'Climatology and Microclimatology', pp. 38–41. UNESCO, Paris.
Denmead, O. T. (1968). *In* 'Agricultural Meteorology' (Proc. WMO Seminar, Melbourne, 1966) Vol. 2, pp. 445–482. Bureau of Meteorology, Melbourne.
Denmead, O. T. (1969). *Agric. Met.* **6**, 357–371.
Denmead, O. T. (1970). *In* 'Prediction and Measurement of Photosynthetic Productivity' (Proc. IBP/PP Tech. Meeting, Trebon, 1969), pp. 149–164. Centre for Agricultural Publishing and Documentation, Wageningen.
Denmead, O. T. (1973). *In* 'Plant Response to Climatic Factors' (Proc. Uppsala Symp., 1970), pp. 505–511. UNESCO, Paris.
Denmead, O. T. and Bradley E. F. (1973). Proc. 1st Australasian Conf. on Heat and Mass Transfer, Monash University, Melbourne, 1973. Monash University, Melbourne.
Denmead, O. T. and McIlroy, I. C. (1970). *Agric. Met.* **7**, 285–302.
Denmead, O. T. and McIlroy, I. C. (1971). *In* 'Plant Photosynthetic Production Manual of Methods' (Z. Sestak, J. Catsky and P. G. Jarvis, eds), pp. 467–516. W. Junk, The Hague.
Denmead, O. T. and Millar, B. D. (1976a). *Agron. J.* **68**, 297–303.
Denmead, O. T. and Millar, B. D. (1976b). *Agron. J.* **68**, 307–311.
Downes, R. W. (1970). *Aust. J. biol. Sci.* **23**, 775–782.
Evans, L. T. and Dunstone, R. L. (1970). *Aust. J. biol. Sci.* **23**, 725–741.
Evans, L. T. and Rawson, H. M. (1970). *Aust. J. biol. Sci.* **23**, 245–254.
Fritschen, L. J. (1966). *Agron. J.* **58**, 339–342.
Fritschen, L. J. (1967). *Agric. Met.* **4**, 55–62.
Gaastra, P. (1969). *In* 'Physiological Aspects of Crop Yield' (J. D. Eastin, F. A. Haskins, C. Y. Sullivan and C. H. M. van Bavel, eds), pp. 140–142. American Society of Agronomy, Madison.
Haugen, D. A., Kaimal, J. C. and Bradley E. F. (1971). *Quart. J. R. met. Soc.* **97**, 168–180.

Impens, I. and Lemeur, R. (1969). *Arch. Met. Geophysk. Bioklimt.* (B) **17**, 261–268.

Kanemasu, E. T., Thurtell, G. W. and Tanner, C. B. (1969). *Pl. Physiol.* **48**, 437–442.

King, R. W. and Evans, L. T. (1967). *Aust. J. biol. Sci.* **20**, 623–635.

Lemon, E. R. (1963). *In* 'Environmental Control of Plant Growth' (L. T. Evans, ed.), pp. 55–78. Academic Press, New York.

Lemon, E. (1969). *In* 'Physiological Aspects of Crop Yield' (J. D. Eastin, F. A. Haskins, C. Y. Sullivan and C. H. M. van Bavel, eds), pp. 117–137. American Society of Agronomy, Madison.

Long, I. F., Monteith, J. L., Penman, H. L. and Szeicz, G. (1964). *Met. Rdsch.* **17**, 97–101.

Millar, B. D. and Denmead, O. T. (1976). *Agron. J.* **68**, 303–307.

Milthorpe, F. L. and Penman, H. L. (1967). *J. exp. Bot.* **18**, 422–457.

Monteith, J. L. (1959). *Quart. J. R. met. Soc.* **85**, 386–392.

Monteith, J. L. (1962). *Neth. J. agric. Sci.* **10**, 334–346.

Monteith, J. L. (1965a). *Ann. Bot. (N.S.)* **29**, 17–37.

Monteith, J. L. (1965b). *Symp. Soc. exp. Biol.* **19**, 205–234.

Monteith, J. L. (1969). *In* 'Physiological Aspects of Crop Yield' (J. D. Eastin, F. A. Haskins, C. Y. Sullivan and C. H. M. van Bavel, eds), pp. 89–111. American Society of Agronomy, Madison.

Monteith, J. L., Szeicz, G. and Yabuki, K. (1964). *J. appl. Ecol.* **1**, 321–337.

Monteith, J. L., Szeicz, G. and Waggoner, P. E. (1965). *J. appl. Ecol.* **2**, 345–355.

Neales, T. F. and Davies, J. A. (1966). *Aust. J. biol. Sci.* **19**, 471–480.

Paltridge, G. W., Dilley, A. C., Garratt, J. R., Pearman, G. I., Shepherd, W. and Connor, D. J. (1972). CSIRO (Aust.). *Div. Atmos. Phys.* Tech. Pap. No. 22.

Parlange, J.-Y., Waggoner, P. E. and Heichel, G. H. (1971). *Pl. Physiol.* **48**, 437–442.

Passioura, J. B. (1972). *Aust. J. agric. Res.* **23**, 745–752.

Pearman, G. I., Weaver, H. L. and Tanner, C. B. (1972). *Agric. Met.* **10**, 83–92.

Penman, H. L. and Long, I. F. (1960). *Quart. J. R. met. Soc.* **86**, 16–50.

Philip, J. R. (1965). *Aust. J. Bot.* **13**, 357–366.

Puckridge, D. W. and Ratkowsky, D. A. (1971). *Aust. J. agric. Res.* **22**, 11–20.

Rider, N. E. (1954). *Quart. J. R. met. Soc.* **80**, 198–211.

Slatyer, R. O. (1967). 'Plant-Water Relationships'. Academic Press, New York.

Stanhill, G., Kafkafi, U., Fuchs, M. and Kagan, Y. (1972). *Israel J. agric. Res.* **22**, 109–118.

Swinbank, W. C. and Dyer, A. J. (1967). *Quart. J. R. met. Soc.* **93**, 494–500.

Tanner, J. W. (1969). *In* 'Physiological Aspects of Crop Yield' (J. D. Eastin, F. A. Haskins, C. Y. Sullivan and C. M. H. van Bavel, eds), pp. 50–51. American Society of Agronomy, Madison.

Thorne, G. N. (1965). *Ann. Bot. (N.S.)* **29** 317–329.

Wardlaw, I. F. (1965). *Aust. J. biol. Sci.* **18**, 269–281.

Wright, J. L. and Brown, K. L. (1967). *Agron. J.* **59**, 427–432.

Wright, J. L. and Lemon, E. R. (1966). *Agron. J.* **58**, 265–268.

Yabuki, K., Aoki, M. and Hamotani, K. (1972). *In* 'Photosynthesis and Utilization of Solar Energy. Level III Experiments' (M. Monsi and T. Saeki, eds), pp. 7–9. Japanese National Subcommittee for PP, Tokyo.

2. Maize and Rice

Z. UCHIJIMA

Division of Meteorology, National Institute of Agricultural Sciences, Nishigahara, Tokyo, Japan

I. INTRODUCTION

According to the FAO Production Yearbook, world acreages of rice and maize planted in 1970 were about 136 million and about 111 million hectares, respectively. The world rice output (unhulled) reached 300 million tons and the maize output was about 270 million tons. Most inhabitants of the Far East and south-east Asian countries live on rice and the rice exported from these districts forms about 90% of the world output. Maize, on the other hand, is a source of food for animals as well as for man and about 70% of the world output is produced in North American and Asian countries.

33

Since maize and rice are such important food resources in the world, much attention in relevant disciplines of agricultural science has recently been directed to the improvement and stabilization of maize and rice production. The growth and the yield of crop plants, through the radiant energy exchange and the supply of nutrients, water and carbon dioxide to the plants, depend closely upon the conditions and constituents of the crop environment. The main purpose of research in agricultural meteorology is to increase yields by more accurate forecasting and controlling of crop environments.

II. RADIATION CHARACTERISTICS OF MAIZE AND RICE FIELDS

A. Radiation Balance of Whole Stands

The energy required for crop growth, water use, and the warming of soil, water and plants is supplied by the sun. Incoming long- and short-wave radiation is absorbed or reflected by the plants or the soil which continuously dissipate energy to the sky through the emission of long-wave radiation. The balance of incoming and outgoing radiant fluxes is termed net radiation.

1. *Radiation Balance*

Net radiation is given by

$$\mathbf{R}_n = (1 - \rho)\mathbf{S}_t - \mathbf{L}_0, \tag{1}$$

where $\mathbf{R}_n$ is the net radiation, ρ the albedo, $\mathbf{S}_t$ the total short-wave radiation, $\mathbf{L}_0$ the net loss of long-wave radiation. The details of the radiation balance sheet are important for understanding interactions between crop plants and the atmosphere. Measurements of radiation balance of rice and maize stands have been made in relation to their heat balance (Tanner and Lemon, 1962; Sato, 1960; Yabuki, 1957) and to dry matter production (Murata *et al.*, 1968). When the vegetation fully covered the soil surface, the following relations were obtained between net radiation and short-wave radiation (Research group of evapotranspiration, 1967a; Franceschini, 1959; Brown and Covey, 1966; Impens and Lemeur, 1969a):

$$\mathbf{R}_n = 0{\cdot}82\mathbf{S}_t - 90{\cdot}5, \qquad \text{for maize}$$
$$\mathbf{R}_n = 0{\cdot}86\mathbf{S}_t - 55{\cdot}7, \qquad \text{for rice} \tag{2}$$

where $\mathbf{R}_n$ and $\mathbf{S}_t$ are expressed in Wm^{-2}. The above relations can be used to determine the net radiation of fully grown crop fields from the data of solar radiation. The net radiation above vegetated surfaces ranges between 50 and 70% of total short-wave radiation depending mainly on the development of the plant canopy as described later.

Figure 1 represents the variation in the partition of total short-wave radiation reached rice and maize fields. Most notable are the very rapid increase in the amount of radiation absorbed by the plant canopy and the very marked decrease of canopy transmission coefficient for total (i.e. whole spectrum) radiation, τ_t. The research group of evapotranspiration (RGE, 1967a) obtained the following balance sheet for the whole growing period of rice crop (in units of MJ m^{-2}):

$$\mathbf{S}_t = 1487\cdot4, \qquad \rho\mathbf{S}_t = 259\cdot8, \qquad (1-\rho)\mathbf{S}_t = 1223\cdot5,$$

$$\mathbf{L}_0 = 305\cdot9, \qquad \mathbf{R}_n = 921\cdot8$$

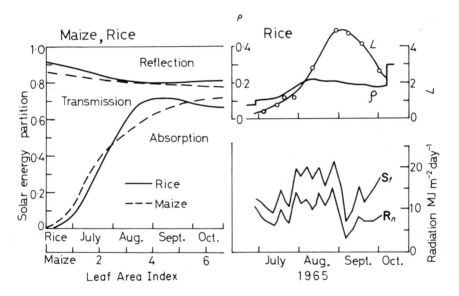

Fig. 1. Seasonal variation of radiation partition above rice and maize fields (Kishida, 1971; Pavlov et al., 1965; RGE, 1967a).

2. Albedo

The energy reflected by canopy surfaces is subtracted from the energy available for plant growth, evaporation and so on. As Fig. 1 shows, the seasonal variation in the values of albedo for a rice canopy is approximately parallel to that of leaf area index L. The dependence of albedo on L was given approximately by (RGE, 1967a)

$$\rho = \rho_p - (\rho_p - \rho_w) \exp(-0.56L), \tag{3}$$

where ρ_p is the albedo of a complete canopy (assumed to be 0.22) and ρ_w the albedo of a shallow water (assumed to be 0.08). More recently, Hayashi (1972) showed that the value of the extinction coefficient varies somewhat among rice varieties, depending on the leaf arrangement in rice canopy.

Figure 2 shows the albedo of fully grown rice canopies as a function of solar elevation (see also p. 3). The albedo of the rice canopy for photosynthetically active radiation (PAR), ρ_{PAR}, was about one fourth of ρ. Tooming (1966) used the following relation to calculate ρ_{PAR} for a maize canopy.

$$\rho = \rho_{PAR}^n c_{PAR} + \rho_{NIR}^n c_{NIR}, \tag{4}$$

where c_{PAR} and c_{NIR} are the fractions of photosynthetically active radiation (PAR) and near infra-red radiation (NIR) in the incident short-wave radiation, respectively, ρ_{NIR} is the albedo for NIR and n is a numerical constant (assumed to be 1.2). From measurements of $\rho(= 0.23)$ and the assumption of $c_{PAR} = 0.52$ and $c_{NIR} = 0.48$, he calculated 0.063 for ρ_{PAR} and 0.405 for ρ_{NIR}. His estimate of ρ_{PAR} for a maize canopy agreed well with the results presented in Fig. 2.

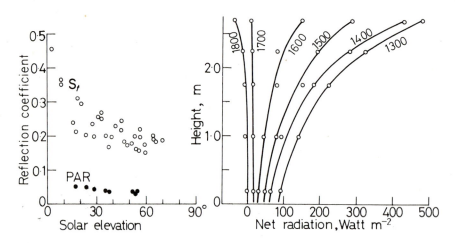

Fig. 2. Left: Relation between solar elevation and reflection coefficient of rice fields (RGE, 1967a; Kishida, 1971). Right: Vertical profiles of net radiation within a maize canopy (Allen *et al.*, 1964).

B. Optical Properties of Leaves

Spectral curves of transmission and reflection of single leaves and of crop canopies are shown in Fig. 3. Characteristics of the reflection spectra of green leaves, $\rho_{s,g}$, are the large values of reflectance at wavelengths longer than 0·7 μm and small values in the visible band. A narrow and relatively small peak of the reflectance at wavelengths around 0·55 μm is considered to arise from the absorption spectrum of chlorophyll in leaves. Yellowing of leaves

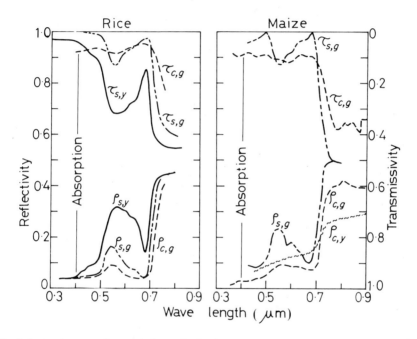

Fig. 3. Spectral curves of transmission and reflection of leaves and canopies (Allen *et al.*, 1964; Kondrat'ev, 1969; Kishida, 1971). Suffixes s and c denote single leaves and canopy, respectively, g and y are green and yellow leaves or canopy.

causes an increase of reflectivity, particularly in the visible band, with a proportionate decrease of absorptivity in this band.

The spectral curves of crop canopies were more flat and lower than those for single leaves. Smoothing and lowering of the spectra were evidently a result of multiple reflection and scattering of radiation by foliage. The mean values of reflectivity at wavelengths of visible and near infra-red radiation were 0·05 and 0·4, respectively, agreeing with those estimated by Tooming (1966).

C. Radiation Transfer Within Canopies

Radiation transfer within plant canopies has a decisive influence both upon the photosynthesis and on the micro-environment of a crop canopy.

1. *Canopy Structure*

The canopy structure is well specified by (a) the vertical distribution function of leaf area density and (b) the orientation function of leaf area, i.e. the inclination and azimuth angle (Ross and Nilson, 1966). In recent years, much attention has been paid to the canopy structure of rice and maize in relation to radiation regimes and canopy photosynthesis (Allen and Brown, 1965; Ito, 1969; Ito *et al.*, 1973; Loomis *et al.*, 1968; Ross, 1964; Tooming, 1966; Tooming and Guljaev, 1967; Udagawa *et al.*, 1968; Williams *et al.*, 1968; Yocum *et al.*, 1964). Figure 4 compares the leaf area orientation function

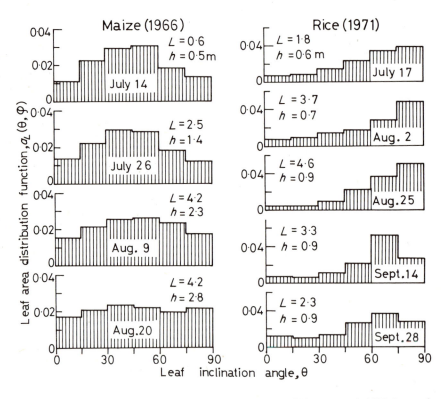

Fig. 4. Leaf area orientation function of rice and maize crops (Udagawa *et al.*, 1968; Ito *et al.*, 1973) (see Vol. 1, p. 25).

between maize and rice crops. Maize canopies appeared to be of plagiophile type (de Wit, 1965) rather than erectophile like rice canopies and to behave like canopies formed from leaves with inclination angles of 40 to 50°. The rice canopy structure shifted clearly from an erectophile type before heading time to a spherical leaf distribution after heading. The main reason for the shift of leaf arrangement is probably the development of the flag leaf and of rice panicles. Analysis of the canopy structure by layers showed that, after heading, a rice canopy consisted of horizontal upper leaves grading to erect ones at low canopy levels. This is contrary to the arrangement which is theoretically most favourable for canopy photosynthesis. In practice, horizontal upper leaves largely shade the lower leaves and reduce the photosynthetic activity.

2. *Radiation Penetration into a Leaf Canopy*

The flux density of total short-wave radiation below a leaf area index of L comprises the following three components:

$$S_t(L) = S_b(L) + S_d(L) + S_{c,d}(L), \tag{5}$$

where S_b and S_d are the direct and diffuse solar radiation flux densities, respectively and $S_{c,d}$ the downward flux of complementary diffuse radiation due to scattering and reflection of radiation by plant elements.

  a. Direct solar radiation. The attenuation of direct solar radiation was measured in rice (Ito, 1969; Ito *et al.*, 1973) and in maize (Ross and Nilson,

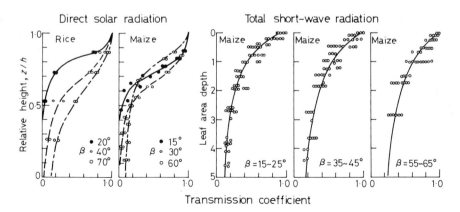

Fig. 5. Left: Direct solar radiation profiles within rice and maize canopies (Ross and Nilson, 1965; Ito *et al.*, 1973). Right: Total short-wave radiation profiles within a maize canopy (Tooming and Ross, 1965).

1965), by using a sunlit measuring bar. Figure 5 presents the penetration of direct solar radiation into the canopies as a function of solar elevation β. When the elevation is low, direct solar radiation is completely absorbed in the upper half of the foliage. With increasing elevation, part of the direct beam can penetrate into deeper layers. At noon when the solar elevation was between 60 and 70°, direct solar radiation penetrating to the ground surface was about 10% of the flux at the top of the canopy.

The extinction of direct solar radiation within a plant canopy can be expressed as follows (Ross and Nilson, 1965):

$$\mathbf{S}_b(z) = \mathbf{S}_b(h) \exp\left(-\mathcal{K}_b \int_z^h a_l(z)\,\mathrm{d}z\right), \qquad z \leqslant h \qquad (6)$$

where $\mathbf{S}_b(h)$ is the flux density of $\mathbf{S}_b$ at the top of canopy and $\mathcal{K}_b$ the extinction coefficient, related to the effective leaf area projection, G, by $\mathcal{K}_b = \mathrm{cosec}\,\beta G$ (where β is the solar elevation). Values of the effective leaf area projection were determined for maize and rice canopies (Ito, 1969; Udagawa et al., 1968, 1974). For a rice canopy G was a decreasing function of β, but for a maize canopy it was an increasing function of the solar elevation. In the top half of a maize canopy, profiles of $\mathbf{S}_b$ calculated from Eq. (6) agreed well with measurements.

   b. *Complementary diffuse radiation.* The downward flux of complementary diffuse radiation is given by the formula

$$\mathbf{S}_{dc}(z) = \mathbf{S}_t(z) - \mathbf{S}_b(z) - \mathbf{S}_d(z),$$

but no measurements of this flux have been made in rice and maize canopies. Solutions of the transfer equation of radiation fluxes within plant canopies have therefore been used to estimate the downward and upward components of this flux (Ross, 1964; Ross and Nilson, 1968; Isobe, 1969; Horie and Udagawa, 1971). The results so obtained can be summarized as follows: $\mathbf{S}_{dc}(z)$ is zero at the top of canopy, increases to a maximum value at a certain depth, and then decreases monotonically toward the ground (see Vol. 1, p. 36). The upward flux of complementary diffuse radiation diminishes also monotonically with depth. More recently, Udagawa et al. (1974) studied the vertical profiles of upward and downward fluxes of complementary diffuse radiation within canopies of two rice varieties.

In a high latitude region or on cloudy days, diffuse sky radiation plays an important role in canopy photosynthesis. The flux density of diffuse sky radiation, $\mathbf{S}_d$, changes with time of day, season, weather and other factors.

Table I shows a relation proposed by Tooming (1965) to estimate the attenuation of diffuse sky radiation within a maize canopy. Using the assumption of a uniform overcast sky for diffuse radiation, Udagawa *et al.* (1974) calculated an extinction coefficient between 0·6 and 0·7 for a rice canopy.

Table I

Transmission function of a maize canopy for diffuse sky radiation, $\tau_d(L, \beta)$ (Tooming and Guljaev, 1967)

L	\multicolumn{5}{c}{Solar elevation, β}
	20°	30°	40°	50°	Cloudy
0·5	0·56	0·59	0·60	0·65	0·63
1·0	0·34	0·38	0·40	0·44	0·42
2·0	0·16	0·18	0·19	0·22	0·21
3·0	0·08	0·08	0·09	0·11	0·10
4·0	0·04	0·04	0·05	0·06	0·06

c. *Extinction of PAR.* The clear difference in optical properties of leaves between PAR and NIR regions has a great influence on the penetration of these fluxes into canopies. Tooming (1966) calculated the transmission coefficient of total short-wave, near infra-red and photosynthetically active radiation in a maize canopy. NIR is able to penetrate through deep layers of maize, but PAR is more absorbed than other two fluxes. He proposed the following values for evaluating τ_{PAR} and τ_{NIR} from the measurements of the transmission of total radiation τ_t:

τ_t	0·9	0·8	0·7	0·6	0·4	0·2
τ_{PAR}	0·88	0·73	0·60	0·48	0·23	0·06
τ_{NIR}	0·92	0·90	0·83	0·75	0·58	0·34

More recently, using measurements of S_t and PAR and applying the Beer–Lambert law, Kishida (1971) studied the relationships between PAR and S_t for a rice canopy. His results can be summarized as follows:

$$\tau_{PAR} = \tau_t^{1·48},$$
$$\alpha_{PAR} = 1 - (1 - \alpha_t)^{1·82}, \tag{7}$$

where α_{PAR} and α_t are respectively the absorption coefficients for NIR and PAR. From the above relations, it appears that $\tau_{PAR} < \tau_t$ and $\alpha_{PAR} > \alpha_t$.

d. *Total short-wave radiation.* Figure 5 shows how total short-wave radiation profiles change in relation to solar elevation. When the elevation is low, total radiation is absorbed mostly by the upper leaf layers of maize. Isobe (1962) showed that the transmission coefficient of a rice canopy depends not only on solar elevation but also on the composition of incident radiation. On a clear day, τ_t, showed a pronounced diurnal change with a maximum of about 0·8 shortly after noon. On a dense overcast day when incident radiation was almost diffuse sky radiation, the transmission coefficient remained approximately constant at a level of about 0·5. Tooming and Guljaev (1965) used the following relation to determine the profile of total radiation in a maize canopy.

$$\tau_t(L, \beta) = \frac{S_b\, e^{-c_1 A}/S_d + \tau_d}{1 + S_b/S_d} + c_2[e^{-c_1 c_3 A} - e^{-c_1 A}], \tag{8}$$

where $A = L/\sin \beta$. The second term on the right-hand side of Eq. (8) is introduced semi-empirically to account for the influence of complementary diffuse radiation on the profiles. Profiles of total radiation (TR, solid curves) calculated from the above equation agree well with measurements. By analysing observations, they determined the values of each numerical constant as follows:

TR	$c_1 = 0\cdot5$	$c_2 = 0\cdot3$	$c_3 = 0\cdot15$,
PAR	$c_1 = 0\cdot5$	$c_2 = 0\cdot04$	$c_3 = 0\cdot15$,
NIR	$c_1 = 0\cdot5$	$c_2 = 0\cdot6$	$c_3 = 0\cdot15$.

e. Net radiation. Since the development of reliable instrumentation, the distribution of net radiation within crop canopies has received considerable attention (e.g. Denmead and Shaw, 1959; Denmead et al., 1962). It was found from the analysis of evapotranspiration structure of rice field that net radiation decreases exponentially with leaf area (Uchijima, 1961; Iwakiri, 1964) i.e.

$$R_n(L) = R_n(0) \exp(-\mathscr{K} L), \tag{9}$$

where $R_n(0)$ and $R_n(L)$ are the net radiation flux above the canopy and beneath a leaf area index L. The value of $\mathscr{K}$ was between 0·45 and 0·65. The fraction of net radiation absorbing the foliage therefore increases curvilinearly with increasing leaf area index. Denmead et al. (1962) found that R_n at the soil surface beneath a maize canopy decreases from 42% of its value above the canopy when $L = 2\cdot5$ to 22% as plants matured. By replacing L by the leaf area depth, Uchijima (1962a) applied the exponential relation for evaluating net radiation profiles within a rice canopy.

Allen et al. (1964) and Brown and Covey (1966) made extensive measurements of net radiation profiles in canopy and some of their results are reproduced in Fig. 2. The relation between net radiation and height changed clearly from a sharp increase of net radiation with height near noon, to almost uniform, but very small, values near sunset, depending on the flux density of total short-wave radiation at the top of canopy. By 1800, the slope of the net radiation profile had become negative. A plot of net radiation against height or leaf area depth showed that the net radiation profiles can be approximated by an exponential relation as in a rice canopy. Brown and Covey (1966) obtained a value of 0·58 for the mean diurnal attenuation coefficient, in good agreement with values obtained for rice canopies. More recently, Impens and Lemeur (1969b) proposed the following empirical relation:

$$\mathbf{R}_n(L) = \mathbf{R}_n(0)\exp\left(-0.622L + 0.055L^2\right). \tag{10}$$

III. WIND REGIME ABOVE AND WITHIN CANOPIES

A. Wind Profile Above and Within Canopies

The maintenance of plant life largely depends on the existence of a vigorous exchange of energy and mass between vegetated surfaces and the air. For this reason, wind profiles above and within canopies have been measured and analysed by many workers (e.g. Inoue, 1960, 1963; Isobe, 1972; Lemon, 1960; Nakagawa, 1956; Saito et al., 1962, Tani, 1963). Aerodynamical properties of crop canopies and the interactions between crop plants and the atmosphere have been studied in these researches.

Figure 6 illustrates typical mean wind profiles above and within a maize canopy. The most obvious feature is that the mean wind velocity, u, decreases with decreasing the height above the ground, particularly near the top of the canopy. If the temperature gradient is small throughout the air layer, the relation between the mean wind velocity and the height above the vegetation is given by

$$u = \frac{u_*}{k}\ln\left(\frac{z - d}{z_0}\right), \qquad z > d + z_0 \tag{11}$$

where $u_* = \sqrt{\tau/\rho}$ is the friction velocity, τ the shearing stress, z_0 the roughness length, d a zero-plane displacement and $k = 0.41$ (see Vol. 1, p. 63).

This relation cannot be applied to air flow *in* the canopy, because of plant elements behaving as a vigorous sink for momentum. Many studies have been made to explore the nature of the wind regime in plant canopies and to deduce a more reliable expression for the mean wind profile in maize and rice canopies. Assuming that the mixing length, l_w, is invariable with height within the foliage and equals to the value $l(h)$ at the top of the canopy, Inoue

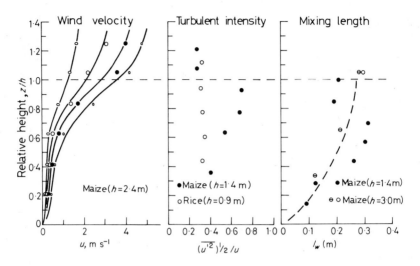

Fig. 6. Wind regime above and within crop canopies (Nakagawa, 1956; Uchijima and Wright, 1964; Wright and Lemon, 1966a; Saito *et al.*, 1970).

(1963) obtained an exponential relation for the canopy wind profile:

$$u = u(h) \exp [-n(h - z)], \qquad z \leqslant h \tag{12}$$

where

$$n = \left(\frac{a(z)C_M}{4l(h)} \right)^{\frac{1}{3}},$$

$u(h)$ is the wind velocity at the top of the canopy and $a(z)$ the leaf area density and C_M the leaf drag coefficient. Saito (1964) and Isobe (1964) also found that the simple exponential formulae fits well wind profiles within the top half of a canopy.

B. Aerodynamic Characteristics of Whole Stands

As described above, plant elements and the ground extract momentum by surface friction and form drag. The frictional drag exerted on the atmosphere by a crop canopy in bulk is given by

$$\tau = \rho K_M \frac{du}{dz} = \frac{u^2 k^2}{[\ln (z - d)/z_0]^2} = \tfrac{1}{2}\rho C_{aM} u(z)^2, \tag{13}$$

where C_{aM} is the surface drag coefficient. This relation shows that the frictional drag can be related to the aerodynamic properties of a crop canopy as specified by z_0 and d.

1. Roughness Length and Zero-plane Displacement

The analysis of wind records obtained in near neutral conditions showed that mean values of d and z_0 for maize and rice crops increase with increasing stand height (Lemon, 1965; Seo and Yamaguchi, 1968). For rice and maize fields this dependence can be approximated by

$$d = 1 \cdot 04 h^{0 \cdot 88},$$
$$z_0 = 0 \cdot 062 h^{1 \cdot 08}. \tag{14}$$

For rice, the large scatter of individual values of d and z_0 concealed the wind dependence presented in Fig. 7, but z_0 for a maize field is clearly an increasing function of wind velocity, while d for a maize field is a decreasing function

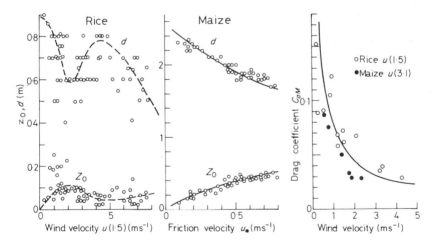

Fig. 7. Wind dependence of z_0, d, and C_{aM} of rice and maize fields (Tani, 1963; Denmead, 1971; Maki, 1969).

of wind velocity. More recently, Denmead (1971) reported the opposite relation between z_0 and u for a maize crop. Tani (1963) indicated the peculiar change of d and z_0 of rice fields with wind velocity illustrated by dotted lines in Fig. 7. He suggested that the behaviour of z_0 and d was closely related to the phenomena of waving plants ('*Honami*'). The discrepancy in wind dependence of aerodynamical properties between rice and maize canopies seems to be primarily related to differences in plant geometry and in the elasticity of plants.

Inoue (1960) suggested that the resonance of vibration of crop plants with the movement of a dominant eddy at the top of the canopy plays an important part in the occurrence of the *Honami*. He studied the physical properties of *Honami* occurring over well grown crop fields. In relation to the lodging of crops, the period of vibration of plants was measured and found to be in the range 0·7 to 1·2.s for maize (Tani and Suzuki, 1966) and between 0·8 and 1·2 s for rice (Tani, 1963; Hitaka, 1968).

The frictional drag imposed on the atmosphere by a crop canopy is uniquely related to a dimensionless drag coefficient, C_{aM}, given in Eq. (13). The values of surface drag coefficient of maize and rice fields decreased rapidly with wind velocity from the order of 0·1 in light winds to the order of 0·01 in strong winds.

2. *Momentum Transfer Within Canopies*

The momentum flux, $\tau(z)$, at a height z *within* a plant canopy is given by

$$\tau(z) = \tau(h) - \frac{\rho}{2} \int_z^h C_M a(z) u^2 \, dz, \tag{15}$$

where $\tau(h)$ is the momentum flux at the top of the canopy, and C_M is the leaf drag coefficient. Applying some reasonable assumptions to wind records, the decrease of momentum flux within an immature dense maize canopy was calculated by Uchijima and Wright (1964). The calculation gave the relationship:

$$\tau(z) = \tau(h) \exp\left[-5\cdot6\left(1 - \frac{z}{h}\right)\right]. \qquad z \leqslant h \tag{16}$$

in good agreement with results obtained by Isobe (1964) in a wheat canopy.

Values of leaf drag coefficient, C_M, were in the range from 0·047 to 0·542 with a mean value of 0·17. Wright and Brown (1967) evaluated C_M for maize leaves and showed that C_M changes not only with the canopy depth but also with wind velocity. Contrary to the decrease of C_{aM} with increasing wind velocity, Wright and Brown (1967) and Saito *et al.* (1970) showed that C_M is an increasing function of wind velocity. The results reported here indicate that the assumption of the constancy of the leaf drag coefficient with height and with wind velocity is invalid (see Vol. 1, p. 67). Direct measurements of momentum flux within crop canopies may make it possible to determine more exactly the relation between leaf drag coefficient, wind velocity and canopy structure.

C. Turbulence Within Canopies

Measurements within rice and maize canopies indicated that the air flow in foliage is fully turbulent. The turbulent intensity ($i = \sqrt{\overline{u'^2}}/u$, where u' is the instantaneous departure from the mean wind velocity, u) is large compared with the intensity in the surface air layer (Nakagawa, 1966; Uchijima and Wright, 1964; Wright and Lemon, 1966a; Saito et al., 1970). Several papers report a decrease of turbulent intensity with depth below the top of the canopy (Uchijima and Wright, 1964; Isobe, 1964), but others do not confirm those conclusions and indicate the presence of a constant turbulent intensity throughout the canopy (e.g., Nakagawa, 1966). More recently, Cionco (1972) reconciled the reported height dependence of turbulent intensity as follows: (a) for nearly ideal canopy with leaves uniformly distributed vertically, $\sqrt{\overline{u'^2}}/u$ is essentially constant with height, (b) for complex canopies, it is closely related to the leaf area distribution in the canopy and reaches a maximum in the layer where the leaf area density is greatest.

Statistical characteristics of turbulence were studied in a rice canopy (Nakagawa, 1956) and in maize canopies (Uchijima and Wright, 1964; Wright and Lemon, 1966a; Saito et al., 1970, Isobe, 1972). These studies showed that the fundamental relationships deduced from similarity theory for isotropic turbulence are also applicable to the analysis of canopy air flows. Isobe (1972) studied the eddy-generating potential of maize stems from the Strouhal number.

1. Mixing Length

On the basis of mixing-length theory, the turbulent transfer coefficient for momentum, K_M, is given by

$$K_M = l_w\sqrt{\overline{w'^2}}, \qquad (17)$$

where l_w is the mixing length and w' the departure of the vertical wind speed from its time average. Uchijima and Wright (1964) calculated profiles of l_w within a maize canopy ($h = 1\cdot4$ m). Saito et al. (1970) also obtained profiles of l_w within a maize ($h = 3\cdot0$ m) on the basis of the statistical theory of turbulence. The results presented in Fig. 6 indicate that l_w in maize is far from constant. In order to confirm such conclusions, however, more observations of the wind regime in crop canopies are required.

2. Turbulent Transfer Coefficient

In the past decade, increasing interest has been directed toward studying the mechanisms which establish vertical profiles of each meteorological quantity within crop canopies. In these studies, both the fundamental concept of heat balance and boundary layer theory have been used successfully. The following relation has been used to calculate the turbulent transfer coefficient within plant canopies (Uchijima, 1962a; Uchijima et al., 1970):

$$K = \frac{\mathbf{R}_n(z) - \mathbf{G}}{-\left[c_p\rho\dfrac{dT}{dz} + \lambda\rho\dfrac{dq}{dz}\right]}, \qquad (18)$$

where K is the turbulent transfer coefficient, $\mathbf{R}_n(z)$ the net radiation flux density at a level of z, $\mathbf{J}$ the rate of change of stored heat, c_p the heat capacity of air at constant pressure, ρ the density of air, T and q the air temperature and specific humidity, and λ the latent heat of vaporization. Figure 8 presents the vertical profiles of K estimated from Eq. (18) and from measurements

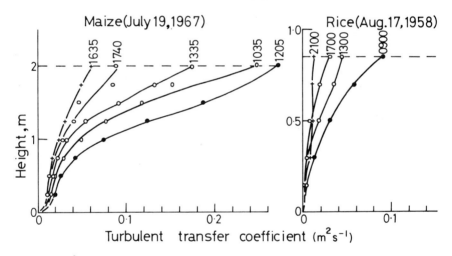

Fig. 8. Vertical profiles of turbulent transfer coefficient within crop canopies (Uchijima, 1962a; Uchijima et al., 1970).

within rice and maize canopies. For the daylight hours, K decreased very rapidly from 10^2 to 10^3 cm^2 s^{-1} at the top of the canopy to value of the order of 10 near the soil surface. Above $0.3\,h$ or so, the profiles of K fit well the equation

$$K(z) = K(h)\exp\left[-n\left(1 - \frac{z}{h}\right)\right], \qquad z \leqslant h \qquad (19)$$

where $K(h)$ is the transfer coefficient at the top of the canopy. Values of n in the range 2·5 to 3·0 were reported for rice and maize canopies. Several workers found that the extinction coefficient showed a significant dependence on the canopy structure and on meteorological conditions particularly the thermal stratification in a canopy. At night when thermal stratification favouring convection is usually observed in a crop canopy, profiles of K were found to be characterized by a maximum in the middle of the canopy (Uchijima, 1962a; Gillespie, 1971; Gillespie and King, 1971). This behaviour may be due to the influence of buoyant convective motion upon the air mixing in the bottom half of the canopy.

IV. HEAT AND WATER BALANCE

A. Heat Balance of Whole Stands

The partition of solar radiation and rainfall that reach vegetative surfaces is of considerable importance because it determines the micro-environment in which crop plants live.

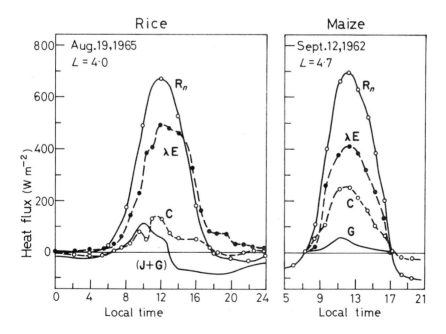

Fig. 9. Heat balance of a rice field and a maize field (RGE, 1967b; Brown and Covey, 1966).

1. *Heat Balance*

Heat balance characteristics of rice and maize fields are presented in Fig. 9. General features of the diurnal energy balance pattern are similar between the two fields. The most notable fact is that the greater amount of radiation surplus is used for evaporation. The daily value of Bowen ratio $\beta = \Sigma C / \Sigma \lambda E$ for the rice field was 0·18, about one third of that for the maize field (0·55). The following heat balance characteristics of a rice field was obtained for the July through September growing season (RGE, 1967b):

$$\mathbf{R}_n = 921 \cdot 8, \qquad \lambda \mathbf{E} = 725 \cdot 7, \qquad \mathbf{C} = 196 \cdot 1 \, \mathrm{MJ \, m^{-2}}, \qquad \beta = 0 \cdot 27.$$

About 80% of the total net radiation was consumed as a source of heat for evaporation (310 mm) and only 20% for heating the air. The Bowen ratio for the rice field was three times as large as for shallow water ($\beta = 0 \cdot 08$). A marked increase of Bowen ratio during the ripening period of crop plants was observed on both fields (Fritschen and Shaw, 1961a; RGE, 1967b).

2. *Evaporation*

The heat balance equation

$$\lambda \mathbf{E} = (\mathbf{R}_n - \mathbf{J})/(1 + \beta)$$

has been successfully applied to determine evaporation from field crops (Tanner and Lemon, 1962; Fritschen and Shaw, 1961a, b; RGE, 1967b). Lourence and Pruitt (1971) reported that evaporation evaluated by the heat balance method was within 2% of the lysimeter for the day. Total water consumption of rice fields in Japan ranges from 3000 to 6000 ton/ha, depending on climatic conditions, particularly on the net radiation income for the growing season.

Over uniform, extensive and well irrigated fields, the following relation is obtained

$$\mathbf{E} = f \cdot \frac{\mathbf{R}_n}{\lambda}, \tag{20}$$

where f is a crop coefficient. Under moist climatic conditions, the value of f can be less than unity, e.g. about 0·8. In dry or arid weather, however, the value of f becomes larger than unity due to the effect of advected heat (Tanner and Lemon, 1962; Lourence and Pruitt, 1971). If W_a and W_{am} are the available and maximum water contents of the soil, then provided $W_a < 0 \cdot 6 \, W_{am}$ it is known that the value of f decreases linearly with decreasing W_a in a root zone.

The following relation was derived to relate the supplementary water requirement, Q_w, for a rice field during the growing season to climatic conditions (Uchijima, 1962b)

$$Q_w = \left[\frac{\mathbf{R}_{nw}}{\lambda P} \cdot \frac{f}{a} \left\{ 1 + \frac{\lambda (Q_p + W_p)}{f \cdot \mathbf{R}_{nw}} \right\} - 1 \right] aP, \tag{21}$$

where $\mathbf{R}_{n,w}/\lambda P$ is the radiation dry index calculated on the basis of net radiation $\mathbf{R}_{n,w}$ over shallow water, P is the rainfall, a the 'effectiveness' of rainfall, Q the percolation, and W the preirrigation for puddling. The supplementary water requirement is proportional to the radiation dry index characterizing the aridity of climate. This relation can be used to evaluate the actual water requirement of large rice fields from climatic data.

Because of the interception of solar radiation and air mixing by plant elements, the transpiring structure of field crops change substantially throughout the growing season (Sato, 1960; Hanyu and Ono, 1960; Uchijima, 1961; Iwakiri, 1965). The relation between evaporation and leaf area index can be expressed as follows:

$$\mathbf{E}_T = \mathbf{E}[1 - \exp(-\mathscr{K}L)], \tag{22}$$

where $\mathbf{E}$ is the total water loss from crop and soil or water, $\mathbf{E}_T$ the transpiration rate, and $\mathscr{K}$ an extinction coefficient. Experiments show that the value of $\mathscr{K}$ is 0·45 to 0·65, nearly equal to the extinction coefficient for net radiation within a rice canopy. A similar change in transpiration with crop development was also observed in a maize field (Tanner and Lemon, 1962; Denmead and Shaw, 1959).

3. *Water Temperature*

The temperature of rice fields and of irrigation water has long been a subject of considerable interest in Japan. Figure 10 shows the difference in temperature between the standing water and the air clearly changed from positive values of 2 to 3°C in the early growing stages to negative and small values in the late growing stage. With the development of the rice canopy the daily range of water temperature in the rice field, ΔT_w, diminished gradually. It was approximated by the exponential relation

$$\Delta T_w = \Delta T_{w,o} \exp(-0·21L),$$

where $\Delta T_{w,o}$ is the daily range of water temperature in open shallow water (Nakayama *et al.*, 1965).

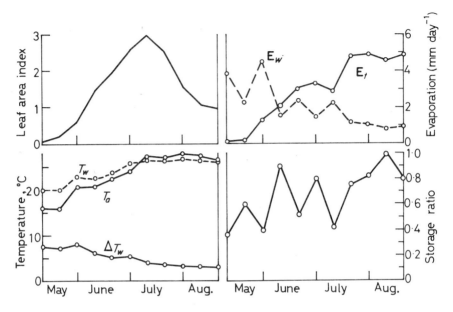

Fig. 10. Seasonal variation of water temperature and heat balance components of a paddy field (Uchijima, 1961).

According to Uchijima (1959), the average temperature of shallow water, T_w, can be estimated for a day, month or other period by the expression

$$T_w = T_a + [(\mathbf{R}_{n,w}/\rho c_p g_a) - 2\delta e]/(1+2\Delta),\qquad(23)$$

where T_a is the air temperature, g_a is the water-to-air exchange (transfer) velocity ($= 1/r_a$), δe is the saturation deficit and Δ is the slope of the curve relating saturation vapour pressure to temperature at T_a. In relation to the problem of preventing damage to the rice crop following the application of cool irrigation water, much research has tried to clarify the mechanism responsible for the warming of irrigation water in specially constructed ponds (Mihara *et al.*, 1959).

B. Turbulent Transfer of Heat and Water Vapour Within Canopies

1. *Sensible and Latent Heat Fluxes*

Since the major site of energy exchange in crop canopies is at the surfaces of leaves, the profiles of sensible and latent heat fluxes are closely related to the leaf distribution in canopies. As Fig. 11 shows, the profiles of vapour

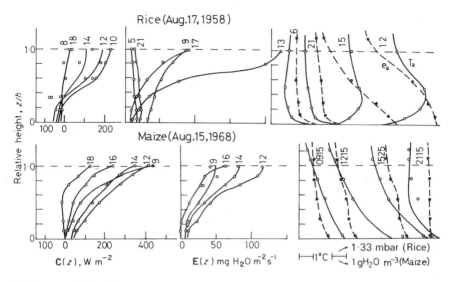

Fig. 11. Vertical profiles of air temperature, humidity, sensible and latent heat fluxes in rice and maize canopies (Franceschini, 1959; Uchijima, 1962a; Groom, 1969).

pressure within rice and maize canopies decreased monotonically throughout the canopy with increasing the height above the ground. Such profiles indicate that the flux of water vapour is upward throughout the day. The temperature profiles in a rice canopy during daylight were characterized by curves with a maximum in a middle layer, but the temperature profiles in a maize canopy were lapse throughout the day. The difference in temperature profile between rice and maize canopies may be accounted for by the very low moisture of the soil on which the maize was grown. Within an immature dense maize canopy, Brown and Covey (1966) observed temperature profiles similar to those obtained in rice (The significance of profile shapes is discussed in Volume 1, p. 77.)

The upward latent heat flux originates mainly in the layer between height $0.5\ h$ and $0.8\ h$. The sensible heat flux in the rice canopy was upward in the upper layer above a level of about $0.4\ h$. In the layer below this level, there was a downward sensible heat flux. The heat transferred downward was used for evaporation from the water surface. In the maize canopy, the sensible heat flux increased monotonically with increasing height above the ground. The decrease of water vapour flux observed at night denotes that part of the water vapour transferred upwards from the bottom layers of the rice canopy condensed on the plant elements.

2. Conductances at Leaf Surfaces

The source or sink intensity of C and λE within plant canopies is given by

$$Q_C = -\rho c_p \frac{d}{dz} K \frac{dT_a}{dz} = \rho c_p a_l g_l (T_l - T_a),$$

$$Q_{\lambda E} = -\lambda \rho \frac{d}{dz} K \frac{dq_a}{dz} = \lambda \rho g_l (q_l - q_a),$$

$$(24)$$

where Q_C and $Q_{\lambda E}$ are the source- or sink-intensity of sensible and latent heat, respectively, T_a and q_a the temperature and specific humidity of the air among plant elements, T_l and q_l the corresponding temperature and specific humidity at leaf surfaces, and g_l the leaf-to-air conductance†. The heat balance method (Uchijima, 1966; Uchijima et al., 1970; Brown and Covey, 1966) and the dry-and-wet-leaf method (Impens et al., 1967) were used to determine the values of g_l assessed by the heat balance method and of the stomatal exchange velocity, $g_s = 1/r_s$, for the maize leaves calculated by the procedure to be described below are presented in a tabular form (Uchijim et al., 1970)

height	250	200	150	100	cm
g_l	2·5–7·0	1·0–4·0	0·8–3·0	0·5–2·0	cm s^{-1}
g_s	0·2–0·6	0·1–0·3		0·1	cm s^{-1}

The stomatal conductance of the maize leaves was calculated by

$$g_s = g_l g_e / (g_l - g_e)$$

where g_e is the effective conductance for transpiration given by

$$g_e = E / \rho (q_w(T_l) - q_a),$$

and $q_w(T_l)$ is the saturation vapour pressure at the leaf temperature. g_s for the top leaves increased curvilinearly with solar radiation (W m^{-2}) and was approximated by

$$g_s = 0·0032 S_l / (1 + 0·0057 S_l).$$

This dependence was due to the stomatal response upon light intensity. The decrease of g_l and g_s with depth into the canopy was approximately an exponential function. The value of extinction coefficient was in the range between 0·5 n and 1·0 n, where n is the extinction coefficient for profiles of K within maize canopy (p. 48).

---

† Sometimes referred to as an 'exchange velocity'.

It was found experimentally that the leaf-to-air conductance is closely associated with the turbulent transfer coefficient in the bulk air (Uchijima, 1966; Uchijima *et al.*, 1970). In spite of the relatively large spread of the results, a tendency is manifest for g_l to be proportional to K. The dependence of g_l on K was approximated by linear relation $g_l = a + bK$. The proportionality constant, b, was between $0.73 . 10^{-3}$ and $1.25 . 10^{-3}$, agreeing with the value of 10^{-3} expected by Philip (1964). The value of a was estimated to be 0.53 cm s^{-1}.

V. CARBON DIOXIDE BALANCE

A. Photosynthetic Behaviour of Leaves

Since photosynthesis by single leaves is the basis for dry matter production, physiological aspects of the photosynthetic behaviour of rice and maize leaves have been studied extensively in recent years.

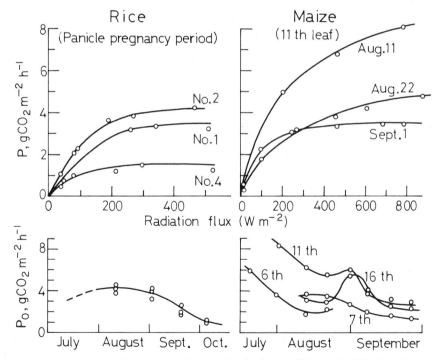

Fig. 12. Photosynthesis–light curves of rice and maize leaves (Akita *et al.*, 1968; Tanaka and Yamaguchi, 1972); developmental phase as defined by Matushima (1966).

1. *Light Response of Photosynthetic Rate*

Figure 12 shows typical light response curves for rice and maize leaves. A fully expanded maize leaf continued to respond to increments of light up to about 750 to 850 W m^{-2}, while the CO_2 uptake rate of a rice leaf reached the plateau at an irradiance of about 300 W m^{-2}.

The rectangular hyperbola $P = bS_t/(1 + aS_t)$ was applied for the light-photosynthesis curve of rice and maize leaves (Murata, 1961; Hesketh and Musgrave, 1962; Chmora, 1966). More recently, Akita *et al.* (1968) showed that the relation $P = cS_t^2/(1 + cS_t^2)$ gave a better fit to measurements on rice. Varietal difference in CO_2 uptake rate were studied for rice (Murata, 1961) and for maize (Heichell and Musgrave, 1969). They showed that the differences range from about 50% in rice to as much as 200% in maize. The value of P_0 denoting the CO_2 uptake rate at the light-saturating point (Fig. 12) was larger for maize than for rice. The mean value of P_0 for rice was about 3 to $4\,g\,CO_2\,m^{-2}\,h^{-1}$, whereas the mean for maize was about 5 to $6\,g\,CO_2\,m^{-2}\,h^{-1}$. Other interesting features of Figure 12 are both the progressive fall in the value of P_0 and the change in the shape of light-photosynthesis curves with the ageing of leaves. Murata (1961) studied the protein nitrogen content in leaves as a likely main cause for the decline of P_0 of rice plants. Chmora (1966) observed a pronounced diurnal change in the light-photosynthesis curve of maize leaves under field conditions. She suggested that this change may be attributed to the variation of chloroplast structure throughout the day.

2. *Temperature Response of Photosynthetic Rate*

Photosynthetic rate and respiration rate of crop plants are affected significantly by leaf or air temperature. Murata (1961) concluded that no pronounced change of CO_2 uptake rate of rice leaves was observed in the temperature range from 20 to 33°C. On either side of this temperature range, however, the CO_2 uptake rate decreased considerably with temperature. No temperature effects were apparent for maize leaves between 24 and 40°C (Hesketh and Musgrave, 1962). The respiration rate of rice leaves, R_l, can be expressed as a function of leaf temperature, T_l, in °C (Munakata, 1969).

$$R_l = a \exp(d \cdot T_l),$$

where the coefficient d was $0.056°C^{-1}$. The value of a ranged from 3·7 $\times 10^{-2}$ to $4.5 \times 10^{-2}\,g\,CO_2\,m^{-2}\,h^{-1}$, depending on the application of nitrogen fertilizer. The CO_2 uptake rate and respiration rate of maize and rice plants are known to increase with increasing nitrogen content of the leaves (Murata, 1961; Tanaka and Yamaguchi, 1972).

B. Carbon Dioxide Balance of Whole Stands

1. *Carbon Dioxide Environment*

In clear weather, the CO_2 content of the air over mature field crops exhibits
a marked diurnal variation with a maximum shortly before the dawn and
a minimum at midday (Inoue *et al.*, 1958; Murata, 1961; Horibe, 1964;
Uchijima *et al.*, 1967; Allen, 1971; Takasu and Kimura, 1971, 1972). On
cloudy days, a slight diurnal variation of CO_2 content was observed over
crops. The diurnal oscillation in CO_2 content was more pronounced when the
canopy was dense, solar radiation was intense, and wind was light (Uchijima
et al., 1967; Uchijima and Inoue, 1970). During daylight, CO_2 content
increases with height above the canopy, rapidly in the lowest layer and more
slowly at greater height. At night, CO_2 content decreases with increasing
height (Karpushkin, 1966). These facts imply that CO_2 flux in the air layer
over crop fields is downward during daylight hours and upward at night.

The clear-weather oscillation of the CO_2 profile within rice and maize
canopies is shown in Fig. 13. Within the maize canopy, the height of the

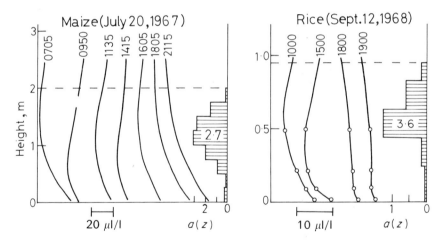

Fig. 13. Vertical profiles of CO_2 concentration in rice and maize canopies (Uchijima *et al.*, 1970;
Ishibashi, 1970).

profile minimum, z_m, moved downward as the forenoon progressed and the
solar elevation increased. After noon, the profile minimum moved upward
(Wright and Lemon, 1966b; Uchijima *et al.*, 1967). Although the level z_m
indicates the characteristic height at which the CO_2 flux direction changes,
it does not correspond to the compensation height.

When photosynthesis was active and the wind was light, the CO_2 content within a maize canopy decreased to a level below about 250 ppm so that the photosynthesis of a good maize crop might be limited by the shortage of CO_2.

2. Carbon Dioxide Flux

The CO_2 flux, F_C, in the surface air layer above crops, roughly corresponding to the photosynthesis or respiration rate of the canopy, is expressed as follows:

$$F_C = g_a(\varphi_1 - \varphi_2), \tag{25}$$

where $g_a = 1/r_a = 1/\int_{z_2}^{z_1} K(z)^{-1} dz$ is the conductance between heights of z_1 and z_2, and φ_1 and φ_2 are the CO_2 concentration of these heights, respectively. Both the logarithmic wind profile and the heat balance method have been utilized to estimate the values of D_a (Inoue et al., 1958, 1968; Lemon, 1960). To avoid laborious data processing and to improve the accuracy of measurements, CO_2 flux-meters have been devised on the basis of the aerodynamic method and eddy correlation techniques and tested under field conditions (Inoue et al., 1969; Ohtaki and Seo, 1972; Yabuki et al., 1972).

The time course of the CO_2 flux at the top of the canopy is in phase with the radiation flux. On a clear windy day, the CO_2 flux at noon reached about $10 \text{ g } CO_2 \text{ m}^{-2} \text{ h}^{-1}$ above a dense maize canopy and about $6 \text{ g } CO_2 \text{ m}^{-2} \text{ h}^{-1}$ above a dense rice canopy. On a calm and sunny day, CO_2 uptake by maize was somewhat diminished due to the low CO_2 content of the air. Lemon (1965) and Yabuki et al. (1972) have reported that the CO_2 uptake rate of both the maize and rice is somewhat affected by wind speed. Model experiments showed that the CO_2 uptake by crop plants increases curvilinearly with wind speed (Uchijima and Inoue, 1970).

Table II
Net fixation of carbon dioxide by maize stand calculated by adding the *downward* flux from the atmosphere ($-F_a$) to the *upward* flux from the soil F_s. Fluxes are quoted in g m^{-2} for the daylight period on each day and the corresponding short wave income is ΣS_t ($MJ \text{ m}^{-2}$). The equivalent efficiency in percent is found from the energy required to fix 1 g of CO_2 or $\mu = 10.6 \text{ kJ/g}$

Date (1966)	ΣS_t	$\Sigma(-F_a)$	ΣF_s	$\Sigma(F_s - F_a)$	$\dfrac{\Sigma F_s}{\Sigma(F_s - F_a)}$	$\dfrac{\mu \Sigma(F_s - F_a)}{\Sigma S_t} \times 100$
July 26	17·1	41·3	8·5	49·8	0·17	3·1
July 27	19·1	48·8	9·2	58·0	0·16	3·3
Aug 8	22·0	57·3	15·6	72·8	0·21	3·5
Aug 9	8·7	54·5	6·4	60·9	0·11	7·4
Aug 19	11·5	60·6	8·2	68·8	0·12	6·3
Aug 20	18·8	70·7	18·3	89·0	0·21	5·0

The carbon dioxide balance of a maize field is summarized in Table II. The ratio of soil CO_2 respiration to the total CO_2 fixation by maize ranges from 10 to 20%, indicating that 80 to 90% of the CO_2 utilized by the plants come from the atmosphere. Carbon dioxide release from the water surface under a rice canopy was measured at rates from 0·02 to 0·2 g CO_2 m^{-2} h^{-1}, depending on the water temperature and the presence of floating weeds.

C. Carbon Dioxide Transfer Within Crop Canopies

Fluxes of CO_2 and CO_2 fixation within a maize canopy were calculated using Eq. (25) and from CO_2 profiles (Wright and Lemon, 1966b; Inoue et al., 1968). Figure 14 shows how rapidly the CO_2 flux decreases within the canopy during daylight, because of assimilation by the upper leaves. Profiles showing the relative contribution of CO_2 fixation at different height intervals to the total net assimilation for the daylight hours clearly demonstrate

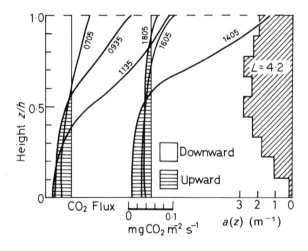

Fig. 14. Profiles of mean CO_2 fluxes, for ten minutes intervals, within a maize canopy (Inoue et al., 1968).

the importance of the upper leaves. The zone from 0·6 h to h accounted for about 70% of the total fixation. However, in a relatively sparse canopy, the contribution of each layer to the total CO_2 fixation was nearly proportional to the ratio of leaf area density to L. Another interesting fact is that the CO_2 assimilated by the leaves in the lowest part of the canopy comes from soil.

D. Dry Matter Production

1. *Canopy Structure and Photosynthesis*

Since Monsi and Saeki's pioneering work (Monsi and Saeki, 1953), considerable attention has been paid to the relationship between canopy structure
and photosynthesis. Tsunoda (1964) and Murata (1961) have pointed out
the importance of canopy structure for increasing photosynthesis and yield.
A number of agricultural practices have been attempted to manipulate the
display of leaves in rice and maize (Tsunoda, 1964; Matsushima, 1966;
Pendleton *et al.*, 1968). This research has led to the 'plant type concept' as
a guide for breeding high yeilding varieties and for establishing better management practices. More recently, Tanaka (1972) demonstrated through
skilful experiments that a rice canopy with leaves held horizontally had a
smaller photosynthesis rate and a plateau-type response to light compared
with an originally erect-leaved rice canopy. The CO_2 uptake rate of the
horizontal-leaved rice canopy at the saturating light intensity was diminished
to about two thirds of that for the erect-leaved rice canopy.

It is well known that most rice varieties bred recently have foliage consisting of more erect and shorter leaves. Simulations with the Nilson (1968),
Kuroiwa (1970), and similar models have revealed that canopies with upper
strata composed of erect leaves becoming more horizontal in display with
increasing depth are more efficient than other leaf arrangements. The simulations also show that the optimum leaf arrangement varies to some extent
with solar elevation and latitude (Tooming, 1967; Nilson, 1968; Kallis, 1969).

2. *Crop Yield and Meteorological Conditions*

Experimental studies showed that below a critical temperature, rice yield
decreased linearly with decreasing mean air temperature over the growing
season. The critical air temperature for rice crops is in the range between
19 and 21°C, depending on varieties and agricultural practices. The critical
water temperature also varies between 22 and 23°C.

Grain yield of rice and maize is strongly affected by the dry matter production for the ripening period (Murata, 1961; Tanaka and Yamagushi, 1972).
On the basis of these results, Murata (1964), Hanyu *et al.* (1966), and Munakata
et al. (1967) have used the solar radiation or sunshine duration and air
temperature for the ripening period (40 days after the heading) as determinants
of rice yield. The relation proposed by Hanyu *et al.* for estimating the rice
productivity, Y, is as follows:

$$Y = D_{sun}\{4\cdot14 - 0\cdot13(21\cdot4 - \overline{T}_a)^2\}. \tag{26}$$

where D_{sun} and $\overline{T}_a$ are the sum of sunshine duration and the mean air temperature for the ripening period, respectively. This analysis shows that the rice yield increases with increasing total of solar radiation during this period. Yield increases fast curvilinearly with mean air temperature, reaches a maximum at a mean air temperature of about 22°C, and then decreases again.

Grain yield of maize appears to be affected not only by air temperature and solar radiation but also by the soil moisture available in the root zone (Baker and Musgrave, 1964; Chirkov, 1968). Chirkov has discussed an agroclimatic method for predicting maize grain yield from climatic data. Dry matter productivity of maize was given by a quadratic function of the mean air temperature for the growing season. The maximum productivity varied linearly with the water available in the root zone. The agroclimatic methods described here could be used to evaluate the productivity of maize or rice crop in given conditions.

VI. CONCLUSIONS

It is apparent from this review of recent investigations on the micrometeorology of maize and rice fields that the introduction of the aerodynamic and the energy balance methods to agricultural meteorology has helped to elucidate the interaction between these crops and their physical environment.

Salient results obtained in the past two decades can be summarized as follows. The energy balance approach has enabled the water consumption by crops to be determined more accurately from climatic data. Information about water consumption has been widely used for planning the supplementary irrigation for field crops and for planning the rational use of water resource. The aerodynamic determination of CO_2 flux above and within crop canopies has made it possible to clarify CO_2 transfer processes in relation to canopy structure. Research on the leaf arrangement and the radiation regime in the canopies has quantitatively revealed that an erect-leaved canopy has a higher photosynthetic activity and can produce a higher grain yield, providing that the concept of plant type is valid as a selection criterion for breeding high yielding varieties of rice and maize.

Although the micrometeorology of rice and maize fields has made many contributions to solving problems in crop production, our knowledge is still too fragmentary to solve complicated relationships of soil-plant-atmosphere systems and much remains to be learned. Synthesizing available fragments of knowledge concerning the micrometeorological processes would greatly increase man's knowledge of the processes involved in forecasting and controlling the crop production.

REFERENCES

Akita, S., Murata, Y. and Miyasaka, A. (1968). *Proc. Crop Sci. Soc. Japan* **37**, 680–684.
Allen, L. H. (1971). *Agric. Met.* **8**, 5–24.
Allen, L. H. and Brown, K. W. (1965). *Agron. J.* **57**, 611–612.
Allen, L. H., Yocum, S. S. and Lemon, E. R. (1964). *Agron. J.* **56**, 253–259.
Baker, D. N. and Musgrave, R. B. (1964). *Crop Sci.* **4**, 249–253.
Brown, K. W. and Covey, W. (1966). *Agric. Met.* **3**, 73–96.
Chirkov, Ju. I. (1969). 'Agroclimatic Conditions and Productivity of Maize Crop', 251 pp. Gidrometeorol. Izd.-vo, Leningrad.
Chmora, S. T. (1966). *In* 'Photosynthetic Systems with High Productivity'. (A. A. Nichiporovich, ed.), pp. 142–148, Izd-vo Nauka, Moscow.
Cionco, R. M. (1972). *Boundary-Layer Meteorol.* **2**, 466–475.
Denmead, O. T. (1971). *In* 'Plant Photosynthetic Production', (A. Sestak, J. Catsky, P. G. Jarvis, eds), pp. 467–516, W. Junk Publishers, Hague.
Denmead, O. T., Fritschen, L. J. and Shaw, R. H. (1962). *Agron. J.* **54**, 505–510.
Denmead, O. T. and Shaw, R. H. (1959). *Agron. J.* **51**, 725–729.
de Wit, C. T. (1965). *Agr. Res. Rep.* (*Wageningen*) **663**, 1–57.
Franceschini, G. A. (1959). Micrometeorological observations over non ideal surfaces. A data report. AFCRC-TR-N-60-254.
Fritschen, L. J. and Shaw, R. H. (1961a). *Agron. J.* **53**, 71–74.
Fritschen, L. J. and Shaw, R. H. (1961b). *Agron. J.* **53**, 149–150.
Gilliespie, T. J. (1971). *Agric. Met.* **8**, 51–58.
Gilliespie, T. J. and King, K. M. (1971). *Agric. Met.* **8**, 59–68.
Groom, M. (1969). *Ann. Rep. Micromet. Inves.* (*Cornell Univ.*) 74–85.
Hanyu, J. and Ono, K. (1960). *J. agric. Met.* (*Tokyo*) **16**, 111–118.
Hanyu, J., Uchijima, T. and Sugawara, S. (1966). *Bull. Tohoku agr. Exp. St.* Part-1 **34**, 27–36.
Hayashi, K. (1972). *Bull. Nat. Inst. Agric. Sci.* D **23**, 1–67.
Heichell, G. H. and Musgrave, R. B. (1969). *Crop Sci.* **9**, 483–486.
Hesketh, J. D. and Musgrave, R. B. (1962). *Crop Sci.* **2**, 311–315.
Hitaka, N. (1968). *Bull. Nat. Inst. Agric. Sci.* A **15**, 1–175.
Horibe, Y. (1964). *J. agric. Met.* (*Tokyo*) **19**, 155–156.
Horie, T. and Udagawa, T. (1971). *Bull. Nat. Inst. Agric. Sci.* A **18**, 1–56.
Impens, I. and Lemeur, R. (1969a). *Arch. Met. Geophysk. Bioklimt.* Ser. B **17**, 261–268.
Impens, I. and Lemeur, R. (1969b). *Arch. Met. Geophysk. Bioklimt.* Ser. B **17**, 403–412.
Impens, I. I., Stewart, D. W., Allen, L. H. and Lemon, E. R. (1967). *Pl. Physiol.* **42**, 99–104.
Inoue, E. (1960). *J. agric. Met.* (*Tokyo*) **16**, 83–84.
Inoue, E. (1963). *J. met. Soc. Japan* **41**, 317–326.
Inoue, E., Tani, N., Imai, K. and Isobe, S. (1958). *J. agric. Met.* (*Tokyo*) **13**, 121–125.
Inoue, E., Uchijima, Z., Saito, T., Isobe, S. and Uemura, K. (1969). *J. agric. Met.* (*Tokyo*) **25**, 165–171.
Inoue, E., Uchijima, Z., Udagawa, T., Horie, T. and Kobayashi, K. (1968). *J. agric. Met.* (*Tokyo*) **23**, 165–176.
Ishibashi, A. (1970). *Ann.* (1969) *Rep. Jap. Ctee, IBP/PP.*, 27–29.
Isobe, S. (1962). *Bull. Nat. Inst. Agric. Sci.* A **9**, 29–67.
Isobe, S. (1964). *Bull. Nat. Inst. Agric. Sci.* A **11**, 1–18.

Isobe, S. (1969). *Bull. Nat. Inst. Agric. Sci.* A **16**, 1–25.

Isobe, S. (1972). *Bull. Nat. Inst. Agric. Sci.* A **19**, 101–113.

Ito, A. (1969). *Proc. Crop Sci. Japan*, **38**, 355–363.

Ito, A., Udagawa, T. and Uchijima, Z. (1973). *Proc. Crop Sci. Japan* **42**, 334–342.

Iwakiri, S. (1964). *J. agric. Met. (Tokyo)* **19**, 89–95.

Iwakiri, S. (1965). *J. agric. Met.* **21**, 15–21.

Kallis, A. (1969). *In* 'Problems of Photosynthetic Efficiency', pp. 44–63, Tartu.

Karpushkin, L. T. (1966). *In* 'Photosynthetic Systems with High Productivity' (A. A. Nichiporovich, ed.), pp. 149–156, Izd-vo Nauka, Moscow.

Kishida, Y. (1971). *In* 'Influence of Climatic Factors on Rice Yield: Studies of Upper Limit of Rice Yield', pp. 239–244, Ministry of Agriculture and Forestry, Japan.

Kondrat'ev, K. Ja. (1969). 'Radiation in the Atmosphere', 912 pp. Academic Press, New York and London.

Kuroiwa, S. (1970). *Proc. IBP/PP Tech. Meet., Trebon*, 79–89.

Lourence, F. J. and Pruitt, W. O. (1971). *Agron. J.* **63**, 827–832.

Lemon, E. R. (1960). *Agron. J.* **52**, 697–703.

Lemon, E. R. (1965). *In* 'Plant Physiology', 4-A 2 (F. C. Stewart, ed.), pp. 203–277. Academic Press, New York and London.

Loomis, R. S., Williams, W. A., Duncan, W. G., Dovrat, A. and Funez, F. A. (1968). *Crop Sci.* **8**, 352–356.

Maki, T. (1969). *J. agric. Met. (Tokyo)* **25**, 13–18.

Matsushima, S. (1966). 'Crop Science in Rice', 365 pp. Fuji Publishing Co., Ltd., Tokyo.

Mihara, Y., Uchijima, Z., Nakamura, S. and Onuma, K. (1959). *Bull. Nat. Inst. Agric. Sci.* A **7**, 1–43.

Monsi, M. and Saeki, T. (1953). *Jap. J. Bot.* **14**, 22–52.

Munakata, K., Kawasaki, I. and Kariya, K. (1967). *Bull. Chogoku Agric. Exp. St.* A **14**, 59–96.

Murata, Y. (1961). *Bull. Nat. Inst. Agric. Sci.* D **9**, 1–169.

Murata, Y. (1964). *Proc. Crop Sci. Soc. Japan* **35**, 59–63.

Murata, Y., Miyasaka, A., Munakata, K. and Akita, S. (1968). *Proc. Crop Sci. Soc. Japan* **37**, 685–691.

Nakagawa, Y. (1956). *J. agric. Met. (Tokyo)* **12**, 61–63.

Nakayama, K., Ichimura, K., Yamamoto, Y. and Matsuda, M. (1965). *J. Agric. Lab.* No. 6, 1–34.

Nilson, T. (1968). *In* 'Radiation Regime Within Plant Canopies', pp. 112–146, Tartu.

Ohtaki, E. and Seo, T. (1972). *Ber. Ohara Insit. Landwirtshaft. Biol.* **15**, 3, 89–110.

Pavlov, A. V. and Ustenko, G. P. (1965). *Izv. A N USSR, Ser. Geog.* No. 6, 47–55.

Pendleton, J. W., Smith, G. E., Winter, S. R. and Johnston, T. J. (1968). *Agron. J.* **60**, 422–425.

Philip, J. (1964). *J. appl. Met.* **3**, 390–395.

Research group of evapotranspiration (RGE). (1967a). *J. agric. Met. (Tokyo)* **22**, 97–102.

Research group of evapotranspiration (RGE). (1967b). *J. agric. Met. (Tokyo)* **22**, 149–157.

Ross, Ju. K. (1964). *In* 'Actinometry and Atmospheric Optics', pp. 251–256, Izd-vo Nauka, Moscow.

Ross, Ju. K. and Nilson, T. (1965). *In* 'Problems of Radiation Regime Within Plant Canopies, pp. 25–64, Izd-vo, Valgus, Tallin.

Ross, Ju. K. and Nilson, T. (1966). *In* 'Photosynthetic Systems with High Productivity', (A. A. Nichiporovich, ed.), pp. 109–125, Izd-vo Nauka, Moscow.

Ross, Ju. K. and Nilson, T. (1968). *In* 'Regime of Solar Radiation Within Plant Canopies', pp. 5–54, Acad. Sci. ESSR, Tartu.

Saito, T. (1964). *Bull. Nat. Inst. Agric. Sci.* A **11**, 67–74.

Saito, T., Inoue, E., Isobe, S. and Horibe, Y. (1962). *J. agric. Met. (Tokyo)* **18**, 11–18.

Saito, T., Isobe, S., Nagai, R. and Horibe, Y. (1970). *J. agric. Met. (Tokyo)* **26**, 177–180.

Sato, S. (1960). *Bull. Kyushu Agric. Exp. St.* **6**, 133–364.

Seo, T. and Yamaguch, N. (1968). *Ber. Ohara Inst. Landwirtshaft. Biol.* **14**, 3, 133–143.

Takasu, K. and Kimura, K. (1970). *Nogaku Kenkyu* **53**, 167–179.

Takasu, K. and Kimura, K. (1971). *Nogaku Kenkyu* **53**, 205–213.

Tanaka, T. (1972). *Bull. Nat. Inst. Agric. Sci.* A **19**, 1–100.

Tanaka, A. and Yamaguchi, J. (1972). *J. Facul. Agric. Hokkaido Univ.* **57**, 71–132.

Tani, N. (1963). *Bull. Nat. Inst. Agric. Sci.* A **10**, 1–99.

Tani, N. and Suzuki, Y. (1966). *Ann. Rep. Kyushu Agric. Exp. St.* 40–47.

Tanner, C. B. and Lemon, E. R. (1962). *Agron. J.* **54**, 207–212.

Tooming, Kh. G. (1966). *In* 'Photosynthetic Systems with High Productivity' (A. A. Nichiporovich, ed.), pp. 126–141, Izd-vo Nauka, Moscow.

Tooming, Kh. G. (1967). *J. Bot. (Leningrad)* **52**, 601–616.

Tooming, Kh. G. and Guljaev, B. I. (1967). 'Methods for Measuring PAR'. 143 pp. Izd-vo Nauka, Moscow.

Tooming, Kh. G. and Ross, Ju. K. (1965). *In* 'Problems of Radiation Regime Within Plant Canopies', pp. 65–72, Izd-vo Valgas, Tallin.

Tsunoda, S. (1964). 'A Developmental Analysis of Yielding Ability in Varieties of Field Crops'. 135 pp. Nihon-Gakujitsu-Shinkokai, Maruzen.

Uchijima, Z. (1959). *Bull. Nat. Inst. Agric. Sci.* A **7**, 131–181.

Uchijima, Z. (1961). *Bull. Nat. Inst. Agric. Sci.* A **8**, 243–265.

Uchijima, Z. (1962a). *J. agric. Met. (Tokyo)* **18**, 1–10.

Uchijima, Z. (1969b). *Bull. Nat. Inst. Agric. Sci.* A **9**, 1–28.

Uchijima, Z. (1966). *Bull. Nat. Inst. Agric. Sci.* A **13**, 81–93.

Uchijima, Z. and Inoue, K. (1970). *J. agric. Met. (Tokyo)* **25**, 5–18.

Uchijima, Z. and Wright, J. L. (1964). *Bull. Nat. Inst. Agric. Sci.* A **11**, 19–66.

Uchijima, Z., Udagawa, T., Horie, T. and Kobayashi, K. (1967). *J. agric. Met. (Tokyo)* **23**, 99–108.

Uchijima, Z., Udagawa, T., Horie, T. and Kobayashi, K. (1970). *J. agric. Met (Tokyo)* **25**, 215–227.

Udagawa, T., Uchijima, Z., Horie, T. and Kobayashi, K. (1968). *Proc. Crop Sci. Japan* **37**, 589–596.

Udagawa, T., Ito, A. and Uchijima, Z. (1974). *Proc. Crop. Sci. Soc. Japan* **43**, 180–195.

Williams, W. A., Loomis, R. S., Duncan, W. G., Dovrat, A. and Nunez, F. A. (1968). *Crop Sci.* **8**, 303–308.

Wright, J. L. and Brown, K. W. (1967). *Agron. J.* **59**, 427–432.

Wright, J. L. and Lemon, E. R. (1966a). *Agron. J.* **58**, 255–261.

Wright, J. L. and Lemon, E. R. (1966b). *Agron. J.* **58**, 265–268.

Yabuki, K. (1957). *Bull. Univ. Osaka Pref.* B **7**, 113–146.

Yabuki, K., Aoki, M. and Hamotani, K. (1972). *Ann. (1971) Rep. Jap. Ctee, IBP/PP.* 7–9.

Yocum, C. S., Allen, L. H. and Lemon, E. R. (1964). *Agron. J.* **56**, 249–253.

3. Sugar Beet and Potatoes

K. W. BROWN

Department of Soil and Crop Sciences, Texas A and M University, College Station, Texas, U.S.A.

I. INTRODUCTION

The micrometeorological and physiological characteristics of sugar beet and potatoes have been investigated by many researchers throughout the world. Most investigations lack the complete set of measurements needed to calculate fluxes of momentum, sensible and latent heat and CO_2, and their dependence on soil- and crop-water relations. There are sufficient observations on these two crops, however, to attempt a complete characterization through a synthesis of the available information.

Sugar beet and potato crops have several common characteristics which may be compared. Both crops canopies are typically 50 to 60 cm tall. Provided the row spacing is 60 cm or closer, the foliage will generally form a complete soil cover by mid-season. Both crops display their leaves at an entire range of angles and have been reported to develop leaf area indexes of six or greater. These factors may lead to similar radiation and aerodynamic properties. One difficulty common to these crops is that pure strains are not generally used in the field. In a sugar beet field, differences in leaf thickness, colour and stomatal densities between plants are common and easily detected. The problem is even more serious in a potato field, where each plant may have a different genetic background resulting in differences in leaf and plant size, number of stomates per unit area and other factors which may influence plant-water relations. Hopefully, however, plants with different characteristics will be located randomly throughout any particular field so that micro-meteorological measurements will be representative and not influenced by small groups of plants which differ from the remainder of the crop. Both crops continue producing new leaves throughout the season.

II. MICRO-CLIMATE

Broadbent (1950) was among the early researchers who investigated the micro-climate of a potato crop. He reported maximum air temperature differences within an irrigated and a non-irrigated crop to be 1·5°C. He also reported some of the earliest profiles of air temperature and vapour pressure within the crop. Since his work, profiles of air temperature, vapour pressure and wind speed have been measured in and above both sugar beet and potato fields. Typical temperature profiles during the day above a sugar beet field at Scottsbluff, Nebraska, U.S.A. are given in Fig. 1A (Brown and Rosenberg, 1971/2). From 1000 to 1300 hours, the profiles were lapse indicating an upward flux of sensible heat. Between 1300 and 1500 hours, the profile passed through a neutral state. At 1500 and 1600 hours, it was inverted indicating a downward flux of sensible heat. These results are typical of records on a clear day with slow wind speeds during the midsummer, in an irrigated valley surrounded by extensive dry area. Figure 1B shows accompanying vapour pressure profiles. The gradients increase throughout the day and are steepest when the temperature profile is inverted. The maximum water vapour pressure gradient on this day was 0·064 mbar cm^{-1}.

Profiles of air temperature, vapour pressure and wind speed in and above a potato crop at Harpenden, Herts, England, reported by Long and Penman (1963) are given in Fig. 2. Here the temperature profiles immediately above the crop are lapse during the day and reveal an inversion only at night. Within the crop, a warm layer develops between 0400 and 0800 hours and becomes more pronounced between 1200 and 1600 hours. The layer is as much as 0·5 C warmer than the air either above or below, indicating both an upward and a downward flux of sensible heat within the crop. This distribution might be attributed to radiation penetrating the partial crop

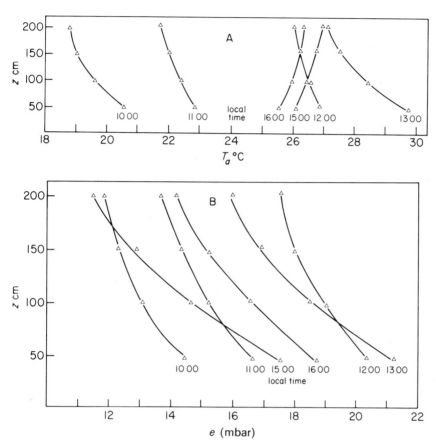

Fig. 1. (A) Air temperature and (B) vapour pressure profiles above a 60-cm tall sugar beet crop at specified times. After Brown and Rosenberg (1971/2).

cover, but similar source-sink relationships have been found within other crops. The profiles of vapour pressure are lapse at all times except for 2000 to 2400 hours, indicating the formation of dew associated with water vapour transport from both above and below the top of the crop. The maximum vapour pressure gradient above this crop was $0.02 \, \text{mbar cm}^{-1}$ or about one-third of the maximum above the sugar beet crop reported above. These differences are most likely attributable to differences in evaporative demand at the two locations, rather than to differences between the crops.

Normalized wind speed profiles are similar for the two crops. The small differences observed between the profiles may be attributed to the influence of wind direction with respect to the rows.

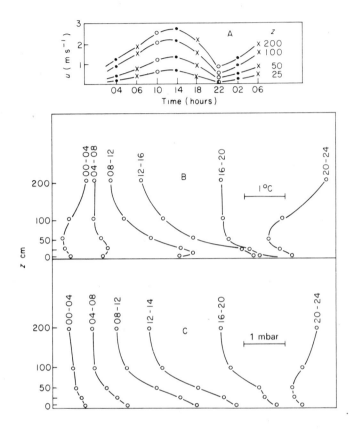

Fig. 2. Mean profiles of wind speed (A), air temperature (B) and vapour pressure (C) profiles in and above a 60-cm tall potato crop at specified times. After Long and Penman (1963).

Diurnal patterns of total incident solar radiation and net radiation above a sugar beet crop for two clear days are shown in Fig. 3. Baumgartner (1970) reported that the short-wave reflection coefficient for both crops varies from 15% to 26%. Szeicz *et al.* (1969) summarized radiation and heat balance components for potato crops grown in California and in southern England. For both locations, they found that percent reflection increases to a maximum as the crop cover develops, and then declines. During the period of maximum leaf area, the average reflection coefficient for both fields was 21%. The soil heat flux was between 5% and 22% (average 12%) of the net radiation above the crop. Changes in the fraction of reflected solar radiation of sugar beet fields have been observed visually during and immediately after furrow irrigation of part of a particular field. This change appears to be associated with a change in leaf angle. The optical properties of leaves may also be altered by changes in leaf-water potential.

III. MOMENTUM BALANCE

Wind speed profiles above a field of sugar beet were used by Monteith and Szeicz (1960) to calculate the turbulent diffusion coefficient above the crop. The average zero plane displacement, determined graphically, was 42·3 cm for their 60-cm tall crop. No values of roughness length were given. More recently, Szeicz et al. (1969) showed that the displacement height of a 60-cm potato crop is 21 cm and independent of wind speed. The roughness length of the same crop varies from 9 cm at a wind speed of 0·5 m s^{-1} to 4 cm at wind speeds of 3 m s^{-1} and greater.

Turbulent diffusion coefficients have also been calculated for the air above sugar beet fields by utilizing vapour pressure profiles and evaporation measured with a lysimeter (Monteith and Szeicz, 1960), and by using measured radiation components, air temperature and vapour pressure profiles in the energy balance (Monteith and Szeicz, 1960; Brown and Rosenberg, 1971a). Reasonable agreement has been found between the two methods, verifying the assumption of equality of turbulent diffusion coefficients for sensible and latent heat. Transfer coefficients above sugar beet crops have generally been linearly related to wind speed. Values for two selected days are given in Fig. 3. The correlation with wind speed can readily be seen by comparing the values before and after a frontal passage at 1335 hours. Whereas turbulent exchange coefficients immediately above this crop were typically 450 cm^2 s^{-1}, values as high as 2,200 cm^2 s^{-1} were observed.

For convenience of comparison with crop resistances, the vertical transport properties of the air may also be characterized by resistance to transfer in the atmosphere (r_a) (Vol. 1, Ch. 3). The dependence of r_a on wind speed is shown in Fig. 4. The curve attributed to Rijtema (1970) is from a table of resistances for a range of crops and wind speeds and is determined from his table based on crop heights alone. Szeicz et al. (1969) calculated values from wind speed profiles while Brown and Rosenberg (1973) made estimates from surface-air temperature differences and the sensible heat flux determined from the energy balance. Although the differences between the curves will alter the energy balance of the crop surface significantly, the differences are within the experimental error and should probably not be attributed to different aerodynamic properties of the two species.

No relevant measurements are available but it might be expected that the boundary-layer resistance of individual sugar beet leaves would be greater that that of potato leaves simply because of the gross difference in size. The hairs on the potato leaves may offset this difference however, at least partially.

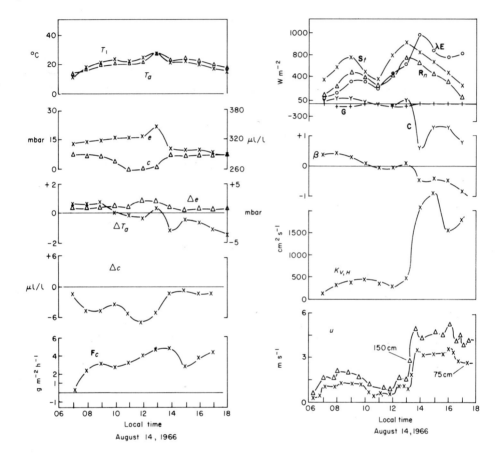

Local time
August 14, 1966

Local time
August 14, 1966

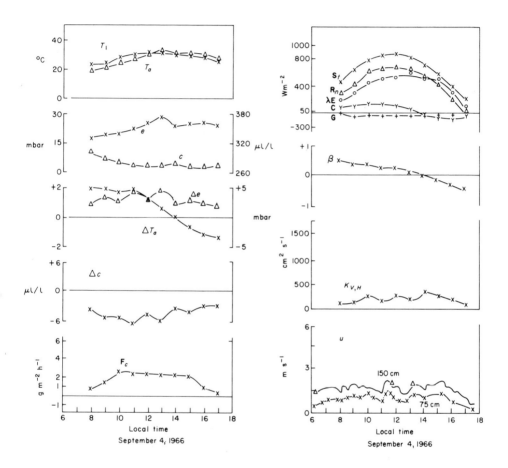

Fig. 3. Diurnal patterns of air and leaf temperature, vapour pressure and CO_2 concentration, air temperature, vapour pressure and CO_2 gradients per 25 cm, CO_2 flux, total solar irradiance, net radiation, soil latent and sensible heat flux, Bowen ratio, turbulent diffusion coefficient for latent and sensible heat, and wind speed above a sugar beet crop on two different days. After Brown and Rosenberg (1971a).

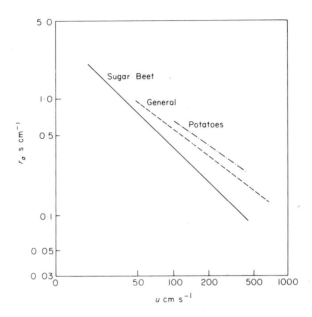

Fig. 4. Calculated resistance to transfer of energy and mass in the atmosphere above 60-cm tall crops of potatoes (after Szeicz *et al.*, 1969) sugar beet (after Brown and Rosenberg, 1973) and a generalized curve (after Rijtema, 1970), wind speed reference at 100 cm above the crop.

IV. HEAT AND WATER BALANCE

A. Stomatal Behaviour

Both plants have stomata on both surfaces of the leaf. Burrows (1969) reported that the ratios of stomatal numbers on the upper and lower surfaces are about 5:8 and 1:3 for sugar beet and potato leaves respectively. While the thicknesses of the guard cells are similar, sugar beet stomata are 20% longer and may open to be 40% wider than potato stomata. From this fact alone, the stomatal resistance per unit leaf area for potatoes would be expected to be larger than for sugar beet exposed to a similar environment. This prediction is borne out by measurements made on field grown plants by Burrows (1969) using both the Alvim type pressure-differences porometer and with the solution-infiltration technique.

Other measurements of stomatal resistance of sugar beet have been reported by Brown and Rosenberg (1970a,b) and for potatoes by Epstein and Grant (1973). Brown and Rosenberg (1970a,b), calculated the resistances from measurements of stomatal densities while Epstein and Grant (1973) used a diffusion type porometer. A good relationship between the two methods was reported in the former reference. Typical minimum resistances reported by both groups of researchers are in the range of 1 s cm^{-1} for both crops, a result somewhat inconsistent with the above discussion of stomatal size. There may still, however, be significant differences between varieties or strains, as borne out by a report of minimum resistance for potatoes of 7 s cm^{-1} reported by Stegman and Nelson (1973).

Diurnal patterns of stomatal diffusion resistance of sugar beet for two clear days are shown in Fig. 5. On August 15, the soil moisture potential was higher

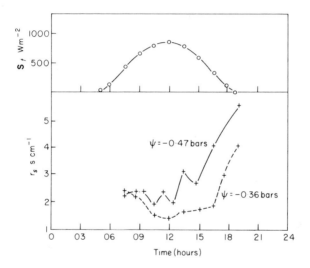

Fig. 5. Total solar irradiance and stomatal resistance of sugar beet leaves for two clear days when the soil-water potential was -0.36 bar and -0.47 bar. Mean air temperatures on both days were 22 and 19°C and mean vapour pressures were 17 and 18 mbar respectively.

and the atmospheric demand was less than on September 4. As discussed earlier for sugar beet, leaf ages were similar for those two dates and the difference could thus be attributed directly to different water relations. Differences in r_s between the 'dry' day (September 4) and 'wet' day (August 15) were small during the morning. By midday, however, the resistance was greater on the dry day and throughout the afternoon the resistance was double or more that of the wet day. No midday stomatal closure occurred, although oscillations from hour to hour were observed on some days. These may be the result of adjustments of the stomatal resistance to compensate for differences between water supply and loss.

The opening and closing of sugar beet stomata corresponds closely with sunrise and sunset. Epstein (personal communications) has observed that stomata of potato leaves reopen two to three hours after sunset. The ecological significance of this behaviour is not yet clear.

Brown and Rosenberg (1973) found that canopy resistance calculated from stomatal resistance on individual leaves was well correlated with the resistance calculated from crop surface temperatures and latent heat flux from the crop. This observation lends support to the proposition that crop resistance is of a purely physiological origin, and is independent of other factors (see Chapter 4).

B. Plant-Water Relations

Fragmentary studies of the water relations of both crops have been reported by various researchers. The influence of soil-water potential on stomatal resistance of sugar beet has been studied by Brown and Rosenberg (1971a). They found that the mean daily stomatal resistance increased from 1·5 cm s^{-1} to 1·9 cm s^{-1} when the water potential decreased from -0.35 to -0.5 bars. Epstein and Grant (1973) reported that when the soil-water potential dropped below -0.25 bars, a potato crop exhibited water stress. Stegman and Nelson (1973) reported that transpiration began to decrease when the relative water content of potato leaves decreased to 80%. Epstein and Grant (1973) found that similar decreases occurred at a relative water content of 88%. These reports show that both crops are sensitive to decreases in soil-water supply. Only occasionally is the atmospheric demand for water strong

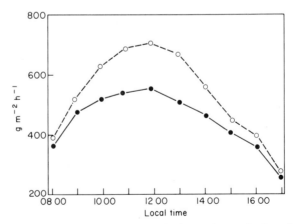

Fig. 6. Diurnal patterns of potential evapotranspiration and transpiration of a potato crop. After Shephard (1972).

enough to override the influence of decreasing soil-water potential on the stomatal response. Diurnal patterns of evaporation for potatoes have been reported by Shephard (1972) (Fig. 6). He concluded that on days of high potential demand the evaporation fell below the potential rate by about 20%. This reduction occurred, however, only on a few days and resulted in a seasonal decrease of only a few percent. Apparently short periods of water stress throughout the season do not affect the final potato yield (Stegman and Nelson, 1973).

Mean monthly values of r_a and r_c for potatoes have been reported by Szeicz et al. (1969) from measurements in southern England and central California and for sugar beet by Brown and Rosenberg (1973) in Nebraska. In units of s cm^{-1}, the values are:

	Potatoes		Sugar beet
	California	England	Nebraska
r_a	0·24	0·45	0·25
r_c	0·90	0·43	0·23

Apparently the wind speeds were lower for the English potato crop than at the other two locations. The potatoes in California may have suffered more soil-water stress than the other two crops. In this case, the influence of the crop resistance on the water balance was greater than that of the air resistance. In the case of the other two crops, the average influence of each resistance was about equal.

Both sugar beet and potatoes can store water in their underground parts. Johnson and Davis (1973) showed that individual sugar beet generally decrease in diameter during the day, and increase their diameter during the night. Once the beet have reached a good size, stored water may be able to supply some of the evaporative demand during the day. There is evidence that water relations of some crops are adversely affected after irrigation. Brown and Rosenberg (1970a) reported that stomatal resistance was greater on the day after an irrigation than would be expected from the soil water potential, thus indicating a decrease in transpiration rate. It is not uncommon to find sugar beet wilting in the field during the first two or three days after irrigation. Johnson and Davis (1973) reported that the diameters of sugar beet roots did not usually begin to increase until the third night after an irrigation. Apparently the lack of oxygen in the soil inhibits the ability of the plant to take up water at these times.

By combining several assumptions, Burrows (1969) was able to develop a curve of the ratio of evaporation to potential evaporation, as given in Fig. 7 for the two crops grown on the same soil. He found that the rate for potatoes fell below the potential rate at water deficits even smaller than those which affected sugar beet indicating that potatoes may be even more sensitive to water deficit than sugar beet.

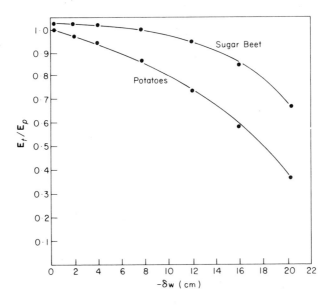

Fig. 7. The relationship between relative evapotranspiration rate and the soil-water deficit. After Burrows (1969).

C. Energy Balance

Profiles similar to those shown in Fig. 1 have been combined with measurements of net radiation R_n and soil heat flux G to evaluate the turbulent transport above sugar beet crops (Brown and Rosenberg 1971a). To check that the measurements were taken within the boundary layer of the crop, the ratio of sensible to latent heat flux was evaluated at height intervals of 50 to 75, 75 to 100, and 100 to 150 cm above the soil. Although the results show some scatter, it was apparent that during portions of certain days the Bowen ratios were identical for all layers and that all measurements were made within the crop boundary layer. On other somewhat windier days, changes in the Bowen ratio with height indicate that advection occurred.

All results given here are therefore based on measurements made within a layer of air not more than 25 cm above the top of the sugar beet canopy corresponding to a minimum fetch of 250 times the height of measurement. Despite this precaution, some measurements do not appear to be representative of the crop surface.

Components of the energy balance and relevant climatic records for one month during 1966 are given in Table I. Relative errors in measurement of energy balance components are greatest during the night. Records are therefore presented only for the period from 0600 to 1800 hours, which closely corresponds to the period of positive net radiation. Most of the days studied were almost cloudless. Only on September 1 did clouds persist for most of the day. Soil heat flux was small throughout the month because the ground was completely shaded by the crop.

Nocturnal energy budgets were less complete. Net radiation and latent heat flux λE determined for several clear nights were therefore averaged and added to the day-time values in order to provide 24 hour estimates of the ratio $\lambda E/(R_n - G)$. Although the fluxes of R_n and λE are much smaller at night than during the day, their omission results in significant underestimation of this ratio.

During most of the days investigated, $\lambda E/(R_n - G)$ did not differ greatly from unity. During such days, winds were light or moderate from a direction providing a fetch of several kilometers over irrigated land. The vapour pressure on these days at crop height was typically near 20 mbar. When strong, dry (vapour pressure of 10 to 14 mbar), southerly winds reached the field after passage over a slightly smaller fetch of irrigated land, they resulted in increased evaporation. On August 29, for example, $\lambda E/(R_n - G)$ was 1·29. Even greater ratios were observed on occasions when northerly winds with a fetch over irrigated fields of only 250 m passed over the field. When northerly winds were light and the vapour pressure about 20 mbar, as on September 5, the ratio $\lambda E/(R_n - G)$ was 1·11.

The range of soil moisture potential was small during this period and did not appear to influence the flux of latent heat. On September 1, a cloudy day, $\lambda E/(R_n - G)$ was 0·86. An even lower ratio was observed, however, on September 2, a nearly clear day.

Detailed energy balance data for selected days are shown in Figure 3. Although the daily average flux of sensible heat from the air was only occasionally negative, the flux was downwards during part of each day. The Bowen ratio often became negative at 1300 or 1400 hours indicating the influence of the advected sensible heat generated outside the experimental field. The influence of water vapour pressure and air temperature on λE can often be seen. When radiation and wind speed were nearly identical,

Table I

Average day-time energy balance (0600 to 1800 hours), climatic conditions, soil-water potential and CO_2 flux data above a sugar beet crop

Data	S_t	R_n	G	C	λE	$\lambda E/(R_n - G)$ (12 h)	$\lambda E/(R_n - G)$ (24 h)	Temperature[a]	Vapour pressure[a]	Wind speed[b]	Predominant wind direction	Soil-water potential	CO_2 flux
	MJ m^{-2} per 12 h							°C	mbar	m s^{-1}		bar	g m^{-2} per 12 h
Aug. 11	26·6	17·5	0·5	+2·4	14·5	0·86	0·97	22·4	21·3	3·1	SE	−0·35	
13	27·2	19·1	0·3	+2·8	16·0	0·85	0·95	15·0	13·8	2·9	SE	−0·40	
14	22·9	14·2	0·3	−6·0	19·9	1·43	1·64	19·5	12·3	2·6	N	−0·43	40·0
15	26·9	18·0	0·6	+2·7	14·7	0·85	0·95	21·5	16·7	1·1	SE	−0·45	
17	26·8	18·6	0·3	−1·7	20·0	1·09	1·21	26·8	18·2	3·5	N	−0·52	34·9
18	24·3	16·6	0·3	+0·8	15·5	0·96	1·07	24·6	20·3	3·3	S	−0·13	22·8
25	27·5	17·3	0·6	+0·7	16·1	0·96	1·07	23·1	20·6	1·5	N and S	−0·22	
27	23·4	14·2	0·4	+1·5	12·3	0·89	1·03	28·4	23·8	1·9	NW	−0·26	24·5
28	25·6	17·0	0·5	+2·2	14·3	0·86	0·98	26·6	24·4	1·5	S	−0·28	22·3
29	25·5	16·4	0·4	−2·4	18·4	1·15	1·29	22·2	12·0	2·8	S	−0·30	31·4
Sept. 1	9·6	5·9	0·0	+1·4	4·5	0·76	0·86	20·2	20·4	2·3	V^c	−0·35	16·8
2	23·6	16·8	−0·1	+4·5	12·3	0·73	0·83	23·4	22·5	1·7	SE	−0·37	18·6
4	25·7	17·8	−0·4	+2·9	14·5	0·84	0·94	24·1	21·2	1·6	NW	−0·41	21·0
5	24·6	17·7	−0·5	+0·3	16·9	0·98	1·11	18·4	12·5	1·5	N	−0·43	26·8
7	24·4	21·4	−0·2	+1·4	18·4	1·08	1·21	19·8	12·7	1·6	V	−0·46	29·6
10	21·6	16·9	−0·4	+0·8	15·7	0·95	1·04	16·2	10·1	1·9	V	−0·51	32·5

[a] At crop height = 60 cm.
[b] At 150 cm.
[c] Variable.

but the vapour pressure and air temperature were greater on one day than on the other, these differences resulted in faster evaporation on the cooler but less humid day. A relationship between wind speed and the exchange co-efficients estimated from the energy balance is evident in each of these figures. Instances can be found, however, when an increase in wind speed was not correlated with deviations from the daily pattern of evapotranspiration found on similar days when wind speed was constant and lower. However, when the increase in wind speed was accompanied by an influx of dry air, as at 1400 on August 14 (Fig. 3a), the evaporation rate was markedly increased, reaching 2·3 times $(\mathbf{R}_n - \mathbf{G})$. The crop wilted at this time but the stomatal resistance was only twice that found earlier in that day. Such examples of atmospheric demand controlling the crop energy balance were infrequent.

A small latent heat loss occurred during most nights. For at least part of many nights a downward flux indicative of dew deposition was observed. Field observations confirmed the occurrence of dew which often persisted until 0700 or later.

IV. CARBON DIOXIDE BALANCE

A. Photosynthesis

Gaastra (1959) reported a series of measurements on the photosynthetic rates of sugar beet leaves exposed to a range of CO_2 concentrations and light intensities at leaf temperatures from 21° to 24°C. The results define rectangular hyperbolic curves with light saturation photosynthetic rates of 2·6 g m^{-2} h^{-1} at 300 μl/l CO_2. Saturation occurred at values of irradiance equivalent to about two-thirds full sunlight in terms of PAR. For greenhouse grown plants, El-Sharkawy and Hesketh (1965) found maximum leaf photosynthetic rates of about 3·0 g m^{-2} h^{-1}, while Kriedeman et al. (1964) reported maximum values of 2·2 g m^{-2} h^{-1}. The small differences between the reported photo-synthetic rates may arise because of different growing conditions or genetic factors. The value reported by Gaastra represents the average of available data.

The influence of temperature on growth rate was investigated by Lundegardh (1927) and Ulrich (1952) for field and growth chamber grown plants respectively. They found the optimum temperature for growth to be between 20° and 25°C. Figure 8 shows their combined results. A point of interest is that growth rate at 0°C is still one-third of that at the optimum temperature.

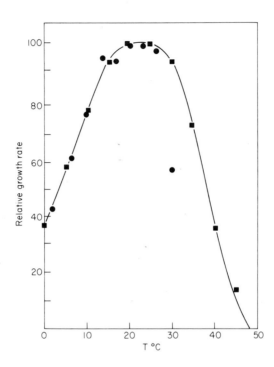

Fig. 8. Dependence of relative growth rate of sugar beet on ambient temperature at atmospheric CO_2 concentrations. Growth-chamber grown plants (●) after Ulrich (1952) and field-grown plants (■) after Lundegardh (1927).

The maximum net photosynthesis of field-grown potato leaves has been reported to range from 4·3 to 5·4 g CO_2 m^{-2} h^{-1} at midday (Buzover 1966). These rates are nearly double those reported for sugar beet leaves. Caution must be exercised in comparing these measurements, however, since differences of this magnitude are found between field and greenhouse plants.

Whether or not differences in leaf photosynthetic rate between varieties might be exploited to increase crop production remains open to question.

The influence of CO_2 concentration of a sugar beet crop enclosed in a field chamber was measured by Thomas and Hill (1949). Their measurements and the response for individual leaves is given in Fig. 9. At normal CO_2 concentrations, it is apparent that the photosynthetic rate for an entire crop is only 50% greater than for individual leaves exposed to a similar environment. Greater rates of field photosynthesis may not be achieved because of shading and the respiratory demand of the large leaf area which sugar beet develops.

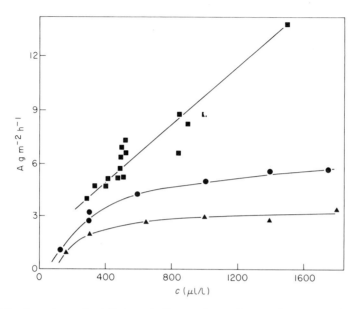

Fig. 9. The dependence of net photosynthetic rate of sugar beet on the concentration of CO_2 in the air c. The net photosynthetic rate of a field-grown crop per unit ground area, presumably at full sunlight (840 W m^{-2}), after Thomas and Hill (1949) (■) and of isolated leaves per unit leaf area at 840 W m^{-2} (●) and at 250 W m^{-2} (▲) after Gaastra (1959).

B. Diurnal Patterns of CO_2 Flux

Figure 3 shows the CO_2 gradients and downward CO_2 flux over a field of sugar beet. The midday gradients immediately above the crop were typically 4 μl/l per 25 cm. Assuming that the turbulent transport coefficient determined for sensible and latent heat is also applicable to carbon dioxide transfer, CO_2 flux to or from the air above the crop can be calculated from CO_2 profiles. The CO_2 flux generally increased during the first 2 or 3 hours after sunrise. The flux then remained nearly constant until about 3 hours before sunset, decreasing thereafter to zero at about 1 hour before sunset. The uptake of CO_2 by the sugar beet crop at midday was about 3 g m^{-2} h^{-1}.

The influence of radiation on the CO_2 uptake by the sugar beet is apparent in the diurnal curves of Fig. 3. Curves of CO_2 flux as a function of radiation, though scattered, indicate that light saturation was about 560 W m^{-2} (total short-wave radiation). Thus, the fixation rate was not regulated by the radiation flux during half of most normally clear days.

Monteith and Szeicz (1960) (Fig. 10) found that the CO_2 flux to sugar beet was closely correlated with the radiation intensity throughout the entire

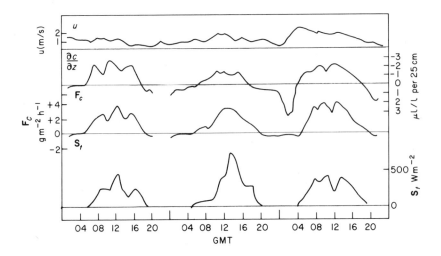

Fig. 10. Diurnal patterns of wind speed at 70 cm, CO_2 gradient between 68 and 93 cm, CO_2 flux and solar radiation above a 60-cm tall sugar beet crop. After Monteith and Szeicz (1960).

day. During the period of their experiments, however, solar radiation only occasionally exceeded $560 \, W \, m^{-2}$ which is about the saturation light intensity for individual sugar beet leaves reported by Gaastra (1959).

There is little evidence in the available records of midday depression in field photosynthetic rate. On occasion, however, the influence of several meteorological factors which might cause this kind of depression were noted. For example, the flux rate decreased slightly on August 14 (Fig. 3) when clouds reduced short-wave radiation to $419 \, W \, m^{-2}$. The strong dry winds which occurred at 1400 hours and thereafter on August 14 also depressed the CO_2 flux rate. This depression probably resulted from the greater stomatal resistances which occurred during this period.

The CO_2 flux from the soil was measured by placing a chamber over the soil surface through which air passed. The flow-rate times the increase in concentration were used to calculate the flux. The records exhibited a distinct diurnal trend closely correlated with soil temperature. The maximum flux was $0.45 \, g \, m^{-2} \, h^{-1}$ at 1300 hours and the minimum of $0.25 \, g \, m^{-2} \, h^{-1}$ occurred at 0300 hours.

On clear days the upward flux of CO_2 at the soil surface averaged 7% of the downward flux from the air above the crop. This does not mean, however, that 7% of the CO_2 fixed by the crop comes directly from the soil atmosphere. Because of the vigorous mixing of the air above and within all but very dense crops, most of the CO_2 evolved from the soil is rapidly dissipated. That which remains affects photosynthetic rates only by increasing the concentration of CO_2 surrounding the plant. Measurements by Monteith (1962) and Lemon (1967) indicate that the concentration near the soil surface exceeds the value at the top of the crop by about 5 $\mu l/l$ in tall open crops and 10 $\mu l/l$ in short dense crops. Gaastra (1959)(Fig. 9) showed that such an increase of concentration would not increase the rate of photosynthesis by more than 2 to 4 %. It appears, therefore, that the flux of CO_2 from the soil beneath the crop has only a very small influence on the photosynthetic rate. Lundegardh's (1927) experiments, in which photosynthetic rate was not increased when CO_2 was released below the crop, support this conclusion.

The question often arises as to whether CO_2 in the field air limits the rate of photosynthesis. For a factor to limit the rate of a process, the rate must be independent of all other factors. Therefore, the CO_2 does not limit the rate of photosynthesis but is *one* of the factors which regulates it. The slope of the line in Fig. 9 indicates that the rate of photosynthesis increases four times for a five-fold increase in CO_2 over a wide range of concentrations.

Assuming the mean atmospheric CO_2 is 318 $\mu l/l$ and using a mean station pressure of 880 mbar, calculations show that all the CO_2 in a layer of air 30 m thick would be required to support the mean photosynthetic rate of 15 g CO_2 m^{-2} day^{-1} which was measured in this field. More realistically, if the concentration is assumed to be reduced uniformly from 318 to 238 $\mu l/l$, a layer of air 276 m thick must be available to the crop daily. Chapman *et al.* (1954) reported that the CO_2 gradient was only 20 $\mu l/l$ between the surface and 152 m. Thus, since CO_2 is not reduced uniformly in the layer of air above the surface, it is evident that the crop extracts CO_2 from a very large volume of air each day. The magnitude of the volume of air involved indicates that the natural vertical turbulent transport is a vigorous process. This argument also suggests that it would be difficult to artificially increase the CO_2 content of field air sufficiently to influence photosynthetic rates significantly, because the released CO_2 would be quickly swept away.

C. Dry Matter Production

Net dry matter production is proportional to the difference between gross photosynthesis and respiration. Respired CO_2 is released from the roots to the soil atmosphere and by the leaves to the surrounding air. It has not been possible to assess these three components of net photosynthesis in the field. Whereas the fluxes from air and soil have been determined in the field, the flux from the roots can be estimated only from laboratory measurements. Net photosynthetic rate may then be calculated as the sum of the flux from the air and from the soil, minus the flux from the roots. This means of calculation may introduce small errors since the release of CO_2 by the roots may not be in equilibrium with the flux from this source at the soil surface, and also since microbial release is not taken into account.

A 24-hour balance sheet of CO_2 fixation for a typical clear day is given in Table II from Brown and Rosenberg (1971a). Day-time and night-time flux

Table II

Calculated diurnal net photosynthetic rate based on the various CO_2 flux components on a typical clear day. All values are given in g equivalent dry matter per m^{-2} per 12 hours. After Brown and Rosenberg (1971a)

Period of flux *from* the atmosphere:	
CO_2 flux from the air	27·8
CO_2 flux from the soil	4·4
Net flux to the above soil portion of the plant	32·2
Estimated root respiration	−2·0
Net photosynthesis during daylight hours	30·2
Period of flux *to* the atmosphere:	
CO_2 flux to the air	−8·3
CO_2 flux from the soil	3·1
Net flux from the above soil portion of the plant	−5·1
Estimated root respiration	−2·0
Net respiration flux during the night	−7·1
Net daily photosynthetic rate	23·1

are calculated separately so that the use of the nocturnal flux from the soil in calculating the leaf respiration may be shown. Although root respiration may differ from day to night, only a crude estimate could be made. The same value was thus used during both periods. The net loss of CO_2 at night was estimated to be 24% of the net day-time fixation rate. Because of higher leaf temperatures and greater biochemical activity, respiration from the leaves should be greater during the day.

The day-time (0600 to 1800 hours) net downward flux of CO_2 is given in Table I. The smallest daily flux occurred on September 1 when the short-wave radiation income was about one-third that on clear days. The day-time CO_2 flux for the remainder of the days, which were all clear or nearly clear, is independent of the incident short-wave radiation but positively related to the latent heat flux.

Dry matter samples consisting of 40 plants selected at random were taken from the same field on August 11 and again on September 2. The dry matter accumulated during this period was 360 ± 34 g m^{-2}. The CO_2 flux measurements were made on half the days during this period and were extrapolated to the missing days. The total calculated fixed CO_2 was converted to dry matter by using the ratio of the molecular weight of $C_{12}H_{22}O_{11}$ to 12 CO_2 molecules. The calculated total of 344 g m^{-2} agrees well with data obtained from dry matter samples. Since no duplicate measurements of flux were made, no standard deviation is available for the CO_2 fixation estimate.

Similar calculations were made on the CO_2 balance of a sugar beet crop by Monteith and Szeicz (1960). They included a longer period and made observations later in the season when the respiratory flux consistently exceeded the normal flux throughout the day. They also found reasonable agreement between CO_2 flux and the equivalent dry matter accumulation. Despite the smaller solar irradiance at their site, their calculated values are about twice those reported here.

Obviously, the most productive use of micrometeorological CO_2 flux measurements is for expanding the knowledge of interactions which occur in short time spans, and not for determining dry matter accumulation. Good agreement between dry matter production determined by plant sampling and by the CO_2 flux calculations, however, demonstrates the reasonable accuracy of micrometeorological techniques. Better techniques of evaluating the exchange coefficients, particularly during the early morning and late evening, and additional information about root respiration rates would improve the reliability of CO_2 balances.

The day-time water vapour and CO_2 fluxes may also be used to calculate the amount of water used per unit dry matter produced. For the data in Table I, greatest efficiency was achieved on September 1, a day during which heavy clouds persisted for all but a few hours. On this day, 171 g of water were used for each gram of dry matter produced. On clear days, an average of 365 g H_2O was used per gram dry matter.

REFERENCES

Baumgartner, A. (1970). *In* 'World H_2O Balance Symposium in Contribution to International Hydrological Decade', pp. 56–65, UNESCO, Paris.
Broadbent, L. (1950). *Quart. J. R. met. Soc.* **76**, 439–454.
Brown, K. W. and Rosenberg, N. J. (1970a). *Agron. J.* **62**, 4–8.
Brown, K. W. and Rosenberg, N. J. (1970b). *Agron. J.* **62**, 20–24.
Brown, K. W. and Rosenberg, N. J. (1971a). *Agron. J.* **63**, 207–213.
Brown, K. W. and Rosenberg, N. J. (1971b). *Agron. J.* **63**, 351–355.

Brown, K. W. and Rosenberg, N. J. (1971/1972). *Agric. Met.* **9**, 241–263.

Brown, K. W. and Rosenberg, N. J. (1973). *Agron. J.* **65**, 341–347.

Burrows, F. J. (1969). *Agric. Met.* **6**, 211–226.

Buzover, F. Y. (1966). *Trudy Khar'kovsk. Sel.-Khoz. Inst.* **57**(94), 57–76.

Chapman, H. W., Gleason, L. S. and Loomis, M. E. (1954). *Pl. Physiol.* **29**, 500–503.

El-Sharkawy, M. and Hesketh, J. (1965). *Crop Sci.* **5**, 517–521.

Epstein, E. and Grant, W. J. (1973). *Agron. J.* **65**, 400–404.

Gaastra, P. (1959). *Meded. Landb. Wageningen,* **59**, 1–68.

Johnson, W. C. and Davis, R. G. (1973). *Agron. J.* **65**, 789–794.

Kriedeman, P. E., Neales, T. F. and Ashton, D. H. (1964). *Aust. J. Biol. Sci.* **17**, 591–600.

Lemon, E. R. (1967). *In* 'Harvesting the Sun' (A. San Pietro, F. A. Green, T. J. Army, eds), pp. 263–290. Academic Press, New York.

Long, I. F. and Penman, H. L. (1963). *In* 'The Growth of the Potato' (J. D. Ivins and F. L. Milthorpe, eds), pp. 183–190. London, Butterworths.

Lundegardh, H. (1927). *Soil Sci.* **23**, 417–452.

Monteith, J. L. and Szeicz, G. (1960). *Quart. J. R. met. Soc.* **86**, 205–214.

Monteith, J. L. (1962). *Netherlands J. Agric. Sci.* **10**, 334–346.

Rijtema, P. E. (1970). *In* 'Plant Response to Climatic Factors', pp. 513–518, UNESCO, Paris.

Shepherd, W. (1972). *Water Resour. Res.* **8**, 1092–1095.

Stegman, E. C. and Nelson, D. C. (1973). North Dakota Res. Report No. 44. 15 pp. Fargo, North Dakota.

Szeicz, G., Endrodi, G., Tajchman, S. (1969). *Water Resour. Res.* **5**, 380–394.

Thomas, M. P. and Hill, G. R. (1949). *In* 'Photosynthesis in Plants' (J. Frank and W. E. Loomis, eds), pp. 19–52, Iowa State College Press.

Ulrich, A. (1952). *Agron. J.* **44**, 66–73.

4. Sunflower

B. SAUGIER

Section d'Eco-Physiologie du C.E.P.E.—Louis Emberger, Centre National de la Recherche Scientifique, Montpellier, France

I. INTRODUCTION

Sunflower (*Helianthus annuus* L.) is a plant native to North America. Introduced in Europe in 1596, it is now grown mainly in the USSR, Argentina,

eastern Europe and Turkey (Table I). The seeds are used mostly for producing a good quality table oil: 3·2 million metric tons were produced in
1969, representing 17 % of the total world table oil production.

Table I

Main producers of sunflower seeds in 1970 (Source: F.A.O. production yearbook, 1970).

Country	Production 10^6 tonnes	%	Yield (100 kg/ha)
U.S.S.R.	6·10	63·2	12·8
Argentina	1·14	11·8	8·5
Roumania	0·63	6·5	13·0
Turkey	0·38	3·9	10·9
Bulgaria	0·35	3·6	15·9
Yugoslavia	0·27	2·8	14·1
Spain	0·15	1·6	10·1
South Africa	0·11	1·1	6·2
Hungary	0·11	1·1	12·8
Others	0·41	4·4	
World total	9·65	100·0	11·7

Several countries have recently taken an interest in sunflower production
because of a shortage in the supply of groundnuts and also because careful
selection of varieties and hybrids have increased the yield up to 3000 kg/ha
(for an irrigated crop) and the oil content of the seeds which went from 30 %
in the 1950s to about 50 % in 1972. As a result, the world seed production
increased from 3·9 million metric tons in 1948–1952 to an average of 10
million tons in 1967–1970.

Sunflower is a plant which thrives in a relatively warm climate and has
moderate requirements for water and nutrients. The yield increases with
plant density up to 3 plants/m^2 and is nearly constant for densities between
3 and 8 plants/m^2. Densities commonly used in agricultural practice lie
between 4 and 8 plants/m^2 so unavoidable gaps in the sowing pattern have
no effect on the final yield. A typical growth period for sunflower growing in
France is shown in Fig. 1. After sowing at the end of March, emergence
follows two weeks later, the flowering period is centred at the end of June,
and harvest is at the end of August. This calendar varies somewhat with the
variety and the climate of the area. The heights of mature sunflowers range
from 1 to 2·5 m or even 3 to 4 m for certain tropical varieties; as for wheat
or corn, there is a trend to breed shorter plants with stronger stems which
tolerate high levels of fertilizers.

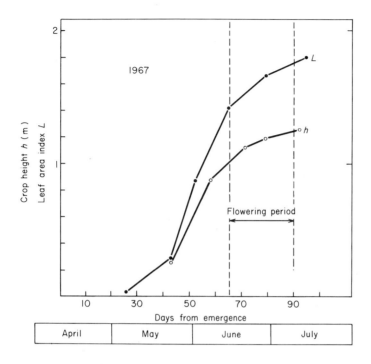

Fig. 1. Increase through the growing season of leaf area index (*L*) and height (*h*) of a crop of *Helianthus annuus* L., var. Peredovik, sown on 24 March 1967 in Montpellier, France. (Reproduced from Eckardt *et al.* 1971, with permission).

II. RADIATION AND CANOPY STRUCTURE

A. Penetration of Radiation

Two types of radiation are of special interest to the micrometeorologist: net radiation, supplying energy mainly for evaporation and heating; and photosynthetically active radiation (PAR) directly related to CO_2 absorption by the leaves.

Figure 2a shows the extinction of net radiation in a sunflower stand under a clear sky. The measurements inside the crop were obtained by averaging, at each level, 20 readings with a net radiometer, using the guiding device of Eckardt and Méthy (1967). Net radiation above the crop was measured with a Gier and Dunkle net radiometer. Measurements lasted from 1000 to 1600, which explains part of the scatter.

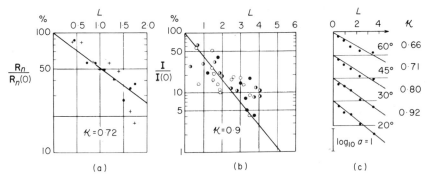

Fig. 2. Semilogarithmic plot of radiation components versus cumulative leaf area index L. (a) Net radiation for two days (redrawn from Eckardt *et al.*, 1971). (b) Visible radiation for several plant densities (reproduced from Hiroi and Monsi, 1966). (c) Sunflecks area for several solar elevations (reproduced from Laisk, 1969, with permission).

Figure 2b represents a similar curve for visible light obtained with a Weston photometer. The profile stays the same for densities ranging from 25 to 100 plants/m², much higher than the usual 4 to 6 plants/m².

Figure 2c shows the decrease in sunfleck area for different solar elevations, as measured by a 'mouse', an ingenious device devised by Laisk (1969). It is a narrow beam, blue-sensitive light receiver which travels in a tube and looks out through each of the many holes bored through its wall. If the 'mouse' sees the sky, the associated electronics send a pulse to a counter; if it sees an obstacle, no pulse is delivered. The number of pulses recorded divided by the number of holes explored is a measure of the canopy transparency, that is, of the percentage of sunfleck area for a solar elevation equal to the elevation of the receiver beam, which can vary from 0° to 90°.

In the three cases shown in Fig. 2, measurements are reasonably linear on a semilogarithmic plot: radiation follows the exponential decrease with leaf area index. But Impens (1973) found his net radiation data were best fitted by an equation of the type: $R_n(L)/R_n(0) = \exp(-k_1 L + k_2 L^2)$ where $R_n(L)$ is the net radiation below a leaf area index L measured from the top of the crop, and k_1 and k_2 are two empirical coefficients. The implied decrease in extinction coefficient going downwards could be caused by a change in the leaf angle distribution: older leaves have a tendency to hang down, becoming more vertical and thus making the canopy as a whole more transparent to downward radiation.

B. Absorption Coefficients and Leaf Arrangement

For relating penetration of radiation to canopy structure, separate consideration is usually given to direct radiation coming from the solar disc,

present in sunflecks only, and diffuse radiation coming from the sky and/or
scattered by the leaves (see Vol. 1, p. 35). The net radiation regime is complicated by thermal radiation from the leaves, but the variation of this component is small compared to direct radiation (Vol. 1, p. 50).

Since most sunflower leaves are nearly horizontal, as Fig. 3 shows, the
extinction coefficient k for any type of radiation is expected to be close to

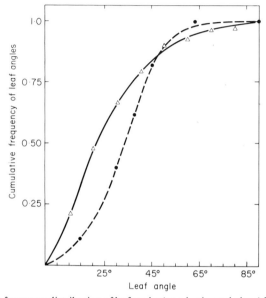

Fig. 3. Cumulative frequency distribution of leaf angles (on a horizontal plane) for two sunflower
canopies; solid line: var. Peredovik (from Eckardt *et al.*, 1971), dashed line, other variety
(redrawn from Ross and Ross, 1969, with permission).

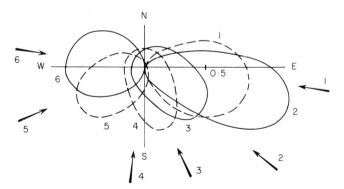

Fig. 4. Azimuthal distribution of young sunflower leaves. Arrows indicate the solar azimuth
(reproduced from Ross and Ross, 1969, with permission).

unity and to depend little on solar elevation. Some variation is expected, depending on the variety used and on the maturation stage of the crop. Young sunflower leaves exhibit heliotropism (Fig. 4) but these movements are confined to immature leaves (Hiroi and Monsi, 1966). Their effect on light interception is thought to be small, except perhaps for young plants with only a few leaves, as observed by Varlet Grancher and Bonhomme (1972) in a crop of *Vigna sinensis*. Light inhibition of growth in leaves exposed to sunlight apparently causes a differential rate of elongation of the two sides of the stem, forcing the leaves to face the sun (Salisbury and Ross, 1969).

Table II summarizes measurements of extinction coefficients for different types of radiation: sunfleck area, diffuse radiation, PAR and net radiation, derived from data of Fig. 2 or from other sources. The value of 0·72 for net radiation was incorrectly quoted as 0·56 by Saugier (1970). Theoretical values have been computed using the formula: $\mathscr{K} = \Sigma (A'/A)_{\alpha\beta} f_{\alpha}/\sin\beta$ where α is the leaf angle and β the solar elevation, f_{α} the fraction of leaves making the angle α on the horizontal plane, and $(A'/A)_{\alpha\beta}/\sin\beta$ the average area of the shadow cast by a leaf of unit area, of random azimuth and of angle α on a horizontal plane. Further details are given in a paper by Duncan *et al.* (1967) which contains a table of the so-called 'Wilson–Reeve ratio' $(A'/A)_{\alpha\beta}$.

Two points are apparent in Table II:

1. theoretical evaluations and measurements agree well although the latter are consistently smaller by 10 to 15 %: the real sunflower crop is thus more transparent to the sun's rays than one would expect if the plant organs were distributed randomly (Laisk, 1969). This 'clumped' distribution is quite common and has to be taken into account in theoretical models (Nilson, 1971);
2. all the values of the extinction coefficients lie between 0·6 and 0·9, reflecting the fact that the leaves are not all horizontal. These values, however, are among the highest quoted for crops (Monteith, 1969).

To summarize, a crude model such as given by the above equation and corrected for the degree of clumping of the vegetation, can give reasonable prediction of radiation penetration, even for PAR or net radiation in which direct radiation is the most rapidly changing component. In particular, there is no need to worry about penumbra effects since sunflecks are relatively large (Norman *et al.*, 1971).

C. Optical Properties of Leaves

The spectral quality of light is modified by absorption in leaves. Near infra-red radiation (0·8 to 1·3 μm) is the least absorbed, while in the visible

Table II

Values of extinction coefficients for different radiation components

Parameter	Method	Solar elevation					L_{max}	h (cm)	Source
		15–20°	30°	45°	60°	—			
{ Sunfleck area { Sunfleck area	theoretical 'mouse'	1·23 0·92	0·96 0·80	0·83 0·71	0·77 0·66		3·9	250	Laisk (1969)
Sky radiation	theoretical	1·06			0·90				Nilson (1968)
{ Diffuse radiation { (overcast)	theoretical					0·88			Anderson (1970)
{ Diffuse radiation	fish-eye camera					0·78	4·1		
Diffuse radiation (overcast)	selenium photocell					0·90	2 to 4		Hiroi and Monsi (1966)
PAR	theoretical	1·27	0·88	0·79	0·74				Ross and Nilson (1968)
{ Sunflecks area { Net radiation	theoretical net radiometer	1·09	0·89	0·86 0·72±0·12			1·8	120	Eckardt et al. (1971)
Net radiation	net radiometer					0·822−0·074L	3·6	220	Impens (1973)

range green light is less absorbed than either blue or red. Figure 5 shows the variation with wavelength of the absorption and reflection coefficients of a sunflower leaf.

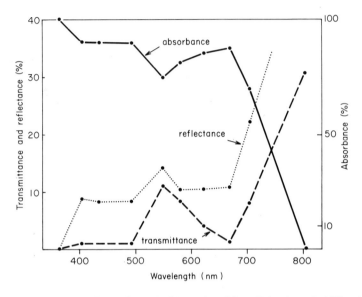

Fig. 5. Optical properties of a sunflower leaf (reproduced from Eckardt *et al.*, 1971, with permission).

The change in light quality may not affect photosynthesis so much as photomorphogenesis through the phytochrome system: the equilibrium between the two forms of this pigment depends on the light absorbed in two spectral regions: red (660 nm) and far red (730 nm) (Salisbury and Ross, 1969). This effect would certainly be more significant in grasses where the meristem is often shaded by the leaves, or in plants growing under a forest, than in sunflower which grows mainly from the top.

III. MOMENTUM

A. Wind Profile in the Canopy

Figure 6a represents a normalized wind profile in a sunflower crop whose leaf area density is given in Fig. 6b. Four anemometers only were available for this study, the sensitive Casella–Sheppard type with a light-bulb photocell device. The basic features of the profile, however, moving downwards

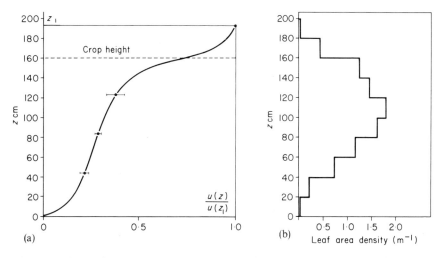

Fig. 6. (a) Wind profile in a sunflower crop normalized with respect to wind speed at height z. (b) Vertical distribution of leaf area density in the same crop (reproduced from Saugier, 1970, with permission).

from the top of the canopy, are as expected:
   (a) a rapid decrease in wind speed initially;
   (b) a more gradual decrease in the bulk of the vegetation;
   (c) a rapid decrease again close to the ground.

B. Momentum Transfer in the Canopy

The momentum flux density decreases in the vegetation according to the equation:

$$d\tau/dz = \rho C_M(z)a(z)u^2(z)$$

with ρ the density of air, $C(z)$ drag coefficient of the vegetation, $a(z)$ leaf area density and $u(z)$ horizontal wind speed (see Vol. 1, p. 66). If one assumes $\tau(0) = 0$, which should be valid for a reasonably dense cover, integration of the above equation gives:

$$\tau(z)/\tau(h) = \int_0^z C_M(z)a(z)u^2(z) \Big/ \int_0^h C_M(z)a(z)u^2(z)$$

where h is the average crop height. This relation allows computation of the profile of τ from data of Fig. 6, using some assumption about the variation of the drag coefficient with wind speed.

This profile has been plotted against leaf area index in Fig. 7, using two different assumptions:

(a) $C_M(z)$ constant (solid line),
(b) $C_M(z) \propto [u(z)]^{\frac{1}{2}}$ as in Thom (1971) (dotted line).

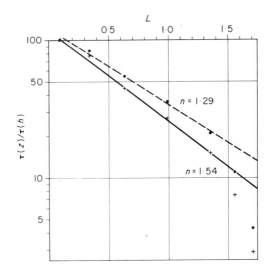

Fig. 7. Logarithm variation with cumulative leaf area index of the normalized momentum flux density, for two assumptions for the drag coefficient (see text).

In the upper half of the canopy, the profile of τ may be approximated by an exponential expression $\tau(z) = \tau(h)\exp(-nL)$. The coefficient n is equal to 1·54 when a constant drag coefficient is assumed, and to 1·29 when $C_M(z)$ is proportional to $[u(z)]^{\frac{1}{2}}$. In the bottom half of the canopy, τ decreases more rapidly than the exponential relation predicts, possibly because the assumption $\tau(0) = 0$ is not valid in a relatively open sunflower crop ($L = 1·8$).

If the flux-gradient relationship $\tau = \rho K_M \, d\bar{u}/dz$ is supposed to be valid in a plant layer, K_M can be computed from profiles of τ and u. A 'kink' in the K_M profile is thus obtained, whatever assumption is chosen for the wind dependence of $C_M(z)$, as Thom (1971) found for a bean crop. The shape of the K_M profile depends on the fact that the wind speed varies very little between $h/4$ and $3h/4$, making the gradient $d\bar{u}/dz$ very small. In some cases the gradient may even be negative if there is a maximum of wind speed in the canopy, as observed by Oliver (1971) in a forest of pine trees where advection is not likely to have occurred. Current models of air flow within plant layers generally assume exponential decays of both $\bar{u}$ and K_M, and thus cannot account for a region of nearly constant wind speed in the bulk of the canopy (Stewart and Lemon, 1969).

C. Boundary-Layer Resistance of Leaves

The boundary-layer resistance for a leaf depends on size, shape and wind speed. According to Raschke (1956), Monteith (1965) and Parlange et al. (1971), the resistance in s cm^{-1} for one side of the leaf may be calculated from the empirical expression $1.3(d/u)^{0.5}$ where d(cm) is an appropriate linear dimension and u(cm s^{-1}) is the wind speed at the length of the leaf.

In the case of the large sunflower leaves, however, Hunt et al. (1968) found very low values for r, ranging from 0.05 to 0.13 s cm^{-1}, much smaller than the above formula predicts. They interpreted these estimates as the result of an enhanced exchange between leaf and air caused by the large size of eddies around the leaf and possibly by irregularities of the leaf which increase its roughness. The values of r were computed from the energy balance equation of the leaf, in which r_s, the epidermal resistance, was estimated with a diffusion porometer. It is suspected that the high values of 3.3 to 4.7 s cm^{-1} reported for r_s may have been too high, thus making r too small.

Raschke (1956) reports measurements by Martin of boundary-layer resistance ranging from 1 s cm^{-1} at a wind speed of 20 cm s^{-1} to 0.3 s cm^{-1} at 260 cm s^{-1}, for a sunflower leaf of maximum dimensions 12.6 × 9.8 cm. These values are only slightly greater than the expression $1.3(d/u)^{0.5}$ would predict. Since Parlange and Waggoner (1972) also found this expression to hold for big reed leaves in the field, it will be used from then on.

IV. HEAT AND WATER

A. Temperature and Humidity Profiles

Figure 8 shows profiles of temperature and water vapour pressure as measured by platinum-resistance aspirated psychrometers (Saugier, 1970) in an irrigated sunflower crop during the flowering period (L = 1.8).

There is an inversion in the early morning (0700); then a maximum of temperature is observed in the upper layer of the crop, till about 1300 when lapse conditions prevail throughout. By 1700 the profile has returned to a slight inversion. A maximum of temperature is often observed within the canopy layer, at a depth depending on the pattern of light absorption and of momentum transport. In vegetation like corn with more erect leaves (Brown and Covey, 1966), or native grassland (p. 382), it is located deeper in the canopy, and is related to a smaller extinction coefficient.

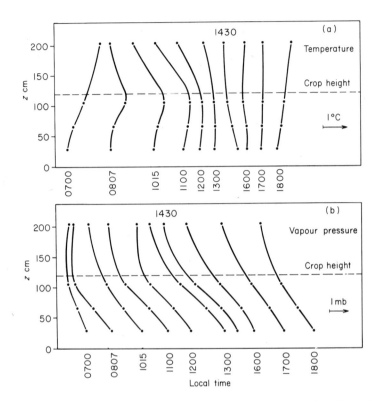

Fig. 8. Temperature and humidity profiles within a sunflower crop (reproduced from Saugier, 1970, with permission).

Profiles of water vapour pressure (Fig. 8) show the usual decrease from the ground upwards. Advection due to an insufficient 'fetch' may have caused the break in the first two morning profiles.

B. Evaporation Measurements by Energy Balance and Cuvette Methods

Sensible and latent heat fluxes in and above the canopy have been computed from profiles of net radiation (Fig. 2a), temperature and humidity (Fig. 8), using the energy balance method. Concurrent measurements of evaporation were made using the closed-circuit cuvette of Eckardt (1967). In this cuvette the water vapour transpired by the plants is condensed on a cooling device and collected in a container whose water level is continuously recorded.

Evaporation rates obtained by the two methods on a clear day are plotted in Fig. 9. There is general agreement between the two, leading to about the

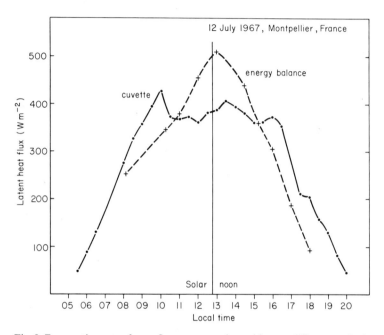

Fig. 9. Evaporation rate of a sunflower crop, estimated by two different methods.

same total evaporation for the day, 5·7 mm, although individual values are different. In particular, the cuvette curve presents midday plateau while the energy balance exhibits a rather narrow peak. In the cuvette, the ground was covered with a plastic film so that the cuvette evaporation represents only plant transpiration whereas the energy budget takes into account the soil evaporation. Moreover, it takes time for the evaporated water to be condensed, collected and measured by the cuvette device and this lag leads to an underestimation of the peak values. Another reason for the discrepancy is the difference in air flow conditions in the cuvette and outside, resulting in different leaf boundary-layer resistances. This is particularly significant for sunflower leaves, where r (see Section III.C) and r_s (Section IV.D), the leaf resistance, are of the same order of magnitude. The boundary-layer resistance may thus limit the evaporation rate. Each method has its limitations, in the energy balance the fetch requirements were not completely met; both methods introduce errors of the order of 20 %.

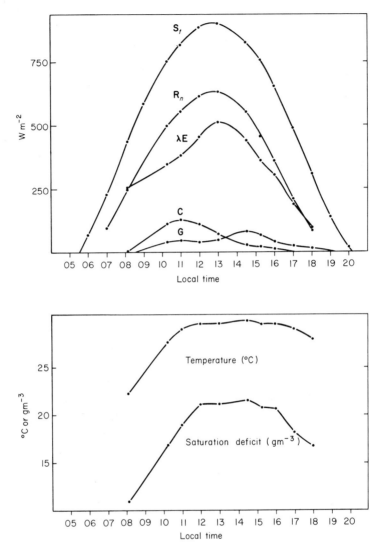

Fig. 10. (Upper) Energy budget components above sunflower, Montpellier, 12 July 1967 (re-drawn from Saugier, 1970, with permission). (Lower) Variation of screen temperature and saturation deficit for the same day.

The various terms of the energy budget are plotted in Fig. 10; λE is measured by the energy balance method and **G**, the soil heat flux density, with a plate buried at 2 cm below the soil surface. The global radiation S_t, measured with an Eppley pyranometer, is added for reference; net radiation R_n was measured with a Gier and Dunkle net radiometer. The sensible heat flux C reaches a peak at about 1100 local time, long before solar noon. A similar trend is apprent in the sunflower measurements of Impens (1973): the evaporation is greater in the afternoon than in the morning for the same net radiation, resulting in a lower sensible heat flux. Such a feature is characteristic of vegetation well supplied with water (Rose *et al.*, 1972). The saturation deficit usually reaches its maximum in the afternoon (Fig. 10) and so will the transpiration if the resistances stay small.

C. Sensible and Latent Heat Fluxes in the Canopy

Figure 11 represents flux densities of sensible and latent heat at several heights in a sunflower crop as computed by Impens (1973) from data similar

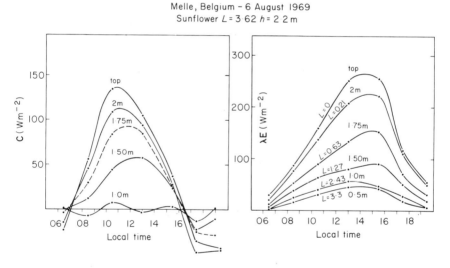

Fig. 11. Variation with time of sensible (C) and latent (λE) heat fluxes at several depths in a sunflower crop (redrawn from Impens, 1973, with permission).

to those of Fig. 8 but with 7 levels of measurements instead of only 4. Only the top half of the crop contributes to the sensible heat flux, since C at 1 m remains quite small throughout the day, probably because there is little radiation at this level. The maximum sensible heat flux occurs later in the day as one goes deeper into the canopy, possibly because direct radiation penetrates more easily at high solar elevations (Table II).

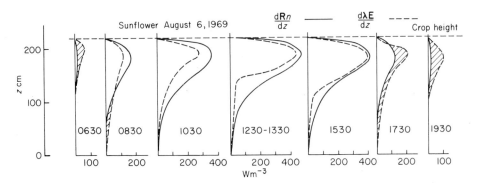

Fig. 12. Distribution of sources and sinks for net radiation and latent heat, in a sunflower crop. In the shaded zones the energy for evaporation comes from the sensible heat flux. (Reproduced from Impens, 1973, with permission).

The distribution of sources and sinks of latent heat and net radiation is given in Fig. 12. The two curves are similar, indicating that a relatively constant percentage of net radiation is used for transpiration. This is not the case in the early morning and in the late afternoon when energy for evaporation is supplied mainly by sensible heat flux.

D. Stomatal Behaviour; Water Status in Plant and Soil

Berger (1971, 1973) in Montpellier made an extensive study of the water relations of sunflower plants. Figure 13 represents some of the results he obtained during two days; A with little water stress ($\psi_{\text{soil}} = -0.8$ bar) and B with high water stress ($\psi_{\text{soil}} = -5.5$ bars). In Figs. 13A and B the water potential measurements are shown in (1); the total water potential ψ is measured on leaf samples in a temperature controlled water-bath with calibrated thermocouple psychrometers (Spanner, 1951); the sum of osmotic and matric potentials $\pi + \tau$ is measured in the same way on previously frozen samples; and the turgor pressure P is calculated as $P = \psi - (\pi + \tau)$. The centre of the figure (2) represents the diurnal change of the leaf resistance to water vapour, estimated from the temperature difference between a freely transpiring leaf and another leaf coated with a plastic film to suppress transpiration, the boundary layer resistance being known as explained in Section III.C. The bottom curves (3) show leaf resistance plotted against global radiation.

In Fig. 13A, ψ stays relatively high during the day, with only a small depression down to -2 bars between 1200 and 1600. For some reason the turgor pressure also decreased after 1100. The corresponding curve of r_s

exhibits the usual U shape (minimum 1.2 s cm^{-1} at 1200) with a slight increase after 1200 that accompanies the decrease in ψ. The radiation dependence of r_s (bottom curve) is more or less hyperbolic, with higher values in the afternoon for the same radiation level.

In Fig. 13B, ψ is very low between 1000 and 1700, causing a negative turgor pressure between 1000 and 1400. The resulting leaf resistances, lowest at 0830 (1.5 s cm^{-1}), increases to 5 s cm^{-1} at 1200 and stays the same until 1500 when it increases again towards the night values. Obviously water stress induced stomatal closure. This is even more obvious in the bottom curve: although in the early morning r_s decreases when radiation increases, this relation does not hold after 0900 and in the afternoon when stomatal aperture is limited by water stress and not by radiation.

The dependence of leaf resistance on total water potential is represented in Fig. 14 for non-limiting radiation of 500 to 850 Wm^{-2}. The resistance r_s increases only slightly from 1.1 to 2.5 s cm^{-1} when ψ decreases from 0 to -12 bars but thereafter the increase is very rapid, since r_s is more than 10 s cm^{-1} when $\psi \sim -15$ bars. The curve is well defined and it may be possible to develop a simple model of r_s as a function of two variables S_t and ψ. The reality is more complex. Curve 14 was plotted using only results obtained around solar noon; at the same water potential, leaf resistances were always lower in the morning and higher in the afternoon. After rejecting a possible effect of temperature on stomatal aperture, Berger (1971) put forward the explanation that differences of potential exist in the different compartments of a transpiring plant: vessels and cell walls, mesophyll cells, epidermal cells, guard cells. The thermocouple psychrometer measures an average of these potentials after equilibrium has been reached, but the stomatal aperture depends on the actual turgor pressure of the guard cells when the plant is transpiring. Reliable *in situ* measurement of ψ would help greatly to solve this matter. On the other hand, a realistic model of stomatal resistance must include the different plant compartments.

V. CARBON DIOXIDE

A. Carbon Dioxide Profiles

Figure 15 represents profiles of CO_2 concentration obtained with a Beckman infra-red gas analyser at the same times as profiles of Fig. 8. There is a persistent minimum of CO_2 concentration at about 65 cm which means that the CO_2 flux is always directed upwards in the bottom half of the canopy. The absolute CO_2 concentration, also recorded at 2 m, usually varies from a minimum value of $300\,\mu$l/l to a maximum of $350\,\mu$l/l; the former value

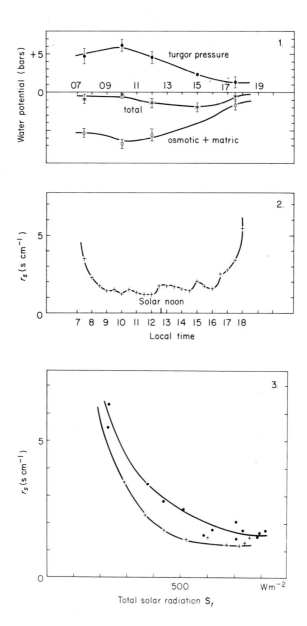

Fig. 13A. 1: diurnal change of water potential components of sunflower leaf on a day with slight water stress; 2: diurnal change of stomatal resistance on same day; 3: relation between stomatal resistance and total solar irradiance. + before noon; · after noon (reproduced, with permission, from Berger, 1971).

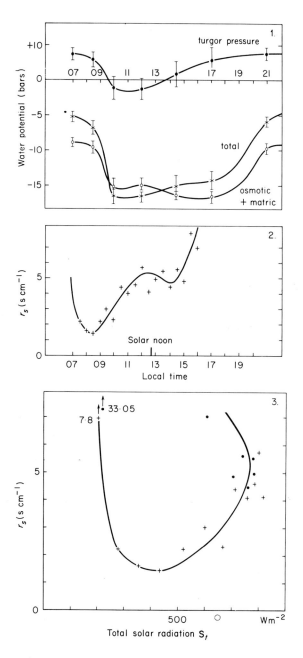

Fig. 13B. As Fig. 13A but on day with severe water stress.

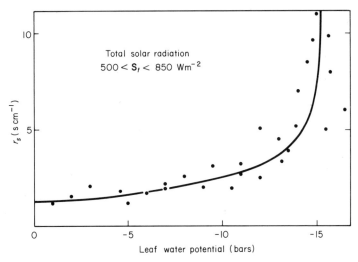

Fig. 14. Variation of sunflower leaf resistance with water potential, when radiation was not limiting (reproduced from Berger, 1971, with permission).

is quite stable between about 0800 and 2200, while the latter is much more variable on account of low and variable wind velocity.

The CO_2 gradients are not large: 5 $\mu l/l$ is the maximum difference observed between the top level and the level of the minimum. So in this relatively open canopy ($L = 1\cdot8$) photosynthesis will not be restricted by a reduction in atmospheric CO_2 concentration. This is also true of the denser crop used by Impens (1973) who found that the maximum CO difference was 8 $\mu l/l$ for a leaf area index of 3·62.

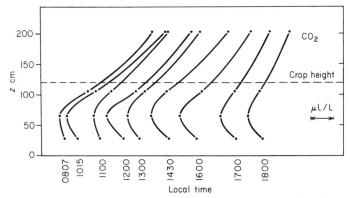

Fig. 15. Carbon dioxide profiles within a sunflower crop in Montpellier, 12 July 1967 (reproduced from Saugier, 1970, with permission).

B. Carbon Dioxide Exchanges of the Whole Vegetation by Profile and Cuvette Methods

Assuming the same diffusivities for carbon dioxide and water vapour, vertical CO_2 fluxes can be computed from CO_2 and H_2O gradients and from the H_2O fluxes given by the energy balance method.

The closed-circuit cuvette used for transpiration has also been used for measuring the CO_2 exchanges of the green parts of four sunflower plants separated from the soil by a plastic film as seen in Section IV B; the net assimilation is equal to the injection rate of pure CO_2 necessary to maintain CO_2 concentration under the cuvette the same than outside; this injection rate is controlled automatically (Eckardt, 1967).

Figure 16 shows CO_2 fluxes obtained by the two methods, and by the mathematical model of de Wit (1965) (Eckardt *et al.*, 1971); in this model the visible radiation load on each leaf is first computed from the geometry of the crop (leaf angles, leaf areas), then photosynthesis of the entire crop is calculated by adding the contributions of each leaf, using as an input the photosynthesis-light curve of Eckardt *et al.* (1971) discussed in Section V.E and presented in Fig. 20.

As in the case of water vapour, the profile method takes into account the soil contribution to the net CO_2 flux F_c i.e. $F_c = A - R_s$, where A is the

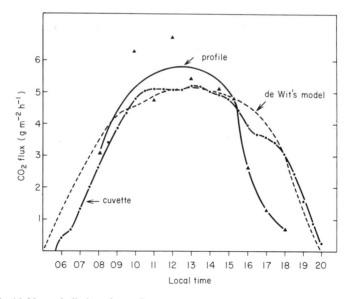

Fig. 16. Net assimilation of a sunflower crop, estimated by three different methods.

net assimilation (gross photosynthesis minus respiration of above-ground organs) and R_s the soil respiration rate. The cuvette method measures A directly, and de Wit's model computes the net photosynthesis of the leaves, equal to A if the respiration of stems and heads is neglected.

There is good agreement between cuvette and model estimation, except in the early morning, where cuvette values are lower. This may be an artefact introduced by the CO_2 control system of the cuvette which sometimes leads to overshooting; it may also be real and correspond to a delay in stomatal response to increasing light intensity.

The profile estimates are much more scattered and a smooth line has been drawn by hand through the points in a somewhat subjective way. These estimates should be lower than the cuvette or model data by an amount equal to the soil respiration. In fact they are higher most of the day except during late afternoon, when presumably soil respiration increased, although not as drastically as shown in Fig. 16.

Taking the areas under the curves and extrapolating when necessary, one can compute the total amount of atmospheric CO_2 absorbed by the plants between sunrise and sunset. Results are presented in Fig. 17, converted in grams of carbon per square meter to allow comparison with biomass

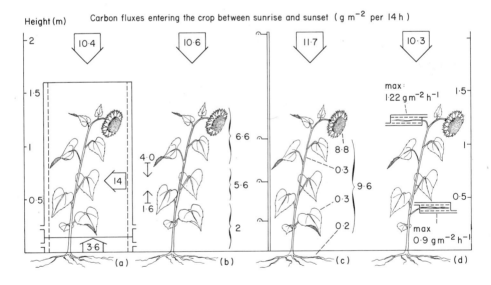

Fig. 17. Input of atmospheric carbon to a sunflower crop, calculated using several techniques. (a) Cuvette. (b) Energy balance. (c) Growth analysis. (d) De Wit's model (reproduced from Eckardt *et al.* (1971), with permission).

data (c). The cuvette measurements (a) give a net input of carbon of 14 gm^{-2} to the green parts; of this amount about three-quarters comes from the atmosphere (10·4 g m^{-2}) and one-quarter from the soil (3·6 g m^{-2}).

In (b) 10·6 g m^{-2} of carbon enter the crop from the atmosphere, as measured using the energy balance method between the top levels. The top third of the crop assimilates 6·6 g m^{-2}, the centre 5·6 g m^{-2} and the bottom third 2 gm^{-2}, assuming again a soil respiration of 3·6 g m^{-2} of carbon. Figure 16(c) represents results obtained by growth analysis (see below). In 24 hours there is a net increase in carbon of 9·6 g m^{-2} in the plants, mainly in the heads (8·8 g m^{-2}). To that value has been added the night respiration of the aerial parts (2·1 g m^{-2}) to give a day-time carbon flux from the atmosphere of 11·7 g m^{-2}. Figure 15(d) gives the results of the model of de Wit (1965), assuming the maximum photosynthesis of individual leaves decreases linearly from the top of the crop downwards. The soil respiration has been subtracted as in (a).

The good agreement observed between the cuvette (a) and the energy balance (b) results is somewhat fortuitous, since neither method could claim an accuracy better than 20%. It is encouraging, however, to see reasonably good agreement of these methods with the growth analysis technique (c). The model (d) gives some confidence in the exercise of extrapolating to the entire crop the photosynthetic behaviour of individual leaves.

C. Dry Matter Production, Carbon and Energy Content

Harvesting plant samples provides measurements of plant biomass at a given time which may then be used to estimate the net production between two consecutive harvests. Figure 18 shows the development of the biomass of several sunflower organs during growth, as measured by G. Heim and reported by Eckardt et al. (1971). Stems and leaves reach a maximum weight at the end of the flowering period, after which all the photosynthates are used for seed production. Moreover, the carbon content of the leaves, which stays constant at a value of 42% until the beginning of the flowering period, decreases thereafter to 36%, while for the heads it increases from 46% to 57% in mature seeds (Eckardt et al., 1971).

Energy content changes in the same way. As a net result, the production of dry matter, and the storage of carbon and chemical energy by the whole plant are proportional to each other only until the end of the flowering period: the fixation of 1 g of carbon is equivalent to the production of 2·35 g of dry matter or to the storage of 39·5 kJ of chemical energy. After flowering is finished, dry matter production is reduced and energy fixation increases in relation to carbon fixation. This behaviour can be interpreted in the following

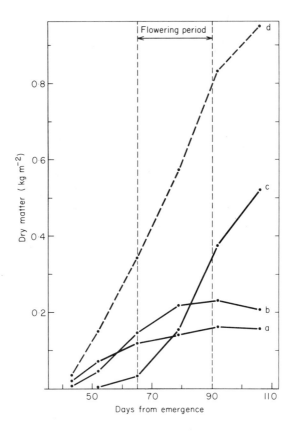

Fig. 18. Development through the growing season of the biomass of different sunflower organs: (a) leaves; (b) stems; (c) heads and (d) total biomass (including roots) (reproduced from Eckardt *et al.* (1971) with permission).

way: to manufacture leaves and stems the plants synthesize mostly carbo-hydrates for which 1 g of carbon is equivalent to 2·5 g of dry matter or 39·0 kJ of chemical energy. The formation of seeds then require the synthesis of lipids for which the carbon and energy contents are much higher: palmitic acid, for example, contains 74% of carbon and 51·9 kJ of energy per gram of carbon. The present varieties of sunflower contain about 50% of lipids in mature seeds.

Two important conclusions are:
1. measurements of CO_2 exchanges of plants refer to carbon production which is not always proportional to dry matter or energy production;

2. even when such proportionality exists the plant may be synthesizing different compounds, like carbohydrates or proteins, which have about the same energy content per gram of carbon.

D. Carbon Dioxide Fluxes Within the Canopy

Figure 19 presents CO_2 fluxes at three levels in a sunflower crop, calculated from data in Figs. 2a, 8 and 15 using the energy balance method. Smooth curves have been drawn by hand through the somewhat scattered points.

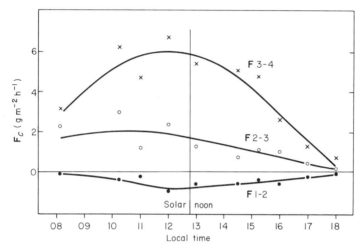

Fig. 19. CO_2 fluxes $\mathbf{F}_c$ at three depths: 46 cm ($\mathbf{F}_{1-2}$); 85 cm ($\mathbf{F}_{2-3}$) and above the crop ($\mathbf{F}_{3-4}$) (reproduced from Saugier, 1970, with permission).

The maximum CO_2 flux is apparently reached before solar noon whereas maximum evaporation occurs after solar noon (Figs. 10, 11b), probably because soil respiration later increases in response to increasing soil temperatures. The same trend is apparent in Impens' (1973) measurements.

$\mathbf{F}_{1-2}$ is the net CO_2 flux at 46 cm; it is equal to the soil respiration minus the photosynthesis of the leaves located below 46 cm. Daily totals of the three fluxes have been calculated and were presented in Fig. 18b.

The accuracy of fluxes measurement within the canopy should not be overestimated: 50% seems to be a reasonable order of magnitude, while 20% would be more appropriate for total fluxes above the crop (Lemon, 1970; Saugier, 1970). This degree of uncertainty may seem discouraging in view of the complexity of the instrumentation required. It may be remembered, however, that measured profiles may be used in two different ways: to derive fluxes as above, or to check the assumptions made in complex micro-meteorological models (Stewart and Lemon, 1969; Lemon *et al.* 1971) by comparing computed and observed profiles. By using the model, fluxes may be computed with a good accuracy, provided computed profiles fit the observations reasonably well. Besides, the interpretation of the results is easier because the assumptions have to be explicit in a mathematical model.

E. Photosynthetic Behaviour of Leaves

1. *Light Response*

The light dependence of photosynthesis in sunflower leaves has been studied by several workers: Pisek and Winkler (1959), Hiroi and Monsi (1966) in the laboratory; Hesketh and Moss (1963), Horie (1968), Eckardt *et al.* (1971) in the field.

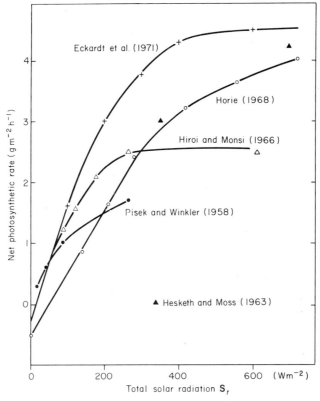

Fig. 20. Light response curves of photosynthesis for sunflower leaves, redrawn from several authors.

Their results have been redrawn in Fig. 20. Radiation units are converted to Wm^{-2} of solar global radiation (0·3 to 3 μm), using when necessary the equivalence 1 W m^{-2} ~ 117 lux; this relation assumes that 47% of the global radiation is photosynthetically active and that 1 W m^{-2} of PAR is equivalent to 250 lux (Gaastra, 1959). Photosynthesis data of Pisek and Winkler, given per gram of dry weight, have been referred to unit area, using a relation given in their paper.

The measurements in the field with the sun as a light source provide the highest maximum photosynthetic rates: 4 to 4·5 g m^{-2} h^{-1}. They are believed to be representative of the behaviour of leaves in a real crop. The laboratory results are lower; this may be due to the use of detached leaves (Hiroi and Monsi, 1966) and for Pisek and Winkler (1959) to the use of a wrong leaf area/dry weight ratio of 330 cm^2/g compared to an average of 110 cm^2/g for sunflower grown in Montpellier.

As shown in Section V.B, the curve of Eckardt et al. (1971) was used as an input in de Wit's model and gave carbon fluxes for the whole crop close to those obtained with other methods (Figs. 17 and 18). The slope at the origin in that curve is $4·5 \times 10^{-3}$ mg/J, giving an efficiency of light energy conversion of 12·8% and a quantum yield of 0·059 mole of CO_2 per Einstein. The energy efficiency is close to values obtained by Gaastra (1959) for sugar beet (13·1%), cucumber (12·2%) and tomato (12·8%). The average quantum yield of 0·059 falls within the range 0·054 to 0·076 found for maximum quantum yield by McCree (1972) for 21 species including sunflower.

2. Temperature and Age Dependence; Photosynthesis and Respiration

Pisek and Winkler (1959) and Hiroi and Monsi (1966) studied the influence of temperature on photosynthesis and respiration. In bright light (30 klux), they found a change in photosynthesis between 15 and 30°C: the photosynthesis temperature curve exhibits a broad peak centred at 20 to 25°C. The peak occurs at lower temperatures for lower intensities, down to 10 to 15°C at 3 klux.

Photosynthesis also depends on leaf age. In broad terms it increases for the first week after unfolding, stays more or less constant for the following two weeks and then decreases quickly to zero during the fourth week (Hiroi and Monsi, 1966; Horie, 1968). This evolution is paralleled by the change in colour of the leaves which turn progressively yellow, then brown. Obviously the duration of each stage will depend on the variety of sunflower and on the climatic conditions.

The respiration rate of a leaf does not seem to vary with leaf age for the first four weeks after which it decreases to zero (Hiroi and Monsi, 1966).

Dark respiration rate, expressed per gram of dry weight, is about the same for leaves, stems or roots of sunflower, from the few measurements available (Hiroi and Monsi, 1966); it has a Q_{10} of about 2, and a value of 2 mg CO_2 $g^{-1} h^{-1}$ at 20°C, for a plant of about 50 days. It is strongly dependent on plant age (Kidd et al., 1921).

3. Water Status

Boyer (1970) studied the effects of moisture stress on photosynthesis, dark respiration and enlargement of sunflower leaves. He found that photosynthesis and respiration rates were not affected until leafwater potential fell below −8 bars, whereas leaf elongation was slowed below −2·5 bars and stopped below −4 bars. Essentially the same results were obtained for corn and soybean. The exact figures would probably be different for a plant grown in the field and adapted to drought, but the conclusion that growth would be inhibited well before photosynthesis would still hold. The reduction in photosynthesis is likely to be the result of stomatal closure (Fig. 14), so as Boyer pointed out, 'there may be considerable differences in the response of stomata and cell enlargement to reduced turgor'.

Eckardt et al. (1971) also reported that photosynthesis per unit area of leaf was about the same in two sunflower crops with different water supply, but leaf area for the stressed stand reached only 1·1 compared with 1·8 in the irrigated stand. If in an early stage of growth, both crops had the same leaf area index, they would fix the same amount of carbon per unit area of ground, but the stressed crop stored a fraction of the photosynthates that could not be used to increase leaf area. As a result the leaf area index of the stressed crop increased less rapidly than that of the irrigated one, because the photosynthates were not used in the same way and because the photosynthesis per unit area of ground became progressively smaller.

4. Resistances to H_2O and CO_2 Transfer

Minimum values of resistances for sunflower leaves (two sides) and for an entire crop are presented in Table III. The boundary-layer resistance is taken for a large leaf (25 cm) at a wind speed of $0·75$ m s^{-1}. Resistances to CO_2 are calculated using the formulae

$$(r)_{CO_2} = (r)_{H_2O} \times (D_{H_2O}/D_{CO_2})^{\frac{2}{3}} \sim 1·5(r)_{H_2O}$$
$$(r_s)_{CO_2} = (r_s)_{H_2O} \times (D_{H_2O}/D_{CO_2}) \sim 1·7(r_s)_{H_2O}$$

where D_{H_2O} and D_{CO_2} are the molecular diffusivities of water vapour and carbon dioxide. The mesophyll resistances were computed from maximum

Table III

Minimum values of leaf and canopy resistances (s cm^{-1})

Type of resistance	Symbol	H$_2$O	CO$_2$	Place	Source
Boundary-layer	r	0·4 (typical)	0·6	field	Berger (unpubl.) Martin (1943) quoted by Raschke (1956b)
leaf	r_l	0·42 0·62 1·1	0·72 1·9	laboratory laboratory field	Holmgren et al. (1965) Laisk et al. (1970) Berger (1971)
Mesophyll	r_m		2·4 2·1 1·9	laboratory laboratory field	Holmgren et al. (1965) Laisk et al. (1970) Eckardt, et al. (1971)
Total	Σr_i	1·5	4·4	field	Eckardt et al. (1971)
Canopy	r_c (L = 1·8) r_c (L = 3·6)	1·1 0·8	3·9 5·8	field field	Eckardt et al. (1971) Impens (1973)

photosynthetic rates and values of boundary-layer and stomatal resistances, assuming the CO_2 concentration in the chloroplasts was zero. The same assumption was made for computing canopy resistances to CO_2.

The values of r_m do not differ appreciably, but the leaf resistance measured in the field differs considerably from the values found in the laboratory. This may be attributed to a difference in behaviour of the leaves in the field, submitted to a much higher evaporative demand than in the usual conditions of the laboratory. The high value of $1 \cdot 1 \, s \, cm^{-1}$ seems to be confirmed by estimations of a canopy resistance to H_2O of $1 \cdot 1$ and $0 \cdot 8$ for a leaf area index of respectively $1 \cdot 8$ and $3 \cdot 6$. The higher canopy resistance to CO_2 found in the denser crop may be caused by soil respiration which was not known, and was not added to the atmospheric CO_2 flux.

VI. CONCLUSION

A certain number of subjects have been covered in this chapter; far more have been left aside, such as photorespiration, in spite of the wealth of available measurements, because it was felt that laboratory evidence could not yet be used to interpret the CO_2 exchange of an entire crop. It may be useful to summarize the few specific and general points which have emerged and to point out the numerous topics that require further research.

The problem of radiation penetration is relatively well understood provided the geometrical structure of a crop can be described adequately, including the clumping effect. No theory exists yet for predicting wind speed u and the eddy diffusivity K within the plant stand in terms of wind action on individual elements, although some progress has been made in that direction as discussed in Chapter 3. Despite this lack, temperature and humidity profiles within the canopy may be predicted with reasonable success when leaf resistance is known (Stewart and Lemon, 1969; Murphy and Knoerr, 1970), by solving the energy balance equation for individual leaves (Raschke, 1956) and assuming that both u and K are exponentially related to height. Boundary-layer resistance of sunflower leaves may be estimated by a classical formula; leaf epiderma resistances are found to be larger in the field than in the laboratory (Table III). The stomatal response to water stress is not clearly understood, possibly because of the inability to measure a relevant water potential in a leaf composed of several compartments.

When there is no shortage of water, crop photosynthesis may be computed from light penetration and the photosynthesis–light response of a single leaf using de Wit's (1965) model. Sunflower has one of the highest photosynthetic rates for a Calvin-cycle plant, i.e. 4 to $4 \cdot 5 \, g \, m^{-2} \, h^{-1}$, although some lower rates have been reported for plants grown indoors. Its photosynthetic rate is not too sensitive to temperature but depends markedly on leaf age. It is influenced by water stress to a lesser extent than extension growth.

Plant respiration is not well understood: it depends on plant age (Kidd *et al.*, 1921), on temperature, on water stress and possibly on previous assimilation, but these factors need to be integrated into a comprehensive picture. Soil respiration is still much more difficult to estimate, specially the contribution from microorganisms. It is, however, a required input for predicting CO_2 profiles within a crop (Stewart and Lemon, 1969).

Sunflower plants share their photosynthates between different organs i.e. roots, stems, leaves and heads in a way which seems closely correlated with phenological stages, the duration of each stage being presumably controlled both by heredity and by environment. During the vegetative stage, the plants attempt to form a closed canopy by growing leaves as quickly as possible, provided there are enough roots to take up water and nutrients and provided stems are strong enough to withstand wind action. After flowering, the plant puts all the new photosynthates in the seeds, but does not grow any new leaves, mainly as a result of selection. Ideally then, canopy closure and hence light interception should be nearly complete when flowering starts, and leaves should remain active as long as possible to maximize final yield (Jones, 1967). A leaf area index of 1·8 produced by a density of four plants/m^2 seems to be sufficient to give an optimal yield of 3300 kg/ha. Having a relatively small leaf area index may help to limit water losses by increasing canopy resistance (Table III), although under dry conditions nearly all the net radiation is used for evaporation (Fig. 10).

The above discussion of growth is very crude, and much more research needs to be done on various topics such as translocation, partitioning of photosynthates and its dependence on water stress, isolation of environmental parameters affecting growth and flowering (Monteith, 1972), photosynthesis dependence on leaf age etc. It is felt, however, that the time is coming when plant physiologists and micrometeorologists may help plant breeders in selecting new varieties on a less empirical basis than in the past, the role of micrometeorologists being to supply relevant information on the environment and to test with their measurements models which predict how a plant grows hour by hour.

ACKNOWLEDGMENTS

Most of the data presented here result from teamwork in Montpellier. Particular thanks are expressed to A. Berger, F. E. Eckardt, G. Heim, M.

Méthy and R. Sauvezon. Technical help of J. Fabreguettes, F. Jardon and L. Sonié is also greatly acknowledged. Drs I. Impens, A. Laisk and J. Ross kindly supplied manuscripts or additional information on their work.

REFERENCES

Anderson, M. C. (1970). Radiation climate, crop architecture and photosynthesis. *In* 'Prediction and Measurement of Photosynthetic Productivity', Proc. IBP/PP Tech. meeting Třeboň. PUDOC. Wageningen, Netherlands.

Berger, A. (1971). La circulation de l'eau dans le système sol-plante. Etude de quelques résistances, en relation avec certains facteurs du milieu. Thèse de doctorat d'état. Montpellier. France.

Berger, A. (1973). Le potentiel hydrique et la résistance à la diffusion dans les stomates, indicateurs de l'état hydrique de la plante. *In* 'Plant Response to Climatic Factors'. Proc. of Uppsala Symp., Unesco, Paris.

Boyer, J. S. (1970). *Pl. Physiol.* **46**, 233–235.

Brown, K. W. and Covey, W. (1966). *Agric. Met.* **3**, 73–96.

Duncan, W. G., Loomis, R. S. Williams, W. A. and Hanau, R. (1967). *Hilgardia* **38**, 181–205.

Eckardt, F. E. (1967). *Oecol. Plant.* **1**, 369–400.

Eckardt, F. E. and Méthy, M. (1967). *Oecol. Plant.* **2**, 163–174.

Eckardt, F. E., Heim, G., Méthy, M., Saugier, B. and Sauvezon, R. (1971). *Oecol. Plant.* **6**, 51–100.

Gaastra, P. (1959). *Meded. Landb. Hogesch. Wageningen* **59**, 1–68.

Hesketh, J. D. and Moss, D. N. (1963). *Crop Sci.* **3**, 107–110.

Hiroi, T. and Monsi, M. (1966). *J. Fac. Sci.* (*Tokyo*) ser. III, **9**, 241–285.

Holmgren, P., Jarvis, P. G. and Jarvis, M. S. (1965). *Physiol.Plant.* **18**, 557–573.

Horie, T. (1968). Vertical distribution of photosynthetic intensity within a sunflower community. *In* 'Photosynthesis and Utilization of Solar Energy. Level III Experiments' (M. Monsi, ed.), University of Tokyo, Japan.

Hunt, L. A., Impens, I. I. and Lemon, E. R. (1968). *Pl. Physiol.* **43**, 522–526.

Impens, I. I. (1973). Daytime distribution of energy sinks and sources and transfer processes within a sunflower canopy. *In* 'Plant Response to Climatic Factors'. Proc. of Uppsala Symp., Unesco, Paris.

Jones, L. H. (1967). *Agric. Prog.* **42**, 32–52.

Kidd, F., West, C. and Briggs, G. E. (1921). *Proc. R. Soc.* B, **92**, 368–384.

Laisk, A. (1969). 'Measurement of Plant Cover Transparency'. pp. 174–185. Institute of Physics and Astronomy, Academy of Sciences of the Estonian S.S.R., Tartu.

Laisk, A., Oja, V. and Rahi, M. (1970). *Sov. Pl. Physiol.* **17**, 40–48.

Lemon, E. R. (1970). Summary section 2—Mass and energy exchange between plant stands and environment. *In* 'Prediction and Measurement of Photosynthetic Productivity', Proc. IBP/PT Tech. meeting Třeboň, PUDOC. Wageningen, Netherlands.

Lemon, E. R., Stewart, D. W. and Shawcroft, R. W. (1971). *Science.* **174**, 371–378.

McCree, K. J. (1972). *Agric. Met.* **9**, 191–216.

Martin, E. V. (1943). Studies of evaporation and transpiration under controlled conditions. Carnegie Inst. Wash. Publ. 550, 48.

Monteith, J. L. (1965). Evaporation and environment. *In* 'The State and Movement of Water in Living Organisms', Symp. Soc. Exp. Biol., Cambridge University Press.

Monteith, J. L. (1969). Light interception and radiative exchange in crop stands. *In* 'Physiological Aspects of Crop Yield', (J. D. Eastin, ed.), American Society of Agronomy, Madison, Wisconsin, U.S.A.

Monteith, J. L. (1972). Weather and the growth of crops. Amos Memorial Lecture, *In* Rep. E. Malling Res. Stn for 1971, pp. 21–34.

Murphy, C. E. and Knoerr, K. R. (1970). Modelling the energy balance process of natural ecosystems. Final research report. School of Forestry. Duke Univ. Durham, N.C., U.S.A.

Nilson, T. (1971) 'The Calculation of Spectral Fluxes of Short-wave Radiation in Plant Communities', pp. 55–80. Institute of Physics and Astronomy. Academy of Sciences of the Estonian S.S.R. (in Russian with English summary). Tartu.

Nilson, T. (1971). *Agric. Met.* **8**, 25–38.

Oliver, H. R. (1971). *Quart. J. R. met. Soc.* **97**, 548–553.

Parlange, J. Y. and Waggoner, P. E. (1971). *Pl. Physiol.* **50**, 60–63.

Parlange, J. Y., Waggoner, P. E. and Heichel, G. H. (1971). *Pl. Physiol.* **48**, 437–442.

Pisek, A. and Winkler, E. (1959). *Planta.* **53**, 532–550.

Raschke, K. (1956). *Planta.* **48**, 200–238.

Rose, C. W., Begg, J. E., Byrne, G. F., Goncz, J. H. and Torssell, B. W. R. (1972). *Agric. Met.* **9**, 385–403.

Ross, J. and Nilson, T. (1968). 'The Calculation of Photosynthetically Active Radiation in Plant Communities'. pp. 5–54. Institute of Physics and Astronomy. Academy of Sciences of Estonian S.S.R. (in Russian with English summary). Tartu.

Ross, J. and Ross, V. (1969). 'Spatial Orientation of Leaves in Crop Stands', pp. 44–59. Institute of Physics and Astronomy. Academy of Sciences of the Estonian S.S.R. (in Russian with English summary). Tartu.

Salisbury, F. B. and Ross, C. (1969). 'Plant Physiology', 767 pp. Wadsworth.

Saugier, B. (1970). *Oecol. Plant.* **5**, 179–223.

Spanner, D. C. (1951). *J. exp. Bot.* **11**, 145–168.

Stewart, D. W. and Lemon, E. R. (1969). The energy budget at the earth's surface: a simulation of net photosynthesis of field corn. U.S. Army ECOM Technical Report 2-68 (U.S. Army Electronics Command, Fort Huachuca, Arizona, U.S.A.).

Thom. A. S. (1971). *Quart. J. R. Met. Soc.* **97**, 414–428.

Varlet Grancher, C. and Bonhomme, R. (1972). *Ann. Agron. (Paris)* **23**, 407–417.

Wit, C. T. de (1965). Photosynthesis of leaf canopies. Agric. Res. Rep. Wageningen, No. 663, 1–57.

5. Cotton

G. STANHILL

Agricultural Research Organization, The Volcani Centre, Bet Dagan, Israel

I. INTRODUCTION

A. Anatomy, Physiology and Phenology

The two most widely cultivated species of cotton, *Gossypium hirsutum* L. and *G. barbadense* are allopolyploids ($n = 26$) of central and South American origin. The two other commercially grown species, *G. herbaceum* and *G. arboreum*, are diploids ($n = 13$), found in Africa and western and central Asia and in India, south-east Asia and the Far East respectively.

All four cultivated species are perennial or biennial in habit but the crop is normally cultivated as an annual. The growth habit of the cotton plant is complex. Apical meristems, which can give rise to either vegetative (monpodia) or flowering (sympodia) branches, develop in the axil of each cotyledon,

Contribution from the Agricultural Research Organization, Volcani Centre, Bet Dagan, Israel. 1973 Series, No. 170–E.

prophyll (the inconspicuous first leaf on each branch), and true leaf. Leaf formation on monopodial branches continues until halted by some stress or deficiency. The leaves are arranged spirally either to the left or right and the usual phyllotaxy is 3:8. The leaves have both stipules and petioles, and the latter, together with the leaves, can be either glabrous or hirsute according to variety. In most varieties, the leaves are palmately lobed, the indentations deepening with age and distance from the internode.

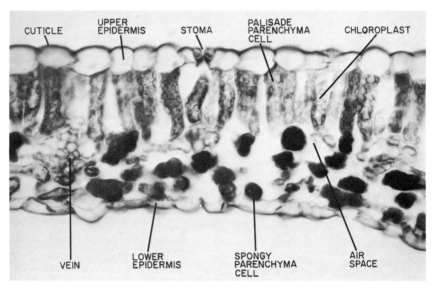

Fig. 1. Photomicrograph of dorsiventral section of cotton leaf (*Gossypium hirsutum* L.) from Schupp *et al.* (1969).

The anatomy of the cotton leaf, illustrated in Fig. 1, is typical of mesophytic crop plants. Stomata occur on both surfaces and densities ranging from 50 to 300 per mm^2 have been reported. The ratio of densities on the upper to lower surface found in six field observations averaged 0·50 and varied by only ±0·05. A wide range of values for the size of a fully open stoma is found in the early literature, a consensus of recent investigations suggests the maximum length of the stoma to be 15 μm with a maximum width of 6 μm.

Sympodial branches terminate in a flower after producing one prophyll and one true leaf, the development of a new branch from an auxiliary bud giving rise to the typical zig-zag appearance of flowering branches. The main or tap root of the cotton plant has a tetrarchal structure which gives rise to four, usually irregularly grouped, rows of lateral roots.

The following brief description of the growth and development of the cotton plant is based on a number of recent field investigations made on plants growing under favourable, semi-commercial conditions. After sowing, cotton seedlings take between 5 and 12 days to emerge above the soil. During the first, purely vegetative growth phase, from seedling emergence until the opening of the first flower 50 to 70 days later, plant dry weight and height increase exponentially with time (Heath, 1937). Over the whole crop life, from emergence till senescence, plant size, expressed as dry matter content on a logarithmic scale, can be described accurately by a second-order polynomial function of time (Hearn, 1969b).

The maximum crop dry weight is reached 100–150 days after emergence and values between 0·8 and 1·2 kg per m^2 soil surface have been reported (Hearn, 1969a, Marani and Aharanov, 1964; Saunt, 1967; Stern, 1965). The maximum rates of dry matter production, measured in the above studies were 20 to 25 $g m^{-2}$ day^{-1}, achieved 80 to 100 days after emergence. The three first references report a rapid subsequent decline in rate, reaching negative values only 10 to 20 days after the maximum rates were recorded and subsequently showing irregular and sometimes large variations above and below zero.

The two determinants of crop growth rate, the efficiency and the size of the photosynthetic apparatus, are both strongly influenced by plant age. Dry matter production per unit leaf area and per unit time, as reported from field measurements by Hearn (1969a) and calculated from measurements by Saunt (1967) declined linearly with time from a maximum of 13 $g m^{-2}$ day^{-1} 30 days after emergence, to zero, 100 days later.

The maximum leaf area index L, is greatly affected by agricultural practice, variety and climate (Hearn, 1969a; Stern, 1965). The maximum values most frequently reported lie between 2·5 and 4·0 and are reached between 90 and 110 days after seedling emergence (Hearn, 1969a; Rijks, 1967; Saunt, 1967; Stern, 1965). The subsequent decline in leaf area index is smooth but less steep than its increase.

There is a strong interaction between vegetative and reproductive growth, the onset of flowering leading to a marked reduction in the growth of both roots and leaves (Eaton, 1931), whilst the number of fruits which develop is in turn largely dependent on the rate of carbohydrate supply by the leaves (Hearn, 1969a). Under favourable conditions, the first flower is differentiated approximately 20 days after seedling emergence. The bracts are visible to the naked eye (the 'square stage') some 10 days later and after another 30 to 35 days the flower opens, i.e. 65 to 75 days after emergence. New flowers continue to appear at a very irregular rate for 50 to 60 days, the maximum rate of 2 per m^2 ground area per day generally coinciding with the middle of the flowering period (Hearn, 1969a; Saunt, 1967). At this peak stage of flowering the weight of flower buds and unopened fruit can comprise one quarter of the total plant dry weight (Marani and Aharanov, 1964).

Each flower is subtended by three large and leafy bracts and consists of five petals, white to creamy yellow in most varieties, with a tendency to pink or redness as the flower develops. The stamens and styles are united in columns and although self-pollution is usual, some cross-pollination by insects does occur. Fertilization is usually completed within one day of the flower opening and the fruit, a leathery capsule approximately 5 mm in diameter known as a 'boll' then starts to grow at the rate of 1 mm diameter per day for a period of 15 to 20 days. Maximum boll diameter, approximately 25 mm, is reached on the 25th day and about three weeks later the ripe boll splits into 4 to 5 loculli or 'locks' each containing 9 seeds. The boll then hardens and is ready to pick within two days, 40 to 50 days after the flower first opens.

It should be noted that a large proportion of the flowers and fruit produced by the cotton plant fail to mature, and in many cases the final yield is determined by the abscision rate of flower buds and bolls. The nutritional and water status of the plant is known to be of great importance in determining these rates, but the precise mechanism of control and the direct influence of environmental conditions is not fully understood.

The cotton seed is of considerable value, the embryo containing 20 to 25% oil. The seed coat is covered by a dense, woolly coat of cellulose hairs of two types. The short hair, or 'fuzz', is white to brown in colour and is inseparable from the seed coat. The long hairs, the cotton of commerce, are known as 'lint' and are usually white in colour, being easily separable from the seed coat. The commercial value of the lint is determined largely by the length and mechanical structure of these unicellular, cellulose hairs.

B. Distribution of Area and Yield

The average area devoted to the cotton crop throughout the world during the decade 1961 to 1970, was 32.6×10^6 ha, 2.3% of the perennially cultivated arable land in the world and identical in area to that cultivated 40 years ago. During the last decade the average world production of cotton per year was 11.0×10^6 tonne of lint and 17.5×10^6 tonne of cotton seed. Approximately one-third of the lint and one-thirtieth of the seed produced entered into international trade forming a major item in the national economies of a large number of countries.

The crop is grown commercially in about 60 countries, ranging from 47°N to 35°S. However, approximately 80% of world production is concentrated in eight countries and details of the area and yield for these centres of production are listed in Table I.

Table I
Area and yields of major centres of cotton production 1960–1969

Country and region	Area[a] (10^6 ha)	Annual yield[a] (tonne/ha) Lint	Seed
U.S.A. Texas, Mississippi, S.E. Coastal plain	5·0	0·54	0·93
U.S.S.R. Central Asia, Aral and Caspian Seas, S. Ukraine	2·5	0·76	1·47
CHINA N. China plain	4·2	0·27	0·54
INDIA Decca plain, Indore	7·9	0·12	0·24
BRAZIL Sao Paulo plain, N.E. coast	2·3	0·23	0·44
MEXICO Rio Grande and Colorado Valleys	0·7	0·68	1·21
EGYPT Nile delta and valley	0·7	0·65	1·18
PAKISTAN Indus valley	1·5	0·31	0·65

[a] From F.A.O. Production Yearbooks, 15–24.

II. TRANSFER PROCESSES

A. Radiation

1. *Characteristics of Individual Leaves*

Data for monochromatic reflectivity $\rho(\lambda)$, transmissivity $\tau(\lambda)$, and absorptivity $\alpha(\lambda)$, of field-grown cotton leaves are available from the following publications: Gausman *et al.* (1971a), McCree (1972), Allen and Richardson

(1968), Allen *et al.* (1970), Allen *et al.* (1971) and Gausman *et al.* (1971b).
Values of the ratio ρ/τ presented in Fig. 2 are taken from the first reference

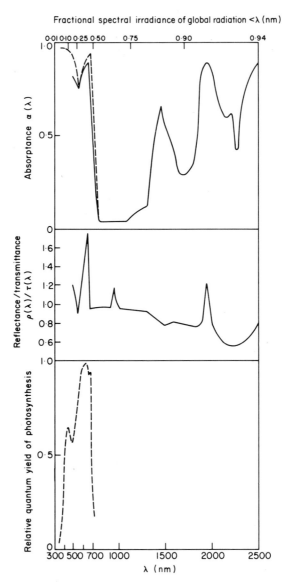

Fig. 2. Spectral absorptance, reflectance to transmission ratio and normalized quantum yield
of cotton leaves. Data from McCree (1972) — — —; and Gausman *et al.* (1971a) ———.

and those of α from the first two references. The same sources were used to compute weighted mean values of ρ, τ and α for photosynthetically active radiation, PAR (375 to 725 nm for cotton leaves, McCree 1972) near infra-red radiation (NIR) and the total solar spectrum (Table II).

Table II
Radiative characteristics of mature cotton leaves and stands

Waveband (nm)	PAR (375–725)	NIR (725–2500)	Whole spectrum (300–2500)
Individual leaves			
ρ	0·066	0·370	0·218
τ	0·060	0·411	0·235
α	0·873	0·218	0·545
Crop stands			
Reflection coefficient	0·073	0·41	0·25

The radiative characteristics of mature, field-grown leaves of cotton do not differ significantly from those of the wide range of mesophytic crop species investigated by Gausman et al. (1971a) and McCree (1972). However, significant differences have been found in cotton leaves of different age and nodal position (Gausman et al., 1971b and 1971c), and salinity (Thomas et al., 1967). The reflectivity for young, saline leaves was up to 30 % greater than for mature, non-saline leaves. These differences have been explained by reference to associated changes in the chlorophyll (Gausman et al., 1970), water content (Thomas et al., 1971) and structure (Gausman et al., 1971b) of the leaves.

Measurements of infra-red emissivity have been reported by Idso et al. (1969) for a wide variety of plant species including two species of cotton. The mean value for G. hirsutum cv. Deltapine 16 was 0·964 ± 0·007 and for G. barbadense cv. Pima S 4 was 0·979 ± 0·008. Both values fall within the range covered by two standard deviations; a range which included more than three-quarters of the 34 plant species examined.

2. Characteristics of Whole Stands

a. Reflectivity. The complexity and rapid rate of change of the canopy structure make it difficult to calculate the reflectivity and transmissivity of whole plant stands from the radiative characteristics of individual plant leaves, although such information is of considerable current interest for use in remote sensing systems.

Gausman *et al.* (1971b) used a very simple model of canopy reflectance to show, both theoretically and experimentally, that reflectance increases to a constant value, termed infinite reflectance, as the number of leaf layers— which they equate with the leaf area index L—increases. They found that in the 500 to 750, 750 to 1350 and 1350 to 2500-nm bands, values of $L = 2, 8$ and 2 respectively are required to reach infinite reflectance. Below 750 nm and above 1800 nm the values so attained are only slightly greater than those of individual leaves whilst in the 750 to 1350-nm band, infinite reflectance is more than 50 % greater than the reflectance from a single leaf. The limited data on the spectral reflectivity of cotton fields available from aircraft measurements (Richardson *et al.*, 1972) is in general agreement with the above predictions.

Table II contains mean daily values of crop reflectivities for the PAR (<690 nm) and near infra-red spectrum (>690 nm), obtained from ground measurements over a mature cotton field in Israel ($h = 1·11$ m, $L = 3·2$).

Detailed studies of reflected radiation in the total solar spectrum have been reported by a number of workers (Zuev, 1956; Fritschen, 1964; Rijks, 1967; Stanhill and Fuchs, 1968; Oguntoyinbo, 1970; Ritchie, 1971). The 54 values of daily solar reflectivity contained in the above references have a median of 0·20, and 90 % of the values fall within a range of $\pm0·05$. Clear evidence of seasonal changes in the daily values of reflectivity was found by Fritschen (1964), Rijk (1967) and Ritchie (1971) whose measurements show an increase of approximately 50 % between the early growth stages and maximum crop development. Much smaller seasonal differences were found by Zuev (1956) and Stanhill and Fuchs (1968).

Changes in the position of the sun and the structure of the crop stand cause diurnal as well as seasonal changes in the reflectivity of cotton stands. Marked diurnal increases in reflectivity at solar elevations below 20° have been reported by Rijks (1967) and Stanhill and Fuchs (1968) and somewhat smaller increases are evident in Zuev's (1956) records. Rijks (1967) also found a diurnal asymmetry in reflectivity with afternoon values generally lower than those measured at equal solar elevations before noon. This difference, as well as the gradual daily decrease noted after each irrigation, were shown experimentally to be due to changes in leaf posture caused by reduced leaf-water content. Evidently the greater gap frequency of the wilted stand, decreased the leaf and increased the soil component of reflected radiation. As leaf reflectivity is greater than that of the soil, stand reflectivity as a whole decreased. A different long-term effect of irrigation on reflectivity can be found in Zuev's (1956) measurements above irrigated and non-irrigated cotton fields in Turkestan where the reflection from irrigated fields was as much as 10 % *less* than that of the non-irrigated fields. The greater cover of young, dark leaves produced by the irrigated crops may explain this difference.

Changes in crop reflectivity induced by changes in plant-water status have been proposed as the basis for methods of controlling the irrigation of cotton stands (Haise and Hagan, 1967).

b. Stand transmissivity and absorptivity. No information on the spectral transmissivity or absorptivity of cotton stands has been found in the literature and even for the broad solar spectrum very few measurements have been reported. Stern (1965) has noted briefly the changing seasonal transmissivity for visible light as measured by a light-meter. Maximum transmissivity of 0·23 coincided with minimum seasonal values of $L \simeq 2$, and minimum transmissivity, 0·05 coincided with maximum values of $L \simeq 5$. Muminov (1968) presented measurements of short-wave absorptivity of cotton stands without specifying details. A curvilinear dependence of absorptivity on L is evident, such that 48 % of incident solar radiation was absorbed when $L = 2$, rising to a maximum of 69 % when $L = 4$.

The effect of stand geometry on short-wave transmissivity was measured by Baker and Meyer (1966) using the time-integrated output of a ground-level pyranometer which automatically transversed four cotton rows. The effect of N–S and E–W row orientations as well as of double and equidistant row spacing, was compared at different growth stages. Transmissivity was inversely related to L and to solar elevation. For a given value of L, the effect of row orientation was relatively small and much less than that of the planting pattern. The double rows transmitted considerably more radiation than the equidistant planting pattern.

Spatially integrated measurements of the radiation balance $\mathbf{R}_n$ at ground level under cloudless noontime conditions have been reported by Ritchie and Burnett (1971) for successive growth stages of a cotton stand. They found that the fraction transmitted by the stand was almost linearly related to the leaf area index : transmitted fractions of $\mathbf{R}_n$ were 80, 60, 45 and 40 % respectively for $L = 1·0, 2·0, 3·0, 3·5$.

3. *Radiation Transfer within Crop Stands*

The absorption of net short-wave radiation $\mathbf{S}_t - \mathbf{S}_r$ was measured by Stanhill (unpublished) within a mature field stand of cotton ($L = 3·2$) using a 3 m-long net pyranometer at four different heights. Daily totals of $\ln(\mathbf{S}_t - \mathbf{S}_r)$ at different heights in the canopy were plotted against L to give a straight line with a slope of -0.64, in almost exact agreement with the value expected for a plane parallel arrangement of leaves whose absorption coefficient $\alpha = 0.545$ (see Table II). Diurnally the extinction coefficient showed the usual dependence on solar elevation, k averaging 0.74 for solar elevations between $70°$ and $80°$ and dropping to 0.51 for elevations between $10°$ and $20°$.

Values of k were 5% greater in the afternoon than in the morning in accordance with the diurnal differences in stand reflectivity noted previously.

Niilisk *et al.* (1970) presented profiles of gap proportion in a dense stand of cotton ($L = 3.5$) and reported the same mean extinction coefficient, $k = 0.65$, already noted for net solar radiation.

4. *Long-wave Characteristics and Radiation Balance of Crop Stands*

Although the net long wave flux density of a crop L_n, can hardly be considered as a purely stand characteristic, a number of investigations have attempted to characterize cotton stands by a heating coefficient, $\beta = -\partial L_n/\partial R_n$ (Monteith and Szeicz, 1962).

For the entire growing season, a mean value of $\beta = 0.07$ was found by Stanhill and Fuchs (1968) to fit their own records as well as that of Fritschen (1964) and the non-irrigated cotton stand measured by Zuev (1956). Rijks (1968) reports a mean value of $\beta = -0.03$ for an irrigated cotton stand in the Sudan and Zuev's data for an irrigated crop yields $\beta = 0.02$. The observations of Fritschen, Stanhill and Fuchs, and Rijks all show a pronounced seasonal decline in β with some indication of an increase at the end of the growing season with the onset of senescence.

Statistical studies in which measurements of L_n over cotton stands were correlated with standard air temperature and vapour pressure measurements have been reported by Fitzpatrick and Stern (1965) for north-western Australia and by Rijks (1965, 1968) for southern Arabia and Sudan, respectively.

Zuev (1956) compared the radiation balance components of an irrigated and non-irrigated stand of cotton in central Asia and showed that R_n increased by as much as 10% as a result of irrigation. The major cause was the lower values of surface temperature of the cooler irrigated surface, the reduction in reflected radiation having only a minor effect. In contrast, Lemon *et al.* (1957) found R_n to be the same for cotton stands receiving different irrigation treatments.

Table III presents the components of the radiation balance as fractions of S_t for the three main growth stages at three sites for which data were available. It can be seen that the values for the short-wave components are similar at the three sites but there are major differences in the size of the long-wave components.

Table III

Radiant energy balance of cotton stands

Site: Latitude and longitude: References:	Central Asia 45°N 45°E Zuev (1956)	Texas 31°N 97°W Ritchie (1971)	Israel 31°N 35°E Stanhill and Fuchs (1968)
VEGETATIVE PHASE: EMERGENCE UNTIL FIRST FLOWER			
Number of days after emergence	1–37	1–35	1–35
S_t (MJm^{-2})	1100	760	990
S_r/S_t	0.19	0.10	0.13
L/S_t	0.25	0.35	0.42
R_n/S_t	0.56	0.55	0.45
FLOWERING PHASE: FIRST FLOWER UNTIL FIRST BOLL MATURE			
Number of days after emergence	38–103	36–90	36–97
S_t (MJm^{-2})	1840	1370	1710
S_r/S_t	0.20	0.17	0.17
L/S_t	0.18	0.24	0.43
R_n/S_t	0.62	0.59	0.38
MATURITY: FIRST BOLL MATURE UNTIL FINAL HARVEST			
Number of days after emergence	104–152	91–115	97–163
S_t (MJm^{-2})	1090	510	1290
S_r/S_t	0.20	0.19	0.17
L/S_t	0.17	0.28	0.44
R_n/S_t	0.63	0.53	0.39
TOTAL GROWING SEASON			
S_t (MJm^{-2})	4030	2650	3990
S_r/S_t	0.20	0.15	0.16
L/S_t	0.19	0.28	0.44
R_n/S_t	0.61	0.57	0.40

B. Momentum

1. *Characteristics of Individual Leaves*

Direct measurements of the drag coefficient of individual cotton leaves do not appear to have been reported in the literature, nor is there sufficient information on their surface characteristics and mechanical properties, or size and orientation to allow this characteristic to be calculated.

2. *Characteristics of Whole Stands*

Roughness lengths, z_0 for three cotton stands at different growth stages have been reported by Stanhill and Fuchs (1968) and Rijks (1971) and an additional value for a stand at its maximum height can be calculated from Jarman's data (1959). A similar pattern emerges from the four stands investigated: z_0 increases rapidly with crop height h, reaching maximum values of approximately 13 cm, afterwards declining slightly with stand maturity and senescence.

The measurements of Rijks (1971) and Jarman (1959) fit the relationship $\log z_0 = 1.476 \log h - 2.022$ derived from a series of measurements by Stanhill and Fuchs (1968), and although this relationship tends to overestimate z_0 somewhat for $50 < h < 80$ cm it does so to a lesser degree than the general relationship calculated by Szeicz *et al.* (1969) from measurements made over a very wide range of plant stands. Stanhill and Fuchs (1968) found that diurnal changes in z_0 averaged $\pm 10\%$ and were unrelated to wind velocity except on one occasion when high velocities coincided with a very open stand (ground cover 0.33). No effect of the wind direction or measurement position relative to that of the row was found.

Zero plane displacement heights, d are reported in the same references for which values of z_0 were quoted. As the height of the cotton stand increases, so does d, but the relationship shows considerable scatter once maximum values of h are attained. Before that stage, Stanhill and Fuchs report $d = 0.77\ h$ whilst Konstantinov (1966) analysed Aizenshtat's measurements over young cotton to yield $d \approx 0.8\ h$. Both these relationships give higher values of d than Rijks (1971) or Jarman's (1959) data suggest. A general relationship derived from measurements made over a very wide range of plant stands (Stanhill, 1969) suggests intermediate values of d for $50 < h < 100$ cm.

Diurnal variations in d reported by Stanhill and Fuchs (1968) and Rijks (1971) were small and unrelated to wind velocity. Unpublished measurements by the author over a cotton stand of $h = 127$ cm demonstrated that individual values of d and z_0 measured during neutral stability varied from their mean values of 70 and 13 cm respectively, by less than $\pm 10\%$ over a three-fold range of wind velocities.

3. *Transfer within Crop Stands*

No measurements of momentum transfer within cotton stands have been found in the literature nor is enough information available about the vertical distribution of density within stands to allow the attenuation of momentum to be calculated from general models. Unpublished data by the author for two cotton stands at maximum vegetative development ($L = 2.8$, 3.2) showed that maximum values of leaf area density, $a(z)$ of 0.035 cm^{-1} were located at relative heights z/h, between 0.45 and 0.70.

C. Water

1. *Characteristics of Individual Leaves*

The total resistance of cotton leaves to water vapour diffusion r_s and its cuticular, stomatal and mesophyll components have been studied extensively under controlled environment conditions. Jarvis and Slatyer (1970) found cuticular resistance to be 71 ± 6 s cm^{-1} for the upper and 191 ± 21 s cm^{-1} for the lower surface of cotton leaves whereas the results of Ehrler and van Bavel (1968) yield 75 s cm^{-1} for the upper and 32 s cm^{-1} for the lower surface of leaves also grown under controlled and constant environment. Measurements on soil-grown plants in a greenhouse (Slatyer and Bierhuizen, 1964a) give a mean cuticular resistance of 32 s cm^{-1} for both surfaces. The first and third reference show that cuticular resistance was independent of air water vapour concentration and temperature respectively.

The above figures represent maximum values of r_s; minimum values show greater agreement. Under constant environment conditions Ehrler and van Bavel (1968) report a minimum of 1.8 s cm^{-1} and Troughton (1969), 1.7 s cm^{-1}. Jarvis and Slatyer (1970) found minimum values of stomatal resistance to be 1.5 s cm^{-1} for the upper leaf surface and 2.5 s cm^{-1} for the lower surface with a minimum mesophyll resistance of 0.5 s cm^{-1}. Under greenhouse conditions even lower minimum values of r_s have been reported. Slatyer and Bierhuizen (1964a, b) reported 0.9 and 0.7 s cm^{-1}. In the second reference a marked increase in r_s with increasing leaf age was demonstrated.

Under favourable leaf-water conditions, r_s varies inversely with incident light intensity, the response of the upper leaf surface being more marked than that of the lower (Ehrler and van Bavel, 1968). Similar responses of stomatal aperture have been found in the field by Dale (1961) and Shimshi (1964). Measurements in a controlled environment by Slatyer and Bierhuizen (1964a) showed that more than 95% of the total decrease in r_s between maximum and minimum values was completed as the illumination increased from zero to 1000 footcandles, corresponding to 84 Wm^{-2} total radiant energy.

Leaf-water status also exerts a marked effect on r_s. Troughton (1969) noted a pronounced decrease in r_s as relative leaf-water content fell below 85% and Slatyer's results, shown by Cowan (1972), indicate a marked inflection in the water content/water potential relation of cotton leaves at 85%. At this same value, Dale (1961) showed that, in the field the stomata began to close in response to decrease in water content. Jarvis and Slatyer (1970) found r_m to be negatively correlated with the relative water content of cotton leaves, increasing from $r_m = 0$ at 100% to $1\,\mathrm{s\,cm}^{-1}$ at 80% and $2\,\mathrm{s\,cm}^{-1}$ at 65%.

Other field investigations have shown changes in stomatal aperture in response to changes in soil–water status (Moreshet, 1964; Shimshi, 1964; Ofir et al., 1968) and the last authors proposed using such changes as an index for irrigation need. The relationship between water status in the soil and in the cotton plant has also been investigated in the field (Weatherley, 1951; Longenecker and Lyerly, 1969, 1971).

The external or boundary-layer resistance to water vapour diffusion r_V has been studied under controlled environment conditions by Slatyer and Bierhuizen (1964a). Using model cotton leaves 100 cm^2 in area they found $r_V \propto u^{-0.47 \pm 0.048}$ where wind velocity u ranged from 0·6 to 3·2 cm s^{-1}. The exponent of u indicates that air flow was laminar over this range. Wind tunnel studies confirmed the expectation that the thickness of the boundary layer and hence r_V, is less for leaves of deeply-lobed cotton varieties (Baker and Myhre, 1969).

Field studies with model cotton leaves exposed at various heights within a crop stand gave the following relationship: $r_V = 2·7(d/u)^{0·5}$ where the characteristic length of the leaf d was 13·5 cm and u was in cm s^{-1} (Stanhill and Moreshet, 1967). The size of r_V calculated with the above equation are somewhat larger than would be expected from previous wind tunnel studies with regularly shaped surfaces parallel to the air flow.

2. *Characteristics of Whole Stands*

The external resistance of whole stands of cotton, r_{aV}, was calculated by Stanhill and Moreshet (1966) from the values of r_{aV} obtained with cotton leaf models exposed at various heights within a large commercial stand using the simple relationship $r_{aV} = r_V/L$. The values so calculated agreed with values of stand resistance to momentum exchange r_{aM}, computed from wind profile measurements corrected for atmospheric stability (Monteith, 1965). The agreement $r_{aV} = r_{aM}$, suggests that in the two cotton stands investigated, the non-transpiring elements of the stand did not contribute significantly to the total drag but a more complex aerodynamic factor may be involved too (p. 197).

The total internal plus external resistance of the stand to vapour diffusion $r_c + r_{aV}$ has been calculated both on a seasonal and on a diurnal basis (Stanhill and Moreshet 1966, 1967) using the relationship $r_c + r_{aV} = \delta\chi/E$ where appropriate values of the absolute humidity difference, $\delta\chi$, were measured and the vapour flux E, was estimated by the energy balance–Bowen ratio method. The internal, leaf resistance of the stand r_C was small although a sharp increase was observed towards the end of the growing season (Stanhill and Moreshet, 1966). In the more reliable diurnal measurements shown in Fig. 3, the minimum value of r_C was 0.17 s cm^{-1} and the

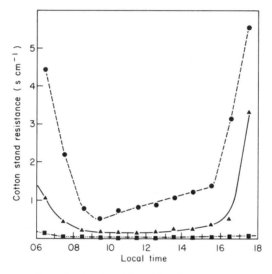

● r_{aH} Aerodynamic resistance to heat flux

■ r_{aV} Aerodynamic resistance to vapour flux

▲ r_C Canopy resistance to vapour flux

Fig. 3. Cotton stand resistance to water vapour and heat diffusion, Bet Dagon, 9.8.1967. $L = 2.5$, $h = 112$ cm. Data from Stanhill and Moreshet (1967).

average for all daylight hours, weighted by S_t was 0.35 s cm^{-1}. When multiplied by the leaf area index, the minimum values of r_C are consistent with previously reported measurements for the individual leaves of greenhouse grown plants.

Evapotranspiration from cotton stands E_t, can be calculated by the combined heat balance and aerodynamic equation (e.g. Monteith, 1965; see also Vol. 1, p. 90 *et seq.*) given the necessary climatic parameters and stand characteristics, including resistances. Figure 4A presents the results of such

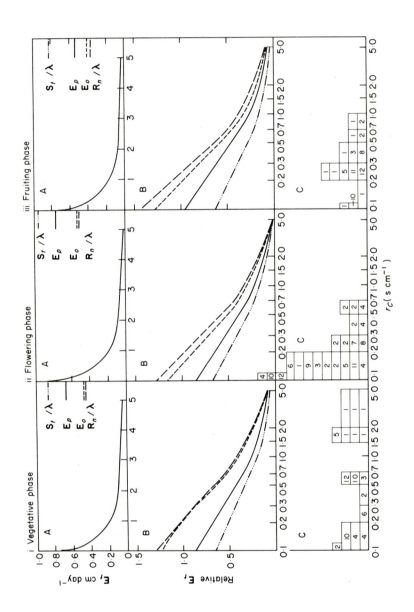

Fig. 4. The effect of cotton canopy resistance to the diffusion of water vapour (r_c) on
(A) absolute rates of stand evaporation;
(B) rates of stand evaporation relative to radiation balance etc. ——— $\mathbf{R}_n/\lambda$, – – – – open water evaporation $\mathbf{E}_0$, ··· — ··· $\mathbf{S}_t/\lambda$, ——— pan evaporation $\mathbf{E}_p$;
(C) Frequency distribution of canopy resistance during three phases of stand development.
   (i) *Vegetative phase*: stand emergence, first flower.
   (ii) *Flowering phase*: first flower, first open boll.
   (iii) *Fruiting phase*: first open boll, final harvest.

A

	Stand climate				Stand characteristics				
	T_a °C	e mbar	u m s^{-1}	$\mathbf{S}_t$ Wm^{-2}	z_0 cm	d cm	$\mathbf{S}_r/\mathbf{S}_t$	β	$\mathbf{L}_n$ (Wm^{-2})
(i)	19·0	14·0	1·8	245	4	30	0·16	0·06	−70
(ii)	25·5	25·3	1·2	266	12	80	0·20	0·02	−70
(iii)	23·5	23·0	1·3	278	10	75	0·24	0·02	−70

C Key to references to measured values of relative evapotranspiration used to estimate canopy resistances.
1. Namken *et al.* (1968). 2. Ritchie (1971a). 3. Zuev (1956). 4. Stern (1967). 5. Jensen and Haise (1963). 6. Carreker (1963). 7. Fuchs and Stanhill (1963). 8. Halkias *et al.* (1955). 9. Fritschen (1966). 10. Rijks (1971). 11. Rijks (1965). 12. Anon (1963).

calculations made for the three main growth stages using different values of r_c including $r_c = 0$, i.e. potential evapotranspiration. The calculated values of E_t are also presented in Fig. 4B as fractions of the water loss from an extensive open water surface E_0, and from a Class A evaporation pan E_p. These two water losses were also computed from the combination equation using appropriate surface characteristics. E_t is also shown in Fig. 4 as a fraction of the latent heat equivalent of the radiation balance of the stand i.e. R_n/λ, and of the incident global radiation S_t/λ. The climatic data used for the calculations were taken from the central Mississippi Valley, a major cotton growing region of the U.S.A. For the calculations of E_t the data was corrected for the microclimate effect of the crop by the relations given by Stanhill and Fuchs (1968).

The only independent test of the calculated values shown in Fig. 4 is a comparison with measurements made on open water surfaces. Values given by Kohler et al. (1959) agree to within 1% with the seasonal totals of E_p and E_0 calculated for the region.

Figure 4A shows E_t to be the same for a given value of r_c during the three stages of development, apparently because the seasonal reduction in the radiative term is compensated for by an increase in the aerodynamic term.

The numerous measurements of relative evapotranspiration from commercial field stands or large irrigation experiments reported in the literature have been used with the relationships given in Fig. 4B to calculate r_c. Although the relationships presented are strictly valid only for the quoted conditions of climate and stand characteristics, the relatively minor differences between the major cotton growing regions are unlikely to cause any major changes in the relationships between r_C and *relative* stand evapotranspiration. Figure 4C shows the distribution of the calculated values of r_C The wide range of surface resistances found during the first, vegetative phase is presumably due to the effect of dry soil surface when ground cover is incomplete and evaporation from bare soil, E_s, is a major component of E_t. During the second, flowering phase when transpiration is the major component of water loss, most values of r_C are less than 0.2 s cm^{-1} and $E_t \geqslant E_0 \geqslant R_n/\lambda$. During the last growth stage, somewhat higher values of r_C are evident, presumably related to the effect of leaf age on r_s already noted.

3. *Transfer within Crop Stands*

Ritchie and Burnett (1971) have described semi-empirical methods for calculating separately the soil, E_s and plant E_c components of cotton stand evaporation, E_t, based on two years of field measurements. E_s was calculated in three stages. In the first, non-water-limiting stage after rainfall or irrigation $E_s = R_n/\lambda$ until $\Sigma(R_n/\lambda) = 4 \text{ mm}$. Thereafter $E_s = 0.37\,(R_n/\lambda)$ until $\Sigma(R_n/\lambda) =$

10 mm after which E_s is independent of R_{ns} and is solely dependent on the water content of the top 3 cm of soil.

The plant component E_c, was then calculated by difference, i.e. $E_c = E_t - E_s$ and related to stand cover and soil–water status. When the soil–water potential in the active root zone was below -4 bars, E_c was found to be solely and linearly dependent on the volumetric soil–water content (Ritchie et al., 1972). Above this critical soil–water potential, E_c was related to leaf area index L and R_n/λ by the following equation:

$$E_c = -0.21 + 0.7L^{0.5}(R_n/\lambda) \quad \text{where } 0.1 \leqslant L \leqslant 2.7$$

The ratio E_c/E_c averaged 0.77 and 0.66 for the two seasons, the lower value being associated with lesser stand development.

4. Effect of Water Status on Evapotranspiration, Growth and Yield

The amount of information available in the literature is so large in relation to the space available here for its review that extended discussion of this important topic is impossible. Instead, reference is made to a recent bibliography which lists 247 items for the period 1955–1969 including a number of review articles (Brouwer and Abell, 1970). The same work lists three Commonwealth Agricultural Bureaux bibliographies on the water requirements of the cotton crops which include 197 items from the period 1927 to 1969. The response to water applied at different growth stages has been reviewed by Salter and Goode (1967). Recently, Framji and Mahajan (1973) have edited the results of a world-wide survey of cotton irrigation whose results are summarized in a 321-page volume containing 603 references.

D. Heat

1. Characteristics of Individual Leaves

The boundary-layer resistance to heat transfer r_H of cotton leaves was measured by Slatyer and Bierhuizen (1964a) under controlled environment conditions. They found $r_H = 0.8$ and 0.7 s cm^{-1} for wind speeds of 1.5 and 3.1 cm s^{-1} respectively; only 0.4 of the values for external resistance to vapour transport r_V, and with less dependence on wind speed. Linacre (1967) demonstrated that the relationship between r_H and r_V depends in a complex way on the values of r_l for the upper and lower surfaces of the leaf.

In the absence of reliable values of r_H it is not possible to calculate with any confidence the leaf-air temperature difference $T_1 - T_a = \Delta T$, to be expected for a given convective heat flux density C, from the Fick law relationship $C = \rho c_p \Delta T / r_H$. However there are a number of extensive field investigations from which the expected range in ΔT can be established. Wiegand and Namken (1966), using infra-red thermometry to measure T_1 of the exposed upper leaves of a cotton stand, found ΔT to be linearly related to global radiation, ambient air temperature and the relative water content of the leaf. Extreme values of ΔT ranged from 9° to $-4°$. A somewhat smaller range in ΔT was found by Palmer (1967) using thermocouple measurements of T_1.

In both investigations, differences of up to 5° were found between the leaf temperatures of wilted and turgid leaves. Similar differences were found in controlled environments during an investigation with chemical transpiration suppressants (Slatyer and Bierhuizen, 1964b). The maximum decrease in transpiration rate, about 80%, increased leaf temperature by 6°.

Palmer's field observations of boll temperature (Palmer, 1967) showed a similar pattern to that of leaf temperature, with some indication of a more clearly marked response to the effect of plant–water status.

2. Characteristics of Whole Stands

The heat transfer resistance r_{aH} of an irrigated, mature field stand of cotton was calculated by Fick's law deriving C and ΔT from energy balance and thermocouple measurements (Stanhill and Moreshet, 1967). The calculated values of r_{aH} presented in Fig. 3. show a clear diurnal trend which was closely and inversely related to that of wind speed, such that $r_{aH} \propto u^{-0.22}$. The mean daily value of r_{aH} weighted by C, was 1.1 s cm^{-1}, two and a half times r_{aV}, the equivalent resistance to vapour flux. Thus the relationship between the external resistance of the field stand to heat and to water vapour is very similar to that noted by Slatyer and Bierhuizen (1964a) for individual leaves.

When the flux of latent heat exceeds that of radiant heat, as it often does during the flowering and fruiting phases of cotton growth, negative values of C are common i.e. sensible heat is transferred from air to ground during the day. The resulting horizontal flux divergence of convective heat ΔC, have been reported as a common phenomenon in a number of large cotton growing areas. Thus Lemon et al. (1957) found C to be negative about 16 km downwind in a cotton growing section of Texas, and Hudson (1964) infers a similar situation from a 16 km-transect of evaporimeter measurements in the Gezira scheme of Sudan.

Rijks (1971) studied the advection of both convective and latent heat in considerable detail on a more local scale within one large cotton stand in Sudan. About 200 to 250 m downwind, ΔC was consistently significant, i.e. $\Delta C = 0.2 \, R_n / \lambda$ during the daylight hours and twice the vertical flux C. Negative values of both ΔC and C were also measured during the night hours. Rijks concludes that on a local scale, advection is of limited significance with fetches exceeding 175 to 200 m, but is of considerable importance as a general phenomenon throughout the entire area. The maximum reduction of air temperature resulting from the divergence of sensible heat flux reaches 3°C (Hudson 1964, Stanhill and Fuchs, 1968). The latter reference shows that the corresponding maximum increase in air vapour pressure is 3 mb.

3. *Transfer within Crop Stands*

Changes in the heat stored within the crop mass are insignificantly small over periods of one day or longer and even for shorter periods seldom reach 1 % of S_t (Rijks 1971; see also Vol. 1, p. 97).

The heat taken up or released by the crop stand in net carbon dioxide exchange **A**, is also a minor term in the heat balance. At the peak growth rate of 25 g m^{-2} day^{-1}, it reaches 4·8 W m^{-2}. Averaged over the period between plant emergence and the attainment of maximum stand dry matter, the net rate of heat storage by photosynthesis is 2 W m^{-2} and a similar value but of opposite sign, represents the mean for the subsequent period until harvest.

The flux of heat into and from the soil surface **G**, is of major significance in the early stages of crop development when ground cover is relatively small. At this stage average net fluxes of between 34 and 41 W m^{-2} have been reported (Zuev, 1956, Fritschen, 1966; Rijks, 1968). As the ground cover develops and the radiant flux at the soil surface decreases so does the soil heat flux. At the stage of maximum crop cover, mean net fluxes into the soil between 12 and 22 W m^{-2} have been measured (Zuev, 1956; Fritschen, 1966; Rijks, 1965, 1968).

Soil and climate characteristics also play a role in determining the magnitude of **G** and these factors may explain the very much lower soil heat fluxes reported for both early and later growth stages by Ritchie (1971) and by Stanhill and Fuchs (1968).

The effect of irrigation on **G** has been studied by a number of scientists. Rijks (1968) showed that with furrow irrigation, there was a clear cycle with maximum fluxes both by day and night, occurring approximately half-way in time between irrigations, the amplitude of the cycle decreasing with stand development. These cyclic changes are explained by the changes in both soil heat capacity and conductivity as the soil–water content varies. Comparisons of **G** measured at the same time under contrasted irrigation treatments shows much smaller differences (Zuev, 1956, Fritschen, 1966).

In Table IV the five terms of the heat balance equation

$$R_n = C + \lambda E + G + \mu A$$

are presented, averaged over the three main growth stages at the three sites for which records were available.

Table IV

Heat flux density balance of cotton stands. Daily average flux densities (W m^{-2}) and seasonal totals (MJ m^{-2})

Site:	Central Asia	Texas	Israel
Latitude and longitude:	45°N, 45°E	31°N, 97°W	31°N, 35°E
References:	Zuev (1956)	Ritchie (1971) Ritchie and Burnett (1971)	Fuchs and Stanhill (1963) Stanhill and Fuchs (1968)
VEGETATIVE PHASE: EMERGENCE UNTIL FIRST FLOWER			
R_n	191	137	146
λE	93	79	126
C	65	54	18
G	33	4	2
μA	—	0.5	0.5
FLOWERING PHASE: FIRST FLOWER UNTIL FIRST BOLL MATURE			
R_n	201	169	121
λE	184	166	146
C	−7	−4	−27
G	—	5	0.5
μA	—	2	2.5
MATURITY: FIRST BOLL MATURE UNTIL FINAL HARVEST			
R_n	163	124	87
λE	144	67	103
C	−7	57	−13
G	26	—	−2
μA	—	0.5	−0.5
TOTAL GROWING SEASON (MJm^{-2})			
R_n	2440	1500	1610
λE	1950	1180	1770
C	140	270	−170
G	350	40	5
μA	—	10	12

4. *Effect of Temperature on Stand Development and Growth*

The rate of development and growth of the cotton plant is strongly influenced by temperature. For emergence in the field, a mean daily soil temperature at sowing depth of at least 18°C is required (Wanjura *et al.*, 1967) but 21°C is needed for acceptably fast rates of germination and of emergence (Lomas and Shashoua, 1969).

The effects of air temperature, during both the night and the day, on rates of leaf and flower production have been studied extensively. The result of a recent investigation in a controlled environment has been presented by Hesketh *et al.* (1972).

E. Carbon Dioxide

1. *Characteristics of Individual Leaves*

The various resistances to carbon dioxide transfer can be divided into two groups: those occurring in the gaseous phase and those in the liquid phase. The first group, which includes boundary-layer, cuticular and stomatal resistances, is formally identical to that previously discussed for water vapour diffusion and may be calculated from them by multiplication with an appropriate function of the molecular diffusion coefficients of the two gases.

Resistances in the liquid phase jointly considered as mesophyll resistance r_m, include the transport resistance within the cell and the resistance to the photochemical carboxylation reaction.

Controlled environment studies with cotton leaves (Troughton and Slatyer, 1969) have shown that in air of normal oxygen concentration, $r_m \simeq 4$ s cm^{-1} and is unaffected by irradiance, leaf temperature over the range 23 to 38°C or relative water content over the range 56 to 92%. The latter finding is surprising because in experiments where resistance to diffusion in the gas phase has been eliminated, linear and inverse relationships have been reported between leaf–water potential and the exchange of CO_2 by cotton leaves, (Boyer, 1965) or the photochemical activity of isolated cotton chloroplasts (Fry, 1972).

Measurements by Slatyer and Bierhuizen (1964b) showed that r_m increased with the age of cotton leaves, but the increase however was one-eighth of the corresponding increase in the gas phase resistance.

The effect of variety and species on r_m is not known but their effect on CO_2 flux has been studied under field conditions by El-Sharkawy *et al.* (1965) and Muramoto *et al.* (1965). They found relatively small and statistically insignificant differences between the mean values of different varieties and species.

The maximum value of CO_2 flux in the above studies was $5 \, g \, m^{-2} \, h^{-1}$ and a large number of the species and varieties had maximum rates of about $4 \, g \, m^{-2} \, h^{-1}$ in an optimum environment. This is similar to the rates found with a number of other, rapid growing crops (El-Sharkawy et al., 1965).

The photosynthetic response of field-grown cotton leaves to the solar spectrum was studied by McCree (1972). The rate of CO_2 assimilation was negligible for light at wavelengths below 350 nm or above 750 nm. Within this waveband the relative quantum yield of the cotton leaves (Fig. 2) exhibited the same triple-peak curve with maxima at 440, 620 and 670 nm as the seven other species examined. Bierhuizen and Slatyer (1964b) and Boyer (1965) report similar, classical response curves for photosynthesis with a saturation flux density of incident light of approximately $100 \, W \, m^{-2}$. A similar response curve was measured in the field by El-Sharkawy et al. (1965). However, a second curve, measured on the leaf of a second cotton plant of identical variety and appearance, on the same day and in the same field and variety, showed no evidence of light saturation even when exposed to intense light at noon and its maximum photosynthetic rate was more than double that of the first leaf examined.

The effect of leaf temperatures between 25° and 40° on net CO_2 exchange is less than 10 % (El-Sharkawy and Hesketh, 1964). As Troughton and Slatyer (1969) report a threefold, linear increase in the photorespiration rate of cotton leaves over the same leaf temperature range, it must be assumed that there is a similar compensating effect on the rate of photosynthesis.

A marked reduction in assimilation rate has been reported for wilting cotton leaves, i.e. those whose water deficits exceeded 10 % (El-Sharkawy and Hesketh, 1964). The response was attributed to a measured decrease in stomatal aperture and hence an increase in resistance. A similar explanation was offered by Pallas et al. (1967) for the reduction in CO_2 flux which followed a decrease in soil–water matric potential. Boyer (1964), however, reported a reduction in CO_2 flux with increasing negative osmotic potential of culture solution during experiments in which the gaseous diffusion resistance did not vary. He also reported a reduction in the dark respiration rate of cotton leaves in the same experiment. Pallas et al. (1967) found that dark respiration rates first decreased and then finally increased with increasing soil–water deficit. They also reported that dark respiration rates decreased with the age of the leaf.

Rates of CO_2 exchange by non-photosynthetic plant parts are also markedly dependent on tissue age. Baker et al. (1972) reported CO_2 fluxes on a unit dry weight basis for cotton bolls exposed to a wide range of air temperatures. The loss of CO_2 by respiration reached a maximum, between 6 and 10 mg CO_2 g^{-1} dry weight h^{-1} (i.e. approximately $1\% h^{-1}$), on the first day of boll development with a reduction to a nearly constant rate of 2 mg CO_2 g^{-1} h^{-1} after ten days. This latter rate is similar to the dark respiration rate of cotton leaves at 25°C. The respiration rates of cotton flowers vary between 7 mg CO_2 g^{-1} h^{-1} at the square stage to 2 mg CO_2 g^{-1} h^{-1} the day after opening, with extremely rapid rates of approximately $4\% h^{-1}$ during the morning of fertilization (Baker, 1969).

The difficulties in applying the results of CO_2 exchange measurements made under controlled environment conditions to the varying climatic conditions found in the field is well illustrated by a recent study of the effect of varying air temperature on the CO_2 exchange of leaves of cotton plants grown under a range of day and night temperatures. Downton and Slatyer (1972) found clear evidence of acclimation in that in every case, the CO_2 exchange rate was fastest and r_m was least when the temperature during measurement corresponded to the daytime temperature at which the plants had been grown.

2. *Characteristics of Whole Stands*

Estimates of carbon dioxide exchange rates for cotton stands growing under undisturbed field conditions are, to date, available only for measurements made by the successive harvesting technique used in growth analysis. Muramoto *et al.* (1965) used this method to show that the dry matter production of cotton stands of different varieties was determined chiefly by varietal differences in rates of leaf area development and that, as with other crops, differences in dry matter production per unit leaf area and per unit time are small. They report that average values during the first two months of growth were within 10% of the mean, 13 g m^{-2} day^{-1}, for the eight varieties examined.

Hearn (1969) also found that variety, spacing and irrigation treatment produced only slight differences of about $\pm 10\%$ in crop growth rate whilst Saunt's (1967) data obtained from successional sowings, showed that the rate was unaffected by climatic factors after the first 50 days of growth.

Measurements of CO_2 exchange have been made on field stands of cotton with a transparent air-tight and air-conditioned system (Baker, 1965). The system was used to study the relationships between CO_2 flux, expressed on a unit ground area basis, and global radiation, air temperature, vapour pressure deficit and CO_2 concentration. The measurements were fitted by a multiple regression analysis which included time of day as a variable as it was found that afternoon rates of gas exchange were considerably less than those measured before noon when the environmental parameters were at the same level.

Baker and Meyer (1966) later used the same equation to show that the effect of different planting patterns and row orientations on CO_2 exchange could be explained by their effect on the flux density of intercepted global radiation, i.e. $S_i = S_t - S_r$.

The CO_2 flux $\mathbf{F}_c$ and intercepted solar radiation flux S_i were measured at 30°C air temperature with soil at field capacity and with a range of planting patterns. The two quantities were related by the simple equation

$$\mathbf{F}_c = a S_i$$

where

$$a = 2 \cdot 8 \times 10^{-6} \text{ g } CO_2/\text{Joule}$$

Baker and Hesketh (1969) adopted the same measurement system to study the effect of air temperature T_a on dark respiration rates. For a mature cotton stand with a total dry weight of $0 \cdot 68$ kg m^{-2} and $L = 3 \cdot 2$, the relation took the form

$$\mathbf{F}_c = n - m T_a$$

where

$$n = 0 \cdot 33 \text{ g } CO_2 \text{ m}^{-2} \text{ h}^{-1}$$

$$m = 0 \cdot 07 \text{ g } CO_2 \text{ m}^{-2} \text{ h}^{-1} \text{ °C}^{-1}$$

and

$$T = \text{temperature in °C } (> 5°C).$$

Stern (1965) used the relationship between $\mathbf{F}_c$ and S_i derived from Bierhuizen and Slatyer's (1964b) measurements made on individual leaves, to calculate the potential dry matter production of cotton stands by De Wit's model (1965). The calculated potential rates were then compared with measured rates of dry matter production throughout the year obtained by successive sowings and harvests. For one third of the year there was a reasonable agreement between the measured and calculated rates after allowing for a 40% respiration loss, but for the rest of the year dry matter production was considerably overestimated. In Israel the maximum rates of dry matter production reported by Marani and Aharanov (1964) agreed with the potential rates estimated by De Wit's method after allowing for 20% loss by respiration.

3. Carbon Dioxide Transfer Between Crop Stands

The importance of the soil beneath cotton stands as a source of CO_2 has not been reported but the combination of high soil temperatures, water and organic matter contents commonly met in cotton fields suggests that it may be significant.

4. *Effect of CO$_2$ Transfer on Crop Yield*

The yield of a cotton crop is only partly determined by the net dry matter yield representing a source. Measurements by Baker and Hesketh (1969) and by Hearn (1969a) show that less than half of the total dry matter produced during the reproductive phase is directed to reproductive growth. The two competing sinks, reproductive and vegetative tissues, are linearly related when plotted on a log scale as long as the growth rates are positive and the limited data available suggest that the slope of this allometric relationship, i.e. the ratio of reproductive to vegetative growth rate, is $= 0.55$, close to that expected between volumetric and predominantly areal growth.

Hearn (1969b) pointed out that the morphology of the cotton plant favours vegetative over reproductive growth and considers the key to increased crop yields is the adoption of practices which will bring the time sequence of dry matter production more closely into phase with that of reproductive sink development. Although it is clear that climatological factors play a major role in determining the size both of the source and of competing sink strengths, neither the relationships nor their interactions have yet been clearly established.

III. CONCLUSIONS

The extensive literature describing the interactions between cotton stands and the atmosphere shows no evidence that the transfer properties of this vegetation are in any way different from those of other mesophytic crop species growing under similar conditions of climate and soil. The transfer of radiation, momentum and heat can be estimated with considerably accuracy from climatological data using standard models of the transfer processes with the addition of a simple term describing the size of the cotton stand. With additional information on the internal resistance to diffusion, water vapour transfer can also be estimated with sufficient accuracy for field applications and there is some evidence that similar methods could be developed to allow the carbon dioxide exchange to be calculated.

A complete, meteorologically-based, model for the estimation of water vapour and carbon dioxide fluxes, which are of great agronomic interest, thus require an understanding of the effect of the climate on the size of the stand and its internal resistances. Whilst the literature contains much information on these relationships they are still almost entirely empirical in nature. A further requirement for the modelling of cotton yields is information on the functional relationships between the development and relative growth of the reproductive and vegetative tissue and the climate.

When cotton is grown under conditions of limited soil-water and nutrients—a common situation the functional relationships between the climate and the development and transfer properties of the root system and its interaction with above-ground growth should be known. The very limited attention devoted to the subject of roots in this chapter should be taken as reflecting the dearth of quantitative data rather than its lack of importance.

REFERENCES

Allen, W. A. and Richardson, A. J. (1968). *J. optic. Soc. Am.* **58**, 1023–1028.

Allen, W. A., Gausman, H. W. and Richardson, A. J. (1970). *J. optic. Soc. Amer.* **60**, 542–547.

Allen, W. A., Gausman, H. W., Richardson, A. J. and Cardenas, R. (1971). *Argon. J.* **63**, 392–394.

Anon, (1963). Bull, 113, Dept. of Water Resources, State of California, Sacramento.

Baker, D. N. (1965). *Crop. Sci.* **5**, 53–56.

Baker, D. N. and Hesketh, J. D. (1969). *Proc. Beltwide Cotton Production Res. Confs.* 60–64.

Baker, D. N. and Meyer, R. E. (1966). *Crop Sci.* **6**, 15–18.

Baker, D. N. and Myrhe, D. L. (1969). *Physiol. Pl.* **22**, 1043–1049.

Baker, D. N., Hesketh, J. D. and Duncan, W. G. (1972). *Crop. Sci.* **12**, 431–435.

Boyer, J. S. (1965). *Pl. Physiol.* **40**, 229–234.

Brouwer, C. J. and Abel, L. F. (1970). Bibliography 8. Int. Inst. Land Reclam. Improv. Wageningen.

Carreker, J. B. (1963). *J. geophys. Res.* **68**, 4731–4741.

Cowan, I. R. (1972). *Planta* **106**, 185–219.

Dale, J. E. (1961). *Ann. Bot. N. S.* **25**, 39–52.

De Wit, C. T. (1965). *Versl. Landb. Onderz.* No. 663.

Downton, J. and Slayter, R. O. (1972). *Pl. Physiol.* **50**, 518–522.

Eaton, F. M. (1931). *J. agric. Res.* **43**, 875–883.

Ehrler, W. L. and van Bavel, C. H. M. (1968). *Pl. Physiol.* **43**, 208–214.

El-Sharkawy, M. A. and Hesketh, J. D. (1964). *Crop Sci.* **4**, 514–518.

El-Sharkawy, M. A., Hesketh, J. D. and Muramoto, H. (1965). *Crop Sci.* **5**, 173–175.

Fitzpatrick, E. A. and Stern, W. R. (1965). *J. appl. Met.* **4**, 649–660.

Framji, K. K. and Mahajan, I. K. (eds) (1973). 'Irrigated Cotton—A World-Wide Survey'. Delhi International Commission on Irrigation and Drainage.

Fritschen, L. J. (1964). Rept. Components of the radiation balance of field crops under irrigated conditions. U.S.D.A. Soil and Water conservation Ariz.-WCL-18 10pps.

Fritschen, L. J. (1966). *Agron. J.* **58**, 339–342.

Fry, K. E. (1972). *Crop Sci.* **12**, 698–701.

Fuchs, M. and Stanhill, G. (1963). *Isr. J. agric. Res.* **13**, 63–78.

Gausman, H. W., Allen, W. A., Myers, V. I. and Cardenas, R. (1969). *Agron. J.* **61**, 374–376.

Gausman, H. W., Allen, W. A., Cardenas, R. and Bowen, R. L. (1970). *Photogrametric Engng* **36**, 454–459.

Gausman, H. W., Allen, W. A., Wiegand, C. L., Escobar, D. E., Rodriguez, R. R. and Richardson, A. J. (1971a). In S.W.C. Research Report 423, June, 1971.

Gausman, H. W., Allen, W. A., Escobar, D. E., Rodriguez, R. R. and Cardenas, R. (1971b). *Agron. J.* **63**, 465–469.

Gausman, H. W., Allen, W. A., Cardenas, R. and Richardson, A. J. (1971c). *Agron. J.* **63**, 87–91.

Haise, H. R. and Hagan, R. M. (1967). In 'Irrigation of Agricultural Lands'. Ch. 30, pp. 577–604 Am. Soc. Agron. Monograph 11. Madison, Wisc.

Halkias, N. A., Veikmeyer, F. J. and Hendrickson, A. H. (1955). *Hilgardia* **24**, 207–233.

Hearn, A. B. (1969a). *J. agric. Sci. Camb.* **73**, 75–86.

Hearn, A. B. (1969b). *J. agric. Sci. Camb.* **73**, 87–97.

Heath, O. V. S. (1937). *Ann. Bot. N. S.* **1**, 515–520.

Hesketh, J. D., Baker, D. N. and Duncan, W. G. (1972). *Crop Sci.* **12**, 436–439.

Hudson, J. P. (1964). *Emp. Cott. Gr. Rev.* **41**, 241–254.

Idso, S. B., Jackson, R. D., Ehrler, W. L. and Mitchell, S. T. (1969). *Ecology* **50**, 899–902.

Jarman, R. T. (1959). *J. Ecol.* **47**, 499–510.

Jarvis, P. G. and Slayter, R. O. (1970). *Planta* **90**, 303–322.

Jensen, M. E. and Haise, H. R. (1963). *J. Irrig. Dr. Div. ASCE* **89**, 1R4, 15–41.

Jones, H. G. and Slayter, R. O. (1972). *Pl. Physiol.* **50**, 283–288.

Kohler, M. A., Nordenson, T. J. and Baker, P. R. (1959). *Tech. Paper* **37**. U.S. Weather Bur. Wash. D.C.

Konstantinov, A. R. (1966). 'Evaporation in Nature, Israel Program for Scientific Translations', p. 128, Jerusalem.

Lemon, E. R., Glaser, R. H. and Satterwhite, L. E. (1957). *Proc. Soil Soc. Sci. Amer.* **21**, 461–463.

Linacre, E. T. (1967). *Pl. Physiol.* **42**, 651–658.

Lomas, J. and Shashoua, Y. (1969). *Cott. Gr. Rev.* **46**, 174–180.

Longenecker, D. E. and Lyerly, P. J. (1969). *Agron. J.* **61**, 687–690.

Longenecker, D. E. and Lyerly, P. J. (1971). *Agron. J.* **63**, 885–886.

Marani, A. and Aharonov, B. (1964). *Israel J. agric. Res.* **14**, 3–9.

McCree, K. J. (1972). *Agric. Met.* **9**, 191–216.

Monteith, J. L. (1965). *In* 'The State and Movement of Water in Living Organisms': *Symp. Soc. Exp. Biol.* Vol. 19, pp. 205–234, Cambridge University Press.

Monteith, J. L. and Szeicz, G. (1962). *Quart. J. R. met. Soc.* **88**, 496–507.

Moreshet, S. (1964). *Israel J. agric. Res.* **14**, 27–30.

Muminov, F. A. (1968). Report of the 6th Interdepartmental Symposium on Actinometry and Atmospheric Optics, June 1966, Tartu, pp. 309–314.

Muramoto, H., Hesketh, J. and El-Sharkawy, M. A. (1965). *Crop Sci.* **5**, !63–166.

Namken, L. N., Gerard, C. J. and Brown, R. G. (1968). *Agron. J.* **60**, 4–7.

Niilisk, H., Nilson, T. and Ross, J. (1970). In 'Prediction and Measurement of Photosynthetic Productivity', pp. 165–177, Pudoc. Wageningen.

Ofir, M., Shmueli, E. and Moreshet, S. (1968). *Expl. Agric.* **4**, 325–333.

Oguntoyinbo, J. S. (1970). *J. geograph Assn. Nigeria.* **13**, 39–54.

Pallas, J. E., Michel, B. E. and Harris, D. G. (1967). *Pl. Physiol.* **42**, 76–88.

Palmer, J. H. (1967). *Agric. Met.* **4**, 39–54.

Pinkas, L. L. H. (1972). *Crop Sci.* **12**, 612–615.

Richardson, A. J., Wiegand, C. L., Leamer, R. W., Peterson, H. D., Gerbermann, A. H. Torline, R. J. (1972). In S.W.C. Research Report 431. U.S.D.A. Weslaco, Texas.

Rijks, D. J. (1965). *J. appl. Ecol.* **2**, 317–343.

Rijks, D. J. (1967). *J. appl. Ecol.* **4**, 561–568.
Rijks, D. J. (1968). *J. appl. Ecol.* **5**, 685–706.
Rijks, D. J. (1971). *J. appl. Ecol.* **8**, 643–663.
Ritchie, J. T. (1971). *Agron. J.* **63**, 51–55.
Ritchie, J. T. and Burnett, E. (1971). *Agron. J.* **63**, 56–62.
Ritchie, J. T., Burnett, E. and Henderson, R. C. (1972). *Agron. J.* **64**, 168–172.
Salter, P. J. and Goode, J. E. (1967). Res. Rev. 2., C.A.B. Farnham Royal, England.
Saunt, J. E. (1967). *Cott. Gr. Rev.* **44**, 2–22.
Schupp, M., Gausmann, H. W. and Cardenas, R. (1969). In S.W.C. Research Report 412, p. 22, U.S.D.A. Weslaco, Texas.
Shimshi, D. (1964). *Israel J. agric. Res.* **14**, 137–143.
Slayter, R. O. and Bierhuizen, J. R. (1964a). *Aust. J. biol. Sci.* **17**, 115–130.
Slayter, R. O. and Bierhuizen, J. R. (1964b). *Aust. J. biol. Sci.* **17**, 131–146.
Stanhill, G. (1969). J. appl. Met. **8**, 509–513.
Stanhill, G. and Fuchs, M. (1968). *Agric. Met.* **5**, 183–202.
Stanhill, G. and Moreshet, S. (1966). In 2nd Ann. Rept. to U.S.D.A. Project AIO-SWC-29 pp. 52–75.
Stanhill, G. and Moreshet, S. (1967). In 3rd Ann. Rept. to U.S.D.A. Project AIO-SWC-29 pp. 39–89.
Stern, W. R. (1965). *Aust. J. agric. Res.* **16**, 347–366.
Stern, W. R. (1967). *Aust. J. agric. Res.* **18**, 259–269.
Szeicz, G., Endrödi, G. and Tajchman, S. (1969). *Water Resources Res.* **5**, 380–394.
Thomas, J. R., Wiegand, G. L. and Myers, V. I. (1967). *Agron. J.* **59**, 551–554.
Thomas, J. R., Namken, L. N., Oerther, G. F. and Brain, R. G. (1971). *Agron. J.* **63**, 845–847.
Troughton, J. H. (1969). *Aust. J. biol. Sci.* **22**, 289–302.
Troughton, J. H. and Slayter, R. O. (1969). *Aust. J. biol. Sci.* **22**, 815–827.
Wanjura, D. F., Hudspeth, E. B., Jr. and Bilbro, J. D., Jr. (1967). *Agron. J.* **59**, 217–219.
Weatherley, P. E. (1951). *New Phytol.* **50**, 36–51.
Wiegand, C. L. and Namken, L. N. (1966). *Agron. J.* **58**, 582–586.
Zuev, M. V. (1956). Formation of the Microclimate of Cotton Field, Hydrometeorol Publ. House, Leningrad. 115 pp. (in Russian).

6. Townsville Stylo (*Stylosanthes humilis* H.B.K.)

C. W. ROSE, J. E. BEGG, and B. W. R. TORSSELL

Division of Land Use Research, CSIRO, Canberra, Australia

I. INTRODUCTION

Townsville stylo, previously known as Townsville lucerne, is a tropical annual legume, native to central and southern America, and adapted to summer rainfall climates. Since its chance introduction to northern Australia in the early 1900s, Townsville stylo (hereafter contracted to T.S.) has become a well-established alien in tropical Australia. Its early establishment, spread, and general agronomic characteristics have been reviewed by Humphreys (1967). In the mid 1960s it had not been widely exploited and Humphreys concluded that, although the basic agronomic information had been available for some time, the financial structure of the beef industry in northern Australia did not favour large-scale investment in pasture improvement.

Since then, economic conditions became more favourable and the total area sown to T.S. in the Northern Territory rose over the last six years from approximately 20,000 to 90,000 ha (Begg, 1972). An assessment of the potential distribution of T.S. in tropical Australia (Begg, 1972) indicates that the total area suitable for its colonization as a naturalized alien is approximately 40 million ha. At present most of this area carries native pasture under Eucalypt woodland and is grazed by beef cattle. Average stocking rates are low—one beast to 20 to 50 ha—and even the best areas of native pasture carry only one beast per 10 ha. A major limitation for beef cattle production is the low nutritive value of the native pastures during the dry season. Thus the year is almost equally divided into a six-month period when the cattle gain weight (December to May) and six months when weight is lost (June to November).

With the introduction of T.S. into the native pasture, or with the provision of high-protein supplement, liveweight gain can be achieved during the dry season too, though cattle still lose weight from the time of the first storm rains of the wet season until edible new growth is produced. The length of this critical period is 6 ± 2 weeks, from late October to early December, and the average loss in liveweight is approximately 10 % (Norman, 1967).

Townsville stylo is being introduced with and without removing the tree component of the Eucalypt woodland (savannah) vegetation. The work described in this chapter refers to the cleared situation where regrowth of tree species and of perennial native grasses has been eliminated by cultivation so that annual grass species are the main competitors. The annual grasses have a low nutritional value at maturity and the liveweight gain of beef cattle in the dry season is related to the amount of T.S. in the pasture (Norman, 1970). Hence invasion by grass is associated with a decline in animal production and one of the aims of management is to check this invasion.

Detailed field environment studies on T.S. have been carried out at Katherine, N.T., Australia (longitude 132·3°E, latitude 14·5°S, altitude 108 m), and most of the quantitative information presented was obtained from this site. The soil type is Tippera clay loam, an extensive lateritic red earth in the Tipperary land system (Stewart, 1956).

At Katherine, temperatures during the growing period (Slatyer, 1960) are unlikely to be low enough to restrain the growth of T.S. significantly. The average Class A pan evaporation for the five-month growing period December to April over a five-year period (1968–72) was 1164 mm (i.e. 7·7 mm day^{-1}), while the average rainfall for the same period was 829 mm (cf. 823 mm average over 91 years). The water holding capacity of the soil (field capacity to wilting point) varies from 70 to 100 mm per metre depth, not all of which is withdrawn by the T.S. pasture. These figures emphasize the dominant importance of the water regime in T.S. production especially when run-off is an appreciable fraction of rainfall, e.g. in the 1971/72 wet season when run-off was 22% of rainfall.

The objectives of this chapter are:

(i) to review micrometeorological studies performed during the main growth period of this crop;

(ii) to summarize and discuss two models based on some of the data collected in (i) which assist in the interpretation and prediction of the very variable seasonal growth of T.S. pastures;

(iii) to outline an approach to population dynamics in the pasture. This approach has arisen in the context of the above studies for interpreting the changes through time in species proportion when the legume is invaded by annual grass—an ecological problem of great concern to the beef cattle industry in northern Australia.

II. STUDIES IN THE MAIN GROWTH PERIOD

The studies described in this section cover the main period of growth, including flowering and early seed setting (Torssell et al., 1968) of ungrazed stands of T.S. In such stands, yields of dry matter for a given level of super-phosphate application differ from one year to another by a factor of three at least, chiefly due to variations in water stress (Norman and Begg, 1973). This great inter-seasonal variability in yield provided the background for the studies of energy exchange and potential growth rates described below.

Kalma and Badham (1972) and Kalma (1972) investigated the radiation characteristics of pure T.S. pastures. Short-wave reflectivity or albedo of the pasture increased during periods of rapid growth, and decreased when severe water stress reduced the crop cover. The seasonal mean value was 0·19. Net long-wave loss from the sward was lower during periods of rapid growth, and after wetting by rainfall. Net radiation R_n (MJ m^{-2} day^{-1}) received by the sward during daylight could be estimated from total solar irradiance S_t (MJ m^{-2} day^{-1}) using a semi-empirical relation ($r = 0.99$):

$$R_n = 0.643S_t - 0.162.$$

Begg and Jarvis (1968) used characteristics of net photosynthesis determined in the laboratory, together with field measurement of radiation above and within the crop to compute maximum rates of net photosynthesis in the field. Comparing these predictions with measured growth rates they concluded that because leaflets were oriented normally to incident radiation, maximum growth rates would be achieved in swards about 20 cm high with a leaf area index L between 2 and 3. Slower growth rates were predicted for taller swards which are common in an ungrazed situation.

Micrometeorological techniques were used in complementary field studies to measure exchanges of water and sensible heat between crop and atmosphere (Rose *et al.*, 1972a), and root and sward morphology were recorded concurrently (Torssell *et al.*, 1968). This combination of micrometeorological and crop data was the basis for the growth models to be described in Section III.

Byrne *et al.* (1971) and Rose *et al.* (1966) described the instrumentation, digital data logging system, weighing lysimeter, and field site used in these studies. Graphical presentation of profiles of air temperature, humidity and net radiation on a computer line-printer allowed visual checking for errors, such as the drying of a wet-bulb, which were difficult to detect by other methods (Rose, 1972; Goncz and Rose, 1972). Following such inspection, data smoothing, interpolation, and analysis were executed using computer techniques. These techniques have the advantages of uniformity and objectivity in analysis, and are required by the large volume of measurements accumulated throughout the growing season of a crop.

Intermittent cloud cover is characteristic of the wet season in the topics. Therefore since conventional micrometeorological transport theory assumes that vertical fluxes are constant, the errors in applying such theory to nonsteady-state conditions had to be determined. This basic question in the application of micrometeorology was investigated by determining latent and sensible heat fluxes in two ways: using a variant of the energy balance or Bowen ratio approach; and independently by weighing lysimeter and fluxatron (Rose *et al.*, 1972a). The measurement of evaporation by lysimeter is indifferent to whether conditions are steady or not.

The energy balance equation was extended to include the rate of change of energy stored within the crop canopy (**J**), a term whose neglect requires validation when there are rapid changes in vegetation temperature. Using symbols defined on pp. ix to xi,

$$\mathbf{R}_n = \lambda \mathbf{E} + \mathbf{C} + \mathbf{G} + \mathbf{J}.$$

The term **J**, computed from measurements of leaf temperature and dry matter (Rose *et al.*, 1972a), was characteristically irregular in time, with peak values up to $50 \; \mathrm{W \; m}^{-2}$, i.e. greater than the soil heat flux in these experiments.

The method of smoothing or averaging raw data is more important when sampling under non-steady than steady conditions. The technique of 'exponential smoothing' (Brown, 1963) employed in this work is a backward-looking moving average, with weight decreasing in proportion to the age of the measurement. It therefore has the advantage over other methods of not introducing an artificial anticipation of future changes into the current mean.

Instead of deriving the Bowen ratio β from differences in air temperature and humidity between two heights above the sward, gradients in these parameters were obtained by fitting a theoretical (logarithmic) curve to three profile measurements using a least-squares criterion. The third above-crop measurement provides the opportunity of detecting significant errors in any one measurement, and fitting a logarithmic curve in no way detracts from the reliable character of the Bowen ratio approach when β is small compared with unity.

Figure 1 illustrates the diurnal pattern in fluxes between a T.S. sward and the atmosphere under a fairly clear sky. In such essentially steady-state conditions, the loss of water from the surface measured by weighing lysimeter, was within $\pm 6\%$ of the flux of water vapour determined by a profile technique (discrepancies were mostly within the limits of experimental error).

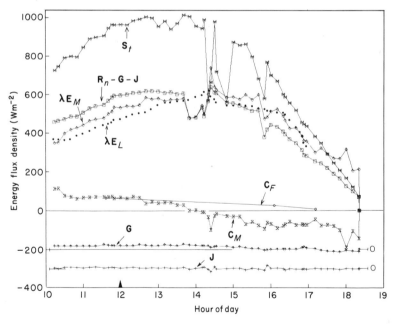

Fig. 1. History of components of the energy balance equation for T.S. pasture on March 20, 1967, under virtually cloud-free conditions. Symbols as in text with subscript M denoting micrometeorologically derived quantities, λE_L latent heat flux density derived from weighing lysimeter record, C_F sensible heat flux density by fluxatron. Origins displaced $-200\ \mathrm{W\ m^{-2}}$ for G and $-300\ \mathrm{W\ m^{-2}}$ for J. (After Rose et al., 1972a.)

In very variable conditions (e.g. Fig. 2) the evaporation from the lysimeter exceeded the flux of atmospheric water vapour by as much as 18 % over periods of nearly a day. Presumably this discrepancy arose because the assumptions of a steady-state were not met, but further clarification is desirable.

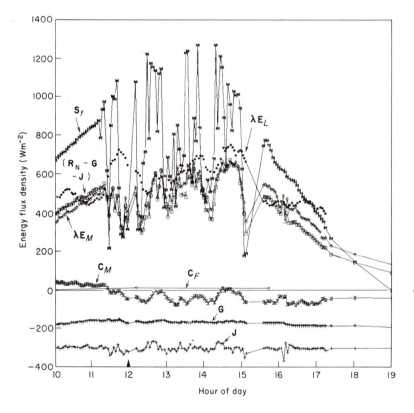

Fig. 2. History of components of the energy balance equation for T.S. pasture on March 30, 1967, with intermittent cloud. Symbols as in Fig. 1. Origins displaced -200 W m^{-2} for G and -300 W m^{-2} for J. (After Rose et al., 1972a.)

Other common assumptions in the application of micrometeorological theory to crops are that turbulence is responsible for transfer only in a vertical direction and that the average vertical wind speed is zero (Webb, 1965). Goncz and Rose (1972) showed that it was not always possible to interpret measurements within a sward without discarding some assumption in this theory. A typical discrepancy was an apparent sink for water vapour within the plant canopy. Using a smoke source and cine-photography in the field, Byrne and Rose (1972) demonstrated that the mean vertical component of wind speed in the sward was not always zero. Difficulties in a steady-state one-dimensional interpretation increased during the experiment as did irregularities in the geometrical characteristics of the stand (Goncz and Rose, 1972).

One conclusion to draw from this experience (which is likely to be relevant to other species) is that failure of the one-dimensional steady-state micrometeorological transport theory is far more likely *within* a plant canopy than *above* it. Such failure is more likely in a pasture known to be spatially heterogeneous in its water environment, commonly associated with patterns of micro-relief (Torssell, in press). It is likely that other workers have collected measurements in crop canopies which are not amenable to a one-dimensional interpretation but the problem has not been widely reported.

III. GROWTH MODELS

Conceptual models have commonly been used as a framework for research. Both in agriculture and ecology the utility of formalized conceptual models in research is now being investigated, and two examples follow. These examples are based on the experience that simple models (e.g. Fitzpatrick and Nix, 1969) are useful provided they concentrate on factors which have a major effect on growth and development. Both models are concerned with T.S. pastures in which other species are a minor component.

A. The Model of Rose et al. (1972b)

Rose *et al.* (1972b) developed a simple model to assist in the interpretation and prediction of the very variable seasonal pattern of dry matter production in swards of undefoliated T.S. Certain empirical relations in the model need to be established for the species and soil type concerned, and the model is designed for climates where extremes of temperature during the growing season are unlikely to retard growth. With these reservations the model should be applicable to other soil types and other pasture species.

There is both theoretical and experimental support (De Wit, 1960) that nutritionally-dependent relationships between accumulated evaporation and dry matter for a given crop are stable enough to be used predictively in seasons other than those in which the relationships were derived. Figure 3 shows these relationships for T.S. The equivalent rate of superphosphate

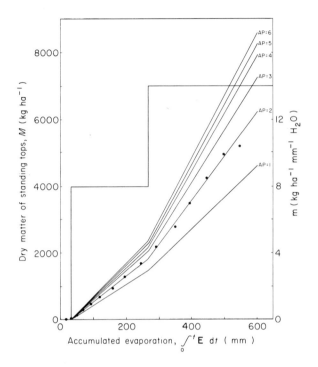

Fig. 3. The family of curves gives the relationships used between dry matter M of T.S. and accumulated evaporation in the model of Rose *et al.* (1972). AP is equivalent rate of super-phosphate applied before the wet season (in cwt/acre units). Points are measurements for AP = 2. Staircase plot gives value of m in Eq. (1) for AP = 2.

applied before the season (AP) is one input required by the model. The constant m also shown in Fig. 3 is defined by

$$\partial M / \partial t = mE \tag{1}$$

where M is dry matter per unit ground area and E is the evaporation rate. Figure 3 shows this relation cumulatively i.e. the standing dry weight M is plotted against $\int_0^t E \, dt$ where t is time after germination. For clarity measurements are plotted for AP = 2 only.

The relationship of Fig. 3 together with those of Fig. 4 (assumed independent of nutrition) are the key elements in the simulation model of Rose *et al.* (1972b). Curve (a) in Fig. 4 is based on data from periods in two seasons when evaporation from the sward was unlikely to be limited by water availability. When $M = 0$ (bare soil), the value of 0·25 for the evaporation ratio (Fig. 4) is the mean of variable ratios, but, as M increases, the value of

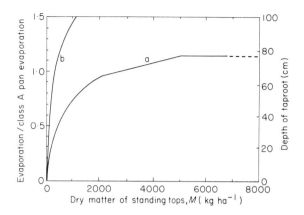

Fig. 4. (a) Stand evaporation/class A pan evaporation ratio if water is not limiting transpiration, as a function of M. (b) Relationship between depth of root system and M. (After Rose et al., 1972b.)

the ratio rapidly becomes more accurate. When water is non-limiting, the value of the evaporation ratio becomes independent of M above about 5000 kg ha^{-1}. It is a matter of computational convenience that the ratio is related to M than to leaf area index. Relating evaporation from a crop to that from an open water surface has proved a procedure of useful accuracy in simple models (e.g. Fitzpatrick and Nix, 1969).

The rudiments of root morphology are also calculated in the model, based on the root and shoot measurements in one season (Torssell et al., 1968). By far the most significant feature of root morphology for model performance was the depth of tap root penetration, shown at b in Fig. 4. Extraction of water by roots from any soil layer was assumed to stop when water content reached the wilting point.

The only inputs required by the model are AP, rainfall, and pan evaporation. Meteorological variables refer to a single day or to any integral number of days, and the distribution of soil-water through the profile is simulated by letting water which exceeds the available storage capacity in any layer of soil percolate to the layer below it. A simple model using daily rainfall to simulate run-off from the pasture has also been developed.

If the availability of water does not limit transpiration, the computation of growth rate proceeds as follows. The evaporation rate for the current value of crop dry matter, calculated from Class A pan evaporation as shown in Fig. 4, is accumulated and inserted in Fig. 3 to give the water-use efficiency-coefficient m governing the rate of increase of dry matter. Integration of this rate gives a new figure for accumulated dry matter. If the availability of water to the root system was inadequate then **E** was reduced, thus reducing $\partial M/\partial T$ [eq. (1)].

This procedure is shown as a flow chart in Fig. 5 using the symbolism of Forrester (1969). (Briefly, system levels are shown by rectangles, and control

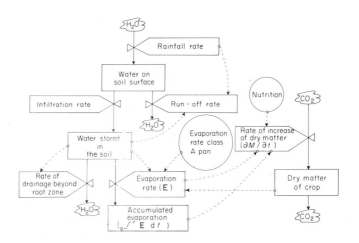

Fig. 5. Generalized continuous-system type presentation of the model used to simulate the growth of ungrazed T.S. pastures.

of rates of flow is indicated by the 'valve' symbol. Material flow is indicated by solid lines and information flow by dashed lines. Sources and sinks outside the system are represented by 'cloud' symbols.)

The growing season was assumed to end when the water content in the top 30 cm of the profile was close to the wilting point for approximately three successive weeks, though the exact period adopted was not critical.

The experimental relationships of Figs 3 and 4 are based on measurements during two seasons. Validity of the model has been tested by comparing predicted and observed growth and maximum yield in six other seasons. Agreement in seasonal dry matter production is good in seasons of average or above-average rainfall, for example an over-estimation of 8 % in maximum yield in a season with well above-average rainfall, as illustrated in Fig. 6.

The different assumptions, whose consequences are plotted with different symbols in Fig. 6, are concerned with a range of possible strategies for the uptake of water by plant roots, and with the range of available soil–water contents. There are apparent differences between the measured and simulated seasonal growth patterns in Fig. 6, and it is not clear why this is so.

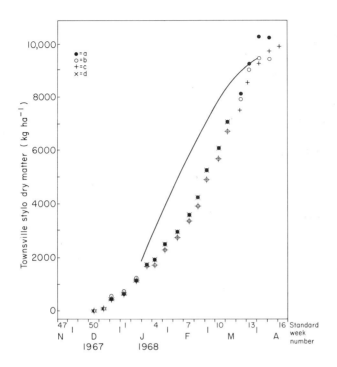

Fig. 6. Line represents experimentally measured growth of T.S. pasture as a function of time in the 1967–68 season. Points indicate simulated values using the model outlined in Fig. 5, different symbols referring to different possible assumptions. (After Rose *et al.*, 1972b.)

Although the model was based on standing dry matter and not total (i.e. standing plus shed) dry matter, there was little difference between these quantities in the two experimental seasons. From the basis of the model (Rose *et al.*, 1972b) which is partly summarized in Eq. (1), it might be expected that M would represent total dry matter rather than standing dry matter (when these are very different). This expectation is somewhat confirmed in seasons with much lower rainfall than those on which the model was based, and in seasons with severe water stress during growth.

This simple model reflects current limitations in our understanding of phosphate- and water availability as well as of other factors believed to be important in production such as environmental and nutritional effects on leaf life. Because the model gives no explicit attention to radiation absorption other approaches may be necessary to consider production of stands with a mixture of species (see Section IV). However, the model has contributed toward the interpretation of a very variable seasonal production, and it provides a method for extrapolating agronomic results to climatically different regions, though the validity of such extrapolation must be checked.

B. The Model of McCown (1973)

The relationship between yield and its probability is of major economic importance, especially in an environment such as tropical Australia where drought imposes the major restraint on agricultural productivity. McCown (1973) has derived such relationships for T.S. swards based on experiments near Townsville, Queensland (latitude 19°26 S, longitude 146° 50 E, altitude 75m) (Fig. 7). He has also shown how such relationships depend on the available soil–water storage capacity (Fig. 7), which can change markedly with soil type.

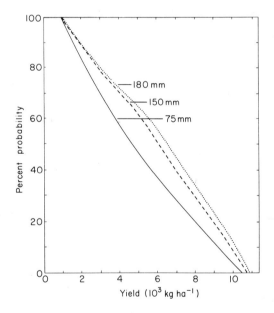

Fig. 7. Empirical probabilities of having more than given yields of T.S. per season at the three available soil-water storage capacities shown labelled in mm. (After McCown, 1973.)

The method used to simulate available soil–water was similar to that already described. Based on this method, the total period of active growth in the wet season was inferred for each of four years of experiments. For any particular water storage capacity, this growth period depended chiefly on rainfall, and not markedly on whether measured or average weekly evaporation was used. Thus, lacking long-term evaporation records, average evaporation was used together with 60 years of rainfall records to predict the long-term probability of any given growth time from a model of the water balance. From the four experimental seasons, relationships between yield and calculated growth period were derived. These relationships, assumed to hold generally, were then combined with those for the probability of a particular growth period to give the results shown in Fig. 7.

It follows from Fig. 7 that variations in yield between seasons can be much greater than those between different soil types in the same season, at least for the range of soils and associated water storage capacities investigated (McCown, 1971). The simulation of growth periods and yields for the 60 years has provided an assessment of the effects of seasonal climatic variations unattainable by the application of statistics to field experiments.

IV. ECOLOGY OF THE PASTURE

Invasion by annual grass weeds is a major management problem in stands of T.S. The main environmental factors determining the performance of this system are water, energy, and phosphorus. Within limits imposed by these environmental factors, significant modification of species composition can usually be exercised through the management of grazing and phosphate supply. Heavy grazing during the wet season usually ensures T.S. dominance, but leaves little fodder for the dry season. With the other extreme of no grazing, T.S. is commonly quickly suppressed. The development of better management strategies between these extremes is being sought through a pasture ecology programme with a strong micro-environmental component.

The principles of T.S. pasture ecology have been studied on cleared land ('The T.S.-annual grass pasture ecosystem of Torssell, in press) and extension of this study to woodland is planned. To describe changes in the proportion of legume to grass in this system, we follow a process identification approach and divide the life cycle of the pasture into three phases:

  (i) germination and establishment;
  (ii) main vegetative growth;
  (iii) seed production.

Changes in species proportion during any of these phases can be related by an expression derived from De Wit (1960):

$$y = \frac{ax}{ax + 1 - x} \qquad (2)$$

where x is the legume proportion at the beginning, and y at the end of the phase under consideration, and a is a constant. Experimental justification for the utility of this relationship for T.S.—annual grass pastures will be discussed in connection with Fig. 8.

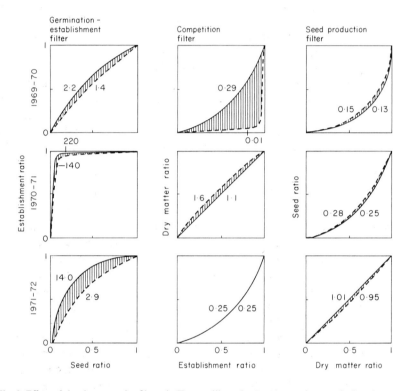

Fig. 8. Effect of the three species filters in Townsville stylo–*Digitaria* mixtures during three consecutive life cycles of the pasture. Note that the y-axis in each filter becomes the x-axis in the succeeding filter. The figures in the diagrams give the measured average values in filter-constant, a. (Low phosphorus = ——; high phosphorus = ------. 1969/70 pasture established under irrigation and mechanically defoliated. 1970/71 and 1971/72 pasture established from self-seeded swards without irrigation, and then grazed). (After Torssell and Rose, unpublished.)

Each phase in the life cycle may be likened to a 'filter' giving rise to changes in species proportion, with the constant a describing its characteristics. If $a = 1$, it follows from Eq. (2) that $y = x$. This equation is the diagonal through the origin in any of the nine replacement diagrams or filters shown in Fig. 8, and it indicates no change in the legume proportion during the period of time covered by the filter. If $a > 1$, the equation describes a curve to the upper left of the diagonal, and the legume proportion is increased during passage through the filter. If $a < 1$ the proportion is decreased, and the magnitude of the species change depends on how different a is from 1.

The life cycle of the pasture may thus be represented as three 'filters' in series representing the three phases of growth already defined: germination, vegetative growth, and seed production. All the ratios y and x in Fig. 8 are expressed as (legume/(legume + grass)). Thus the seed ratio is defined as the ratio of the number of viable T.S. seeds to the total number of viable seeds measured on the same area. The establishment ratio is similarly defined in terms of established plants, and the dry matter ratio is based on dry matter sampled at the end of the growing season. The filter constant, a in Eq. (2), has been determined as follows for each phase in three consecutive years with contrasting management and weather. For each filter the experimental data is converted to the ratio form already described, and each pair of values of x and y (which corresponds to the input and output of the filter) then define a point in the diagram. The set of such points corresponding to the range of experimental data are then fitted using Eq. (2) to give the best value of a. This procedure was carried out separately for a low and a high phosphorus treatment in each filter (Fig. 8), the two curves are each based on eight initially different legume proportions and six replicates. Equation (2) accounts for 88 % (range 63 to 99 %) of the variation in y. Figure 8 illustrates the considerable differences between phases in values of a in any year, and the effect of season and management on a in any given phase.

This approach emphasizes the need for micrometeorological information in studies of pasture dynamics and this need will be briefly discussed with reference to each filter.

Most seeds are dispersed in the upper 3 mm of soil. Changes in species ratio during the *germination and establishment* phase depend on rapidly changing water and temperature regimes in this surface layer.

During the dry season there is an increase in the percentage of 'soft' seed (i.e. viable seed which germinates when given favourable conditions) in response to diurnal temperature fluctuations. Before the main germination there is a larger decrease in viable seeds of *Digitaria ciliaris*, a common annual grass competitor to T.S., than in T.S. itself, associated with 'false starts' in response to early storms. Therefore it is not only the main germination event which is important, but also the changes in the physiological properties of the seed population since it was shed in the early dry season.

At temperatures commonly recorded near the soil surface, the rate of germination of T.S. in the field is greater than that of *Digitaria*. Temperature optima for germination and early growth are 25 to 27°C for T.S. and close to 35°C for *Digitaria*. When the soil is wet enough for germination its surface temperature remains between 25 and 30°C, but surface temperature may reach 60°C when the soil dries. If the wet season starts with an isolated thunderstorm there will be a greater germination of T.S. than *Digitaria*. A series of thunderstorms on consecutive days would result in a longer period when the soil surface was wet and a major germination of all species, including the slower establishing grasses so that the competitive advantage of T.S. would be lost.

Grass seedlings are more susceptible to water stress during germination and establishment than legume seedlings. For example in 1969/70 when swards were sown and established under irrigation, germination and establishment in the two species were similar (i.e. a close to 1, Fig. 8). With self-seeding under natural rainfall in the following two seasons, conditions were relatively unfavourable for grass ($a \gg 1$, Fig. 8). Intermittent germination occurs throughout the wet season, but many plants from later germinations do not survive or else remain very small. Though prolific in seed setting, the carry-over of seeds of *Digitaria* from the end of one wet season to the next is very small—only about 0·01% of the initial viable seed population. The hard seed carry-over in T.S. may amount to approximately 10%.

Since the water and temperature history of soil surface layers in a pasture depend largely on weather, there appears little opportunity for management to influence the a value of the germination and establishment filter. The value of a for this filter can vary over at least two orders of magnitude (Fig. 8), and its ecological significance depends on the long-term relative frequency of a values. Obtaining such information solely from repeated experimental determination for a sufficiently large number of seasons is not feasible. A more promising approach lies in developing a model of the germination filter, validating it for the experimental seasons, and then simulating a values in the large number of seasons for which standard meteorological records are available. This approach is now being developed.

The character of the *competition filter* during the main growth period depends on soil fertility and defoliation or grazing received. High fertility and no defoliation favours the grass through competition for light, resulting in a low value for a. Mechanical defoliation at 10-day intervals is less effective than grazing as is evident from the 1969/70 season $a \ll 1$ Fig. 8). In 1970/71, swards were grazed continuously at a high stocking rate, resulting in successful legume competition during the main growth phase ($a > 1$, Fig. 8). In 1971/72 grazing appeared to be less effective in controlling the grass. However, the lower a value (0·25) is more likely due to drought in January, since this caused early flower initiation in the legume, thus reducing its response to late rains in its vegetative growth. In a given season, therefore, the weather can apparently limit the effectiveness of manipulation through management.

The operation of the competition filter is being sought on the basis of growth rate for each species in relation to radiation absorption, inter-species shading, water availability, and the effects of defoliation or grazing.

The *seed production filter* generally favours the grass when swards are ungrazed (1969/70, $a \ll 1$, Fig. 8). If carefully planned, grazing may have a strong effect on seed setting of the grass, as in 1971/72. Townsville stylo flowers in response to daylength and water stress (Cameron, 1967), and *Digitaria ciliaris* may respond similarly. The effect of environment on seed setting after flowering has been little studied in either species, nor is the specific effect of defoliation on flowering and seed setting sufficiently understood.

V. MICROMETEOROLOGY AND ECOLOGICAL OBJECTIVES

The micro-environmental and physiological studies reviewed in Section II were dominantly concerned with processes and their interpretation in terms of physical and physiological concepts. This work initiated the broader study of pasture stability in a more variable agro-ecological setting (Section IV). The emphasis on identifying and interpreting key processes which is characteristic of micro-environmental work is proving very useful in this broader study of the species changes in pastures, in which an effective integration is sought between the relevant biological and environmental factors.

Another basic ingredient in the approach to pasture dynamics developed in this work is the explicit recognition of continuity in the pasture life cycle, and the possibility of subdividing each cycle into a number of phases, each of which is considered in terms of its micro-climatic and physiological components. The summation of a series of complex processes into one single constant (the a value) for each filter provides a powerful simplification of a variable and complex system. Thus variations in a reflect oscillations in the legume proportion through the annual cycle, and the germination filter is seen as an important natural mechanism in legume persistence.

The variation of a in response to weather and management (Section IV) indicates the dynamic nature of the system, and where management experiments are limited, conclusive results may not be obtained, even if experiments continue over many years. The present approach may shorten the experimental period by identifying the weather-dependent processes and indicating how *macro*-climatological probabilities may be related to the significant plant *micro*-climatic probabilities.

In horticultural crops and cereals, for example, the character of the plant community obtained is largely determined by man through relatively intensive management. The application of micrometeorology in this context (Smith, 1970) is thus concerned with optimizing productivity through management techniques which minimize adverse production factors, for example, by supplementary irrigation, frost protection, or pest and disease control. It can also be applied to broader problems such as matching genotype to climate, or to other decision-making activities.

In an ecological setting, however, the species components and their changes in space and time are much more the result of interaction with a range of primary and resultant micrometeorological variables. In this age of concern with ecological stability, do not such questions provide an even greater challenge to a micrometeorologist? However, to respond adequately to this challenge he should be a member of a suitable multidisciplinary team.

ACKNOWLEDGMENT

The assistance of Dr J. D. Kalma and Dr I. R. Cowan in improving the presentation of this chapter is gratefully acknowledged.

REFERENCES

Begg, J. E. (1972). *J. Aust. Inst. agric. Sci.* **38**, 158–162.
Begg, J. E. and Jarvis, P. G. (1968). *Agric. Met.* **5**, 91–109.
Brown, R. G. (1963). 'Smoothing Forecasting and Prediction of Discrete Time Series'. Prentice-Hall, New York.
Byrne, G. F. and Rose, C. W. (1972). *Agric. Met.* **10**, 13–17.
Byrne, G. F., Rose, C. W., Begg, J. E., Torssell, B. W. R. and McPherson, H. G. (1971). *CSIRO Aust. Div. Land Res. Tech. Pap.* No. 32.
Cameron, D. F. (1967). *Aust. J. exp. Agric. Anim. Husb.* **7**, 489–494.
De Wit, C. T. (1960). *Versl. landbouwk. Onderz. Ned.* No. 66. **8**, 1–82.
Fitzpatrick, E. A. and Nix, H. A. (1969). *Agric. Met.* **6**, 303–319.
Forrester, J. W. (1969). 'Principles of Systems'. Wright–Allen Press, Cambridge, Mass.
Goncz, J. H. and Rose, C. W. (1972). *Agric. Met.* **9**, 405–419.
Humphreys, L. R. (1967). *J. Aust. Inst. agric. Sci.* **33**, 3–13.
Kalma, J. D. (1972). *Agric. Met.* **10**, 261–275.
Kalma, J. D. and Badham, R. (1972). *Agric. Met.* **10**, 251–259.
McCown, R. L. (1971). *Aust. J. exp. Agric. Anim. Husb.* **11**, 343–348.
McCown, R. L. (1973). *Agric. Met.* **11**, 53–63.
Norman, M. J. T. (1967). *J. Aust. Inst. agric. Sci.* **33**, 130–132.
Norman, M. J. T. (1970). *Proc. XIth Inter. Grassld Congr.* pp. 829–832.

Norman, M. J. T. and Begg, J. E. (1973). *CSIRO Aust. Div. Land Res. Tech. Pap.* No. 33.

Rose, C. W. (1972). *Proc. Aust. Soc. Anim. Prod.* **9**, 112–115.

Rose, C. W., Byrne, G. F. and Begg, J. E. (1966). *CSIRO Aust. Div. Land. Res. Tech. Pap.* No. 27.

Rose, C. W., Begg, J. E., Byrne, G. F., Goncz, J. H. and Torssell, B. W. R. (1972a). *Agric. Met.* **9**, 385–403.

Rose, C. W., Begg, J. E., Byrne, G. F., Torssell, B. W. R. and Goncz, J. H. (1972b). *Agric. Met.* **10**, 161–183.

Slatyer, R. O. (1960). *CSIRO Aust. Div. Land Res. Surv. Tech. Pap. No.* 13.

Smith, L. P. (1970). *Agric. Met.* **7**, 193–196.

Stewart, G. A. (1956). *CSIRO Aust. Soil Publ.* No. 6.

Torssell, B. W. R. (1973). *J. appl. Ecol.* **10**, 463–478.

Torssell, B. W. R., Begg, J. E., Rose, C. W. and Byrne, G. F. (1968). *Aust. J. exp. Agric. Anim. Husb.* **8**, 533–543.

Webb, E. K. (1965). *Met. Monogr.* **6**, 27–53.

7. Coniferous Forest

P. G. JARVIS, G. B. JAMES and J. J. LANDSBERG

Department of Botany, University of Aberdeen, Aberdeen, Scotland

I. INTRODUCTION

There have been many descriptive studies of the micrometeorology and environmental physiology of coniferous forests, well summarized in Geiger's (1965) book and in recent reviews by Reifsnyder and Lull (1965) and

Reifsnyder (1967) and Rutter (1968). However, the number of comprehensive, process-oriented studies on coniferous forest is extremely small. Recently, a number of comprehensive studies have been initiated or strongly supported by the International Biological Programme and some are beginning to produce results. Thus a much larger body of information will soon be available on which to base an assessment of the properties of coniferous forest. We are deeply grateful for the unpublished records from these and other studies which have been made available to us in preliminary form. We look forward to their full analysis and publication. In the meantime, they have helped us considerably to build up a broad picture of the properties of coniferous forest from a rather patchy literature.

As far as possible, the micrometeorological records and derived fluxes presented here are based on mean hourly or half-hourly values. Instantaneous, mean daily, or other values are indicated when they have been used. Characteristics of the sites and stands from which data appear in a number of sections are given in the Appendix.

It is evident from the rather small amount of work on conifers that they have not, in general, been popular objects for study. Some of the reasons are obvious. The plant scientist, for example, finds the canopy difficult to describe and to work in, the leaf area difficult to measure, the leaves physiologically sluggish and the stomata hard to find. The micrometeorologist finds it difficult, if not impossible, to get a 'good' site which is sufficiently flat, uniform and extensive for the unequivocal application of theory and which is also representative of normal forestry management practice. Consequently, conifers and coniferous forest are not ideal subjects for the study of fundamental processes. In this review we try to point out how work on conifers has contributed to our basic understanding of plant–atmosphere interactions, and in what respects conifers and coniferous forest are different from other plants and communities.

II. RADIATION

The radiant energy available to drive processes in the canopy is largely dependent on the macro-climate. However, the properties of the canopy itself influence the exchanges of energy by reflection and emission and thereby affect the amount of energy available for canopy processes. The small, in many cases needle-like, leaves of conifers, and the large leaf area indexes frequently found, result in radiant energy exchange characteristics of coniferous forests which are significantly different from those of field crops.

A. Exchanges by Canopies

1. *Net Radiation Flux Density*

The net radiation flux density, or net radiation, R_n, of a stand, can be empirically related to the total solar irradiance on a horizontal surface, S_t, by the linear equation:

$$R_n = a + bS_t. \tag{1}$$

The constants of this simple relationship, a and b, are partly determined by the interaction between canopy structure and radiation exchange. The information available does not justify the use of a less empirical equation, e.g., one containing a long-wave exchange coefficient (Gay, 1971a), for the comparisons between stands made here. Studies bearing on this relationship have been based on various time periods of measurement or integration (e.g. instantaneous, hourly means, daily means) (see Linacre, 1968). Idso *et al.* (1969) compared equations based on daily totals with equations based on hourly means and concluded that the latter were more reliable, since the relatively small proportion of negative net radiation on long days, compared with that on short days, may bias the slopes of equations based on daily values. Tabulated regression coefficients are therefore based on hourly records only. In many cases, the complete regressions are not available and discussion must be restricted to an approximation for the slope, b, the ratio R_n/S_t.

Most of the estimates of R_n/S_t for coniferous stands fall within the range 0·7 to 0·9 with an overall mean of 0·80. Different regression constants would be expected for different species under overcast and sunny conditions, and at different times of the year in relation to solar elevation, canopy temperature and flushing of new leaves, but the observations are generally inadequate to show such effects. Using mean hourly values of R_n and S_t to calculate linear regressions and pooling data from six field crops, Fritschen (1967) obtained a mean value of b of 0·73, and still lower values were obtained for field crops by Tajchman (1967). The higher values of b for conifers are the result of the lower solar reflection coefficients (Section II.A.2) and probably also of the needle-like shape of the leaves, the temperature of which does not deviate much from air temperature (Section IV.B.2). Fritschen obtained a mean value of a for the six field crops of 82 W m^{-2} which is higher than many of the values for a in Table I. This difference is related to lower thermal fluxes from the forest surfaces which are relatively cooler than the surfaces of farm crops.

Table I

Summertime values of the solar reflection coefficient (ρ) and the net radiation flux density ($\mathbf{R}_n$) as a function of solar irradiance ($\mathbf{S}_t$)

Species	Site ref. no.	Reflection coefficient ρ — mean hourly — overcast[a]	mean hourly — sunny[a]	mean hourly — diurnal[b]	mean daily	a(Eq. 1)[c] W m^{-2}	b(Eq. 1)[c] or $\mathbf{R}_n/\mathbf{S}_t$[d]	Reference
Pinus								
radiata	1	—	—	—	0.10	—	—	Denmead (1969)
halepensis	2a	—	—	—	0.12–0.13	—	—	Stanhill *et al.* (1966)
halepensis	2b	—	—	—	0.14–0.20	—	—	Stanhill (1970)
taeda	3a	—	0.13	—	0.11	−110	0.87	Gay and Knoerr (1970), Gay (1971a)
taeda	3b	—	0.15	—	—	—	0.8	Sinclair *et al.* (unpublished)
contorta	7	—	—	—	0.09	—	—	Gay (1971b)
resinosa	6	0.10	0.08	0.08–0.13	—	−126	0.94	Leonard (1968)
resinosa/strobus	9a	0.06–0.08	0.08–0.11	0.08–0.27	—	—	—	Mukammal (1971)
strobus	8	—	0.13–0.19	—	—	—	—	Federer (1968)
sylvestris	12	—	0.12	0.11–0.20	—	—	0.77	Tajchman (1972b)
sylvestris	17	0.09	0.09	0.08–0.17	0.08	−38, −63	0.91, 0.91	Stewart (1971), Gay and Stewart (1973)
sylvestris	18a	—	0.14	0.14–0.18	—	—	0.86	Lutzke (1966)
sylvestris	21b	—	0.10–0.14	0.13–0.18	—	—	0.71	Odin and Perttu (1966), Perttu (1970)
Pseudotsuga								
menziesii	10	—	—	—	0.09	—	0.80	Gay (1972), Gay and Stewart (1973)
Picea								
abies	13	—	0.04	0.04–0.07	0.05–0.06	—	0.75	Tajchman (1967, 1972a)
abies	16	—	0.12	—	—	—	—	Barry and Chambers (1966)
glauca	9b	—	0.15	0.14–0.27	—	—	—	Mukammal (1971)
sitchensis	19	0.15	—	—	—	−27	0.87	Landsberg *et al.* (1973)
sitchensis		0.12–0.14	0.10–0.14	0.10–0.24	—	−66, −6	0.86, 0.76	Jarvis *et al.* (unpublished)

[a] Values for middle of the day periods.
[b] Diurnal range on sunny days.
[c] From regressions of mean hourly values under sunny or overcast (second figure) conditions.
[d] Mean hourly values for middle of the day periods.

2. *Reflection of Solar Radiation*

The solar reflection coefficient ρ (or S_r/S_t, where S_r is the reflected solar irradiance), often called albedo, is given for a number of coniferous forest stands in Table I. Some of the considerable variation in the figures undoubtedly results from variation in the instruments used (e.g., Kipp, Eppley and Lintronic solarimeters). For example, Stanhill *et al.* (1971) demonstrated that Lintronic solarimeters, which were used by Stanhill (1970), Barry and Chambers (1966) and Landsberg *et al.* (1973), gave reflection coefficients several percent more than values given by Kipp solarimeters.

In general, the mean daily and the midday reflection coefficients for pine species and Douglas fir lie in the range 0·08 to 0·14. Values for spruce are more widely spread and range from 0·04 to 0·06 for *Picea abies* in the Ebersberger forest, to 0·15 for *Picea sitchensis* and *Picea glauca* from Fetteresso and Petawawa forests, respectively. These higher values may result in part from differences in needle reflectivity, both *Picea sitchensis* and *Picea glauca* being more glaucous than *Picea abies*, but may also result from differences in the height and density of the stands.

The proportion of incident solar radiation which is reflected from a tall stand depends to a considerable extent on the proportion which is trapped within and below the canopy by multiple scattering. Stanhill (1970) found that the reflection coefficient of tall vegetation was systematically less than that of short vegetation, partly explaining the relatively small reflection coefficients over the stand at Ebersberger, which is much taller than the stand at Fetteresso (see Appendix). Mukammal (1971) suggested that the larger reflection coefficient (0·15) of the *Picea glauca* stand compared with the *Pinus resinosa/Pinus strobus* mixture (0·08) at Petawawa might result from the greater density of the spruce stand. A very dense stand or a canopy with a high leaf area density may behave as if it were shorter because more reflection occurs at the upper levels and there is less opportunity for multiple scattering lower down. Thus the reflection coefficients of shorter, denser stands may be larger than those of taller more open stands, explaining many of the differences found in Table I.

In general, midday reflection coefficients are similar under clear and overcast conditions, values under overcast conditions being no more than about 0·01 larger. Some species show a much wider diurnal range under clear skies than under overcast skies (reaching a maximum of 0·27, see Table I and Fig. 1), but much larger diurnal variations are reported for field crops (cf. Monteith and Szeicz, 1961). The probable reason for this contrast is the very rough surface of a coniferous canopy, which in no way approaches a plane surface, so that even at low solar elevations multiple scattering traps a large proportion of the incident radiant flux density. This is probably also why the diurnal variation in the reflection coefficient seems to be less over spruce than over pine: when the sun is near the horizon, the spire-shaped crowns of the spruce present a rougher surface to its rays than the more rounded pine crowns.

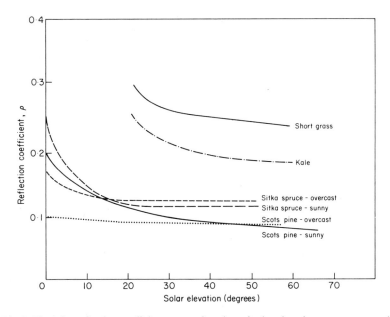

Fig. 1. The solar reflection coefficient, ρ, as a function of solar elevation on overcast and sunny days over Scots pine at Thetford (Site ref. no. 17) (from Stewart, 1971) and Sitka spruce at Fetteresso (Site ref. no. 19). For comparison, curves for short grass and kale on sunny days taken from Monteith and Szeicz (1961) are shown.

Larger values of ρ observed on damp days and when the canopy is wet almost certainly result from the presence of free water droplets on and between the needles.

3. Emission of Thermal Radiation

The thermal emissivity of canopies has not been measured but is usually taken as unity. Dirmhirn (1964) gives a value for coniferous forest of 0·97, compared with 0·96 and 0·95 for meadow and arable lands respectively. These differences are small, but taken in conjunction with the lower temperature of coniferous foliage (Section IV.B.2) can lead to significant differences in the emission of thermal radiation (Rutter, 1968).

B. Exchanges within the Canopy

The distribution of radiant energy within the canopy has an important influence on the spatial distributions of carbon dioxide, heat and water vapour fluxes. The average net radiation flux density is of primary importance since it defines the amount of energy available within the canopy for partitioning into latent and sensible heat fluxes. Photosynthesis depends on the flux density of quanta or of short-wave radiant energy, within the wavelength limits of about 380 to 720 nm. Furthermore, since the rate of photosynthesis is a non-linear function of irradiance or quantum flux, the average flux density at a level within the canopy is less relevant than the area distribution of radiant energy within the canopy.

1. *Extinction of Average Irradiance*

The average irradiance, $S(z)$, at any level, z, in a canopy may be related to the irradiance above the canopy, $S(0)$, an extinction coefficient, $\mathscr{K}$, and the leaf area index, $L(z)$, cumulated from the top of the canopy down to level z,

$$S(z) = S(0)\,e^{-\mathscr{K}L(z)}. \tag{2}$$

Thus a simple extinction coefficient, $\mathscr{K}$, which is a function of wavelength and solar elevation, is defined by

$$\mathscr{K} = \frac{\ln\,[S(0)/S(z)]}{L(z)} = \frac{\ln\,[1/\tau]}{L(z)} \tag{3}$$

where $\tau = S(z)/S(0)$ is the mean transmissivity measured at level z.

Few extinction coefficients have been reported for forests (Table II) because of the difficulties in obtaining good estimates of leaf area index and in adequately sampling to determine the mean irradiance. In fact, very few studies of processes in forest canopies include an adequate description of the amount and distribution of the leaves. The extinction coefficients listed in Table II give a general idea of the size of $\mathscr{K}$ in coniferous forest canopies but must be regarded as approximate because of the difficulties already mentioned and because $\mathscr{K}$ is not constant with height in the canopy, as Table III shows. This increase in $\mathscr{K}$ with increasing depth in the canopy, also found by Impens and Lemeur (1969) in farm crops, may result at least in part from the increasing amounts of absorbing plant structure not included in the estimate of leaf area index (namely twigs, branches and trunks) and, in the case of net radiation, from the increasing magnitude of the downward long-wave radiation flux with increasing depth in the canopy (Landsberg *et al.*, 1973).

Under a clear sky the extinction coefficient is a function of the elevation β of the sun above the horizontal (Anderson, 1966) and would be expected to be linearly proportional to (cosec $\beta - 1$) for random foliage (Monteith,

Table II

Comparative extinction coefficients $\mathcal{K}$ of various stands during the middle of the day

Species	Site ref. no.	L^a		Type of radiation	Condition[b]		Reference
Pinus resinosa	4	2.6	June–July	visible	o	0.35–0.45	Waggoner and Turner (1971)
Pinus resinosa/strobus	9a	3.1[d]	May–Sept.	short-wave	s	0.28–0.57	Mukammal (1971)
					o	0.39–0.45	Mukammal (1971)
Pinus sylvestris	17	4.3	May–Sept.	short-wave	s/o	0.46	Stewart et al. (unpublished)
Pinus sp.		2.7[d]	—	net	s	0.41	Rauner in Tajchman (1972a)
Pseudotsuga menziesii	—	5.5[d]	—	net	s/o	0.42	Kinerson (1973)
				visible	s/o	0.79	
Picea abies	13	8.4[d]	Summer	short-wave	s	0.28	Baumgartner (unpublished)
Picea sitchensis	19	9.8	June–July	visible[e]	s	0.56	Norman and Jarvis (1974)
				visible	o	0.49	Norman and Jarvis (1974)
				near infra-red	s	0.53	Norman and Jarvis (1974)
				near infra-red	o	0.48	Norman and Jarvis (1974)
			June–August	net	s[c]	0.55	Landsberg et al. (1973)
				net	o	0.58	Landsberg et al. (1973)

[a] Leaf area index of the stand on a projected leaf area basis.

[b] s, o and s/o are sunny, overcast and mixed conditions respectively.

[c] Sun overhead.

[d] Projected area taken as total area $\times\ 0.389$: $(\pi/2 + 1)^{-1} = 0.389$.

[e] See Table III, footnote a.

Table III
Measured extinction coefficients of visible, near infra-red and net radiation for several
layers of foliage in Sitka spruce at Fetteresso forest

Leaf area index interval	Visible[a]			Near infra-red		Net radiation[b]	
	overcast	$\beta = 42°$	$\beta = 25°$	overcast	$\beta = 42°$	overcast	$\beta \approx 42°$
0–2·0	0·58	0·67	0·91	0·46	0·53	0·60	0·98
2·0–4·2	0·40	0·58	0·41	0·33	0·45	0·69	1·2
4·2–7·0	0·43	0·50	0·58	0·35	0·37	1·1	—
7·0–8·3	0·85	0·80	0·80	0·48	0·57	4·3	—

[a] Taken from Norman and Jarvis (1974); visible $\lambda \approx 480$ to 600 nm, near infra-red $\lambda \approx 700$ to 1100 nm.
[b] Taken from Landsberg et al. (1973).

1969). Landsberg et al. (1973) found that the extinction coefficient for net radiation in the upper part of the Sitka spruce canopy at Fetteresso increased linearly with increasing (cosec $\beta - 1$) from a minimum of about 0·5 in the middle of the day to values of about 2·5 to 3·0 when the sun is near the horizon in the early morning and evening. The values of $\mathscr{K}$ for Sitka spruce under overcast conditions of 0·4 to 0·6 (Table III) are similar to those of broad-leaved field crops with foliage of intermediate inclination (Monteith, 1969) thus suggesting that the effective foliage units in coniferous canopies are the twigs rather than the individual needles. The effect on radiation penetration in a spruce canopy of the grouping of needles into twigs has recently been examined by Norman and Jarvis (1974, 1975).

2. Area Distribution of Irradiance

The problems of sampling radiation within coniferous forest canopies are undoubtedly severe. A satisfactory solution seems to be to traverse sensors along transects at a number of levels under different sky conditions (Sinclair and Knoerr, 1972; Brown, 1973; Norman and Jarvis, 1974). Figure 2 shows an example of the area distribution of photosynthetically useful radiant energy below a leaf area index of 4·2. The area distribution of irradiance at any depth in a canopy depends not only on the overlying leaf area index but also on the solar elevation and the proportion of diffuse to direct solar irradiance above the canopy (Fig. 3).

We have found no other descriptions of the distribution of radiant energy within a coniferous forest canopy, yet such descriptions are indispensable for the realistic modelling of photosynthesis in different parts of the canopy. Several recent reports describe the variability of short-wave radiation on the floor of coniferous forests (e.g. Vézina, 1964; Vézina and Péch, 1964; Gay et al., 1971; Reifsnyder et al., 1971). Whilst these analyses are of considerable relevance to the production of the ground flora, they provide only limited

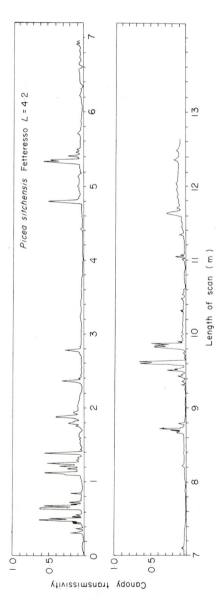

Fig. 2. The transmissivity of radiant energy τ ($\lambda \approx 480$ to 600 nm) along a 14-m transect through the Sitka spruce canopy at Fetteresso below a leaf area index (L) of 4·2. The 2-mm diameter sensor was moved at a speed of 1·8 cm s^{-1} and the irradiance recorded on a data logger every 7 mm along the transect.

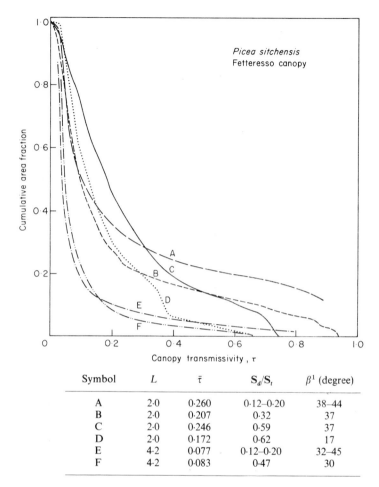

Symbol	L	$\bar{\tau}$	S_d/S_t	β^1 (degree)
A	2·0	0·260	0·12–0·20	38–44
B	2·0	0·207	0·32	37
C	2·0	0·246	0·59	37
D	2·0	0·172	0·62	17
E	4·2	0·077	0·12–0·20	32–45
F	4·2	0·083	0·47	30

Fig. 3. The fraction of horizontal area occupied by transmissivities greater than τ as a function of transmissivity, τ, to show effects of the leaf area index, L, the proportion of diffuse radiant energy above the canopy, S_d/S_t, and the solar elevation relative to the slope, β^1, on the penetration of radiant energy ($\lambda \approx 480$ to $600\,nm$) into the Sitka spruce canopy at Fetteresso (from Norman and Jarvis, 1974).

information on the influence of canopy structure on radiation penetration, and on the availability of radiation for processes within the canopy.

3. Spectral Composition

Measurements of the spectral distribution of diffuse light made below coniferous canopies on clear days show a marked depletion in the blue region of the spectrum (400 to 450 nm), a slight enrichment in the red (650 to 700 nm) and pronounced enrichment in the near infra-red (700 to 750 nm) relative to the composition of diffuse skylight above the canopy (Federer and Tanner, 1966; Freyman, 1968). Between about 450 and 650 nm, the rate of decline of energy was almost uniform and closely followed the rate of decline of energy in the skylight spectrum. There were only small differences amongst the canopies of *Pinus strobus, Pinus resinosa, Pinus banksiana, Pinus taeda* and *Pseudotsuga menziesii,* but there was a slight peak in transmissivity of *Picea* sp. at 550 nm. Thus, over this part of the spectrum, the coniferous canopy behaves like a neutral filter (Fig. 4). The spectral distribution of direct sunlight beneath a canopy of mixed *Pseudotsuga menziesii, Abies concolor* and *Pinus ponderosa* contained a much higher proportion of red light (650 to 700 nm) (Atzet and Waring, 1970). The depletion in the blue is smaller than that generally found beneath broad-leaved plants. Such plants

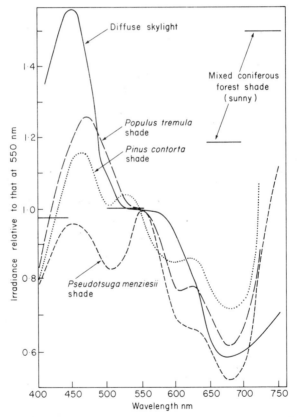

Fig. 4. The spectral distribution of diffuse skylight irradiance and of the shade irradiance beneath three forest canopies on overcast days (from Freyman, 1968) and beneath a mixed canopy of *Pseudotsuga menziesii, Abies concolor* and *Pinus ponderosa* on a sunny day (from Atzet and Waring, 1970).

generally exhibit depletion in the red, in contrast to the enhancement below conifers.

The blue depletion accords with the low reflectivity and high absorptivity of needles in the blue region (see Figs 6a and b), and the near infra-red enhancement agrees with the high near infra-red reflectivity and low absorptivity. The slight red enhancement is more difficult to explain since needles have high absorptivity between 630 to 680 nm.

C. Exchanges by Leaves

The optical properties of microphylls and needles are poorly known because of the difficulties in making measurements with conventional spectrophotometric equipment. Gates *et al.* (1965a) assumed zero transmissivity and found values of α ($= 1 - \rho$) of 0·88 to 0·89 for needle mosaics of *Pinus strobus* and *Thuja occidentalis* in wavebands corresponding to sunny and cloudy daylight conditions. In the visible waveband, absorptivities were as high as 0·97, falling to 0·58 in the near infra-red. These absorption coefficients are much higher than the values of 0·4 to 0·7 found for herbs and broad-leaved trees (Gates *et al.*, 1965a ; Birkebak and Birkebak, 1964).

Using a miniature radiation sensor (diameter 2 mm) sensitive to two wavebands (Norman, 1971 ; Norman and Jarvis, 1974), the reflectivity of Sitka spruce needles was measured using branch mosaics (Table IV).

Table IV
Reflectivity of Sitka spruce needles attached to branches removed from different heights in the canopy when viewed normally from above (top) and below (bottom)[a]

| Relative height of branches (z/h) | Reflectivity | | | |
| | Visible $\lambda \approx 480$ to 600 nm | | Near infra-red $\lambda \approx 700$ to 1100 nm | |
	top	bottom	top	bottom
0·7–1·0	0·06	0·10	0·41	0·45
0·5–0·7	0·04	0·06	0·24	0·28

[a] From Norman and Jarvis (1974).

Transmissivity, measured by placing a mosaic of needles on the sensor (with exhaustive checks to ensure no entry of untransmitted radiation), is shown in Fig. 5. As the transmission was appreciable, especially in the near infra-red, the assumption of Gates *et al.* (1965a) that transmission is zero in conifers is not necessarily correct, and hence their large absorption coefficients

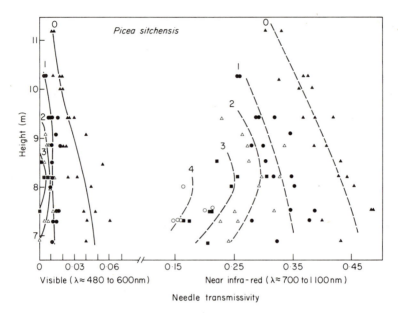

Fig. 5. Needle transmissivity for visible ($\lambda \approx 480$ to 600 nm) and near infra-red ($\lambda \approx 700$ to 1,100 nm) radiant energy as a function of needle age and height in the Sitka spruce canopy at Fetteresso. The figures by the curves are the ages of the needles in years (from Norman and Jarvis, 1974).

may well be overestimates. The decline in transmissivity with increasing needle age can be attributed to the growth of algae and fungi on the surface of the needles.

The diffuse spectral reflectivity of needles of *Pinus resinosa* and *Pseudotsuga menziesii* was measured using integrating spheres by Egan (1970) and Wooley (1971), respectively. The curves (Fig. 6a) are very similar to those for other plants (cf. Gates *et al.*, 1965a) with a low reflectivity through the visible, a high reflectivity in the near infra-red and the characteristic water absorption bands beyond 1400 nm. More detail in the photosynthetically useful region is shown in the spectral reflectivity curves of several conifers by Clark and Lister (1975b). The reflectivity in the blue waveband is particularly large in the spruces, especially in *Picea pungens*, the blue or Colorado spruce, because the needles are made extremely glaucous by a thick covering of fibrillar wax (Hanover and Reicosky, 1971). Because the visible transmissivity is small, the absorption spectrum is similar to the spectrum of $(1 - \rho(\lambda))$ (see Fig. 6b).

The action spectrum of net photosynthesis in Scots pine corresponds closely with the absorption spectrum between 550 and 700 nm but in the

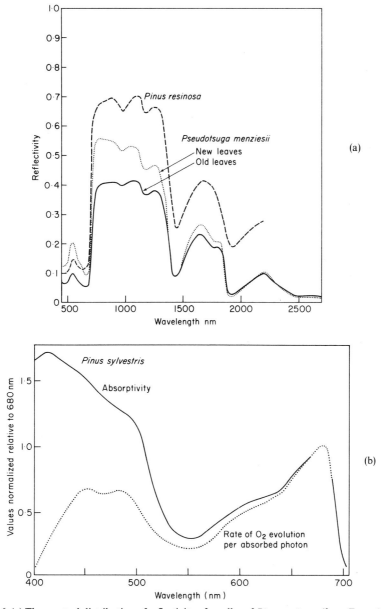

Fig. 6. (a) The spectral distribution of reflectivity of needles of *Pinus resinosa* (from Egan, 1970) and *Pseudotsuga menziesii* (from Woolley, 1971). (b) Absorptivity and photosynthetic action spectrum of a current-year needle from *Pinus sylvestris* normalized relative to values at 680 nm (from Linder, 1971).

blue waveband the photosynthetic effect is low compared with the absorption (Fig. 6b). This could result from photorespiration which is known to be stimulated by blue light, or from strong absorption by photosynthetically inactive carotenoids (Linder, 1971). Clark and Lister (1975a) also found a much smaller photosynthetic effectiveness of blue light than red light of the same *incident* irradiance in *Pseudotsuga menziesii, Picea sitchensis* and *Picea pungens*. In particular, the blue spruce had negative net photosynthesis at wavelengths below 500 nm, but much of the small effect of the blue light in these experiments must be attributed to the high blue reflectivity.

We have found no measurements of the emissivity of coniferous needles. Gates and his associates (Gates and Benedict, 1963; Tibbals *et al.*, 1964; Gates *et al.*, 1965b) used values of 0·95 and 0·97 in energy balance studies of needles of *Pinus ponderosa, Abies concolor* and *Picea pungens*, by analogy with the measured emissivity of a number of broad-leaved trees and herbs (Gates and Tantraporn, 1952).

III. MOMENTUM

A. Aerodynamic Characteristics of Canopies

1. *Properties of the Wind Profile*

The exchanges of water vapour, carbon dioxide, heat and momentum between a canopy and the bulk air above it, depend upon the turbulent exchange properties of the wind profile generated in the boundary layer above the canopy. These are determined by wind speed and by properties of the canopy.

Under near-neutral conditions the wind-speed, u, at a height z is given by the well-known logarithmic equation in terms of the friction velocity u_*, von Karman's constant $k = 0.41$, the zero plane displacement d, and the roughness length z_0 (see Vol. 1, p. 62). The values of d and z_0 may be regarded as characteristics of the canopy. Rutter (1968) summarized some early and rather irregular data for forests. More recent results obtained from wind profile measurements are summarized in Table V, where both parameters are presented as non-dimensional fractions of median canopy height, h.

In most of the studies listed, the values of z_0 and d were determined over a fairly wide range of unstable and stable conditions. Only in the Velen study and that by Belt (1969) were conventional stability corrections applied to all estimations of z_0 and d. In the Thetford study (Oliver, 1971), d was estimated in neutral conditions only and the resulting value taken as a constant which was used in the derivation of z_0 over a range of stabilities. Oliver (1971) found that z_0 at Thetford decreased as conditions became more unstable. The relation between z_0 (m) and Richardson numbers Ri (see Vol. 1, p. 81) was

$$z_0 = 0.91 + 3.2Ri. \tag{4}$$

Tajchman (1967, 1972a) obtained a constant value of d independent of wind speed and stability (although with considerable scatter) and found that z_0 declined with increasing wind speed, measured at a reference height of 30.9 m ($h = 27.2$ m):

$$z_0 = -0.31u^3 + 2.2u^2 - 6.38u + 9.9 \quad \text{(m)} \quad (u \leqslant 2.5 \text{ m s}^{-1}) \tag{5}$$

$$z_0 = 0.15u^2 - 1.4u + 5.5 \quad \text{(m)} \quad (u > 2.5 \text{ m s}^{-1}). \tag{6}$$

However, Landsberg and Jarvis (1973) analysed their records grouped according to wind speed or Richardson number and found no dependence of z_0 or d on either. Although the apparent constancy of z_0 and d may result from the small range of conditions covered by the profiles which they analysed (see Table V), the range is typical for day-time conditions over many forests. It is pertinent to point out that, because of the height and roughness of forest canopies, the daytime temperature gradients immediately above the canopy surface are small (Section IV.A.2) and the wind speed gradients large, in comparison with farm crops. Consequently, Richardson numbers generally remain fairly close to zero. Furthermore, although Ri becomes larger with height above the canopy surface, as shown by Thom et al (1975), profile measurements above forest are generally made as close to the surface as possible, in the region of smallest Ri, because of the difficulty in satisfying the fetch requirements at greater heights. Under normal, unstable, daytime conditions and close to the forest surface, at Fetteresso Ri varies over a relatively small range, say from -0.001 to -0.03. Taking into account the inevitable spread in wind profile parameters (Oliver, 1971; Landsberg and Jarvis, 1973), it is questionable whether any real improvement is obtained in the estimation of z_0 and d for forest by the application of empirical stability corrections obtained for completely different kinds of surface such as grass and stubble (see discussion by Thom et al., 1975).

The values of z_0 and d determined from a logarithmic wind profile are not independent; when d is large, z_0 is small and $vice$ $versa$. Referring to the data compiled in Table V, the relatively large value of $d = 0.9h$ for $Pinus$ $taeda$ and the $Pinus$ $resinosa/strobus$ stand is unusual and must await confirmation. The value of $d = 0.84h$ found with Sitka spruce is sufficiently large to be unusual but Landsberg and Jarvis (1973) used their knowledge of leaf area distribution in the canopy to make a check calculation which gave $d = 0.81h$ to $0.84h$, using Thom's (1971) 'centre of pressure' approach. They attributed the large value of d/h to the very dense layer of foliage

Table V
Aerodynamic exchange characteristics of canopies

Species	Site ref. no.[a]	Median height h (m)	L[b]	No. of observations	Wind speed[d] u (m s^{-1})	Richardson number Ri[e] (or stability)	z_0/h	d/h	η[f]	g_{aM}[g] (m s^{-1})	r_{aM}[g] (s m^{-1})	Reference
Larix leptolepis	5	10.4	3.6[c]	18 × 10 min	1.7–5.8	—	0.11 (0.005–0.26)	0.61 (0.30–0.84)	0.25	0.12	8.0	Allen (1968)
Pinus resinosa	6	11.8	—	30 × 15 min	1.8–5.7	−0.05 to 0	0.07	0.81	0.40	0.32	3.1	Leonard and Federer (1973)
Pinus contorta	3c	10.0	2.8	15 × 7 min	<2 ($h + 6.5$)	near neutral	0.05	0.76	0.26	0.14	7.3	Bergen (1971)
Pinus resinosa/strobus	9a	22.0	3.1	436 × 1h	>1 ($h + 39$)	—	0.03	0.90	0.34	0.24	4.2	Martin (1971)
Pinus taeda	3b	14.0	2.6[c]	—	—	—	0.02	0.92	0.34	0.23	4.3	Sinclair et al. (unpublished)
	—	23.4	—	—	—	—	0.02	0.89	0.24	0.12	8.7	Belt (1969)
Pinus sylvestris	17	15.5	4.3	13 × 20 min	0.3–1.8	neutral	0.06[h]	0.76 (0.75–0.77)	0.30	0.18	5.5	Oliver (1971)

Pinus											
densiflora	3d	4·5	0·8	17 × 5 min	1·0–4·0	near neutral	0·10	0·67	0·34	0·23	4·3	Kondo (1971)
	3e	23·0	1·7	26 × 5 min	1·0–5·1	near neutral	0·05	0·83	0·32	0·20	4·9	Kondo (1971)
Pseudotsuga											
menziesii	10	28·0	—	—	—	near neutral	0·14	0·75	0·73	1·06	1·0	Gay and Stewart (1973)
Picea abies	13	27·2	8·4[c]	ca800 × 1 h	0·5–2·5	wide range	0·26–0·11[i]	0·72	7·50– 0·42	0·56	1·8	Tajchman (1967, 1972a)
					2·5–5·0 (h + 3·7)	wide range	0·11–0·08[i]		0·42– 0·33			Tajchman (1967, 1972a)
		27·5	8·4[c]	—	—		0·11	0·83	0·94	1·76	0·6	Baumgartner and Alfreit (unpublished)
	20	22·0	—	73 × 1 h	3·1–9·5 (h + 32)	−0·66 to +0·05 (h + 22)	0·07 (0·004–0·21)	0·68 (0·36–1·09)	0·27	0·15	6·8	Berggren et al. (1971)
Picea											
sitchensis	19	11·5	9·6	66 × 1 h	0·8–4·7	−0·03 to 0 (h to h + 4)	0·03 (0·002–0·14)	0·84 (0·68–0·91)	0·29	0·17	5·9	Landsberg and Jarvis (1973)

[a] See Appendix.

[b] Leaf area index on a projected area basis.

[c] Total leaf area × 0.389.

[d] $u = u(h)$ unless otherwise indicated.

[e] Richardson number $(Ri) = g(\Delta\theta/\Delta z)/[\theta(\Delta u/\Delta z)^2]$, where g is gravitational acceleration and θ is potential temperature.

[f] $\eta = u_*/u(h) = k/\ln[(h - d)/z_0]$ (see p. 190).

[g] $g_{aM} = 1/r_{aM} = \eta^2 u(h)$; $u(h) = 2$ m s^{-1}.

[h] See Eq. (4).

[i] See Eqs. (5) and (6).

between 6 to 8 m above the ground which effectively reduces the height of the roughness elements of the surface. To some extent, this interpretation is borne out by the two estimates of $d = 0.83h$ and $d = 0.72h$ for the Ebersberger stand of Norway spruce which also has a large leaf area index ($L > 8$ on a projected area basis). The Thetford value of $d = 0.76h$ is also larger than the normally quoted figure of $d = 0.7h$ for farm crops. The smallest value of $d = 0.61h$ for Japanese larch (*Larix leptolepis*), is associated with the smallest leaf area index (excluding *Pinus taeda*). It is, of course, reasonable to expect d to approach h and z_0 to approach zero as leaf area density increases, since, in the limiting situation of a solid canopy, d would equal h and the profile would be generated from the surface upwards, characterized by a very small value of z_0 depending on surface roughness only. There is insufficient variation amongst the heights of the stands to confirm the dependence of z_0 on canopy height as shown by Rutter (1968) and Szeicz *et al.* (1969) for a range of crops.

2. *Exchange of Momentum*

The influence of the canopy characteristics z_0 and d on the process of momentum exchange between the canopy and bulk air above can be combined into a canopy boundary-layer resistance for momentum (r_{aM}), or the corresponding eddy conductance ($g_{aM} = 1/r_{aM}$) (Monteith, 1963, 1965).

The shearing stress, or flux of momentum, transmitted to the canopy, $\tau(h)$, is

$$\tau(h) = \rho u_*^2 = \frac{\rho u(z)}{r_{aM}} \tag{7}$$

and therefore

$$g_{aM} = \frac{1}{r_{aM}} = \frac{u_*^2}{u(z)}. \tag{8}$$

With median canopy height (h) taken as a convenient reference level, the logarithmic wind profile equation can be written in the form

$$\frac{u_*}{u(h)} = \frac{k}{\ln\left[(h-d)/z_0\right]} = \eta. \tag{9}$$

Substituting from Eq. (9) into Eq. (8) for u_*, with $u(z) = u(h)$ gives:

$$g_{aM} = 1/r_{aM} = C_{aM}u(h) = \eta^2 u(h) \tag{10}$$

where

$$C_{aM} = [u_*/u(h)]^2 = \eta^2$$

is a low-level drag coefficient derived close to the canopy surface. It was suggested by Deacon and Swinbank (1958) and demonstrated by Bradley (1971/72) over wheat stubble that C_{aM} is effectively constant over a wide range of instabilities because the large wind shear near the surface reduces the distortion of the wind profile by buoyancy effects. Close to the surface, therefore, the eddy conductivity for momentum may be taken as linearly proportional to wind speed, and to a drag coefficient C_{aM} which depends upon the adequate definition of z_0 and d over a range of conditions. Landsberg and Jarvis (1973) found η for Sitka spruce forest to be 0·29, in general agreement with values for some other types of forest. A wider range of values is apparent in Table V, but most fall within the range 0·25 to 0·35, giving the convenient approximation that, for coniferous forest, $C_{aM} \approx 0·1$. It follows that

$$C_{aM} \approx 0·1u(h) \quad \text{and} \quad r_{aM} \approx 10u^{-1}(h). \tag{11}$$

If, however, z_0 is a function of u or Ri (Eqs 4, 5 and 6), then η must also be treated as a variable. Using Tajchman's (1967) data, with its dependence of z_0 on u (Eqs 5 and 6), Szeicz et al. (1969, Fig. 3) obtained a parabolic relationship between $1/r_{aM}$ and u, with maximum values of $1/r_{aM} = 0·35 \text{ m s}^{-1}$ at wind speeds of 1 and 4 m s^{-1} and a minimum of $1/r_{aM} = 0·29 \text{ m s}^{-1}$ at $u = 2·8 \text{ m s}^{-1}$.

The values of canopy boundary-layer resistance summarized in Table V are compatible with the estimates assembled by Rutter (1968) using much less adequate data. Much of the scatter in the values is likely to be the result of errors in the derivation of z_0 and d, and cannot be taken to reflect real differences amongst the canopies in their drag coefficients.

From Eqs (7 to 10), the shearing stress transmitted to the canopy can be conveniently derived as

$$\tau(h) = \rho u^2(h) . C_{aM} \tag{12}$$

and, if $\eta \approx 0.3$ for coniferous forest (Table V) and $\rho \approx 1 \text{ kg m}^{-3}$, to a good approximation

$$\tau(h) = 0·1u^2(h). \tag{13}$$

Since $u(h)$ generally lies between 0·5 to 5 m s^{-1}, $\tau(h)$ has a usual range of 0·025 to $2·5 \text{ N m}^{-2}$, as found by Landsberg and Jarvis (1973) for Sitka spruce (Table VII).

The eddy diffusivity or momentum exchange coefficient, K_M, may be defined in terms of τ and the wind speed gradient as

$$K_M = \frac{\tau}{\rho \, du/dz} \tag{14}$$

where under adiabatic conditions

$$\frac{du}{dz} = \frac{u_*}{k(z-d)}.$$ (15)

Combining Eqs (10), (12), (14) and (15) gives

$$K_M = \eta k(z-d)u(h).$$ (16)

Thus, at the top of a coniferous forest canopy, for which $z = h = 10$ m and $d/h \approx 0.8$, K_M can be approximated by

$$(0.3 \times 0.41 \times 0.2 \times 10)u(h) \approx 0.25u(h)$$ (17)

and thus may be expected generally to lie between 0·1 and 1·3 $m^2\ s^{-1}$ (Table VII).

Thus C_{aM} (or η) is a highly descriptive canopy characteristic which leads to ready estimates of the shearing stress, the eddy transport coefficient for momentum and the canopy boundary-layer resistance. It seems to have a value of about 0·1 for coniferous forest stands, but its exact value for a particular forest, and hence its usefulness as a descriptive aerodynamic exchange parameter, is limited by the uncertainties inherent in the determination of the wind profile parameters d and z_0.

B. Aerodynamic Characteristics of Shoots

When the appropriate gradient is constant, the flux of mass, heat or momentum across the boundary layer of a leaf is inversely proportional to the boundary-layer resistance of the leaf, r, and proportional to the boundary-layer conductance ($g = 1/r$). This leaf characteristic is defined and ways of determining it are described by Jarvis (1971) together with certain empirical equations for the boundary-layer resistance of broad leaves in relation to leaf dimension and wind speed.

Determination of the boundary-layer resistance of coniferous leaves presents particular problems. The dimensions of needles and microphylls lie outside the range in which most empirical transfer coefficients apply. Furthermore, the needles are grouped together into shoots in such a way that they interfere with each other in the exchange of mass, heat and momentum. Consequently, the boundary-layer resistance, or conductance, of representative shoots has to be determined for different species and shoot structures. So far, determinations have been too few to warrant generalization.

Three methods of determining the boundary-layer resistance of coniferous needles have been used. The boundary-layer resistance to sensible or convective heat transfer, r_H, has been measured from the energy budget of silver replicas of shoots by Tibbals *et al.* (1964), and Gates *et al.* (1965b). The boundary layer-resistance to water vapour transfer, r_V, has been measured by coating shoots with a thin layer of gypsum (Landsberg and Ludlow, 1970; Landsberg and Thom, 1971), and the boundary-layer resistance to momentum transfer, r_M, has been determined, together with the drag coefficient, by measuring the drag force, F, on shoots with a moment balance in a wind tunnel by Landsberg and Thom (1971).

Estimates of boundary-layer resistance by these three methods are shown in Table VI. Much of the variation between species can probably be explained by variation in the orientation and density of the needles on the shoots. This is apparent from the influence of the direction of air flow across white fir (*Abies concolor*) shoots and from the dependence of the resistance on needle density, σ, shown by the experimental removal of needles by Landsberg and Thom (1971). The density of needles on a shoot is conveniently described by the coefficient σ which is defined as

$$\sigma = \frac{A_n + A_t}{A_s} \tag{18}$$

where A_n is the total projected area of the needles when laid out separately. A_t is the total projected area of the twig or shoot axis and A_s is the projected or frontal area of the complete shoot in a particular orientation, usually taken as normal to the direction of air flow. In the Sitka spruce canopy at Fetteresso, σ varies from about 0·5 at the base of the canopy to 1·8 at the top. When needles are crowded together, as on dense shoots, mutual aerodynamic interference increases the resistance to transfer across the boundary layer. A measure of the interference, the shoot shelter factor, p_s, is defined by the ratios

$$p_s = \frac{C_{M,n}}{C_{M,s}} = \frac{r_{M,s}}{r_{M,n}} = \frac{g_{M,n}}{g_{M,s}} \tag{19}$$

where C_M, r_M and g_M are the drag coefficient, boundary-layer resistance for momentum transfer and boundary-layer conductance, respectively ($C_M \cdot u = g_M$). The property of an individual needle is indicated by the subscript n and the property of a needle when grouped together with others into a shoot by the subscript s. Landsberg and Thom (1971) found that for shoots of Sitka spruce in a turbulent air stream:

$$p_{s,M} = p_{s,V} = 1·67\sigma^{0·43}. \tag{20}$$

The shelter factor for shoots of typical density in the canopy ($\sigma = 1·5$) is 2·0, that is to say, the boundary-layer resistance of a needle grouped together

Table VI

Boundary-layer conductance, g, and resistance, r, of coniferous shoots in forced convection

Species	Wind speed (cm s^{-1})	Flow direction	Equation for g[b] (m s^{-1})	r ($= 1/g$) ($u = 1$ m s^{-1}) (s m^{-1})	r_v[d] ($u = 1$ m s^{-1}) (s m^{-1})	Reference
Pinus ponderosa	$0 < u < 120$	cross, parallel	$g_H = (10{\cdot}33 + 0{\cdot}408u) \times 10^{-3}$	19·6	6·7	Gates et al. (1965b)
Abies concolor	$35 < u < 150$	cross	$g_H = (11{\cdot}71 + 1{\cdot}00u^{0{\cdot}75}) \times 10^{-3}$	23·1	7·9	Tibbals et al. (1964)
		parallel	$g_H = (11{\cdot}71 + 1{\cdot}60u^{0{\cdot}75}) \times 10^{-3}$	16·0	5·5	Tibbals et al. (1964)
Picea pungens	$35 < u < 150$	cross, parallel	$g_H = (11{\cdot}71 + 0{\cdot}115u^{0{\cdot}97}) \times 10^{-3}$	46·0	15·8	Tibbals et al. (1964)
Picea sitchensis	$35 < u < 220$	stirred[a]	$g_V = 8{\cdot}3u^{0{\cdot}5} \times 10^{-3}$	12·0	4·7	Landsberg and Ludlow (1970)
	$150 < u < 520$	cross	$g_V = 8{\cdot}9\sigma^{-0{\cdot}38}u^{0{\cdot}58} \times 10^{-3}$	9·6[c]	9·6	Landsberg and Thom (1971)
	$150 < u < 520$	cross	$g_M = 3{\cdot}2\sigma^{-0{\cdot}42}u \times 10^{-3}$	3·9[c]	2·8	Landsberg and Thom (1971)

[a] In assimilation chamber; others in wind tunnels.

[b] $g_H = [h/c_p\rho]0{\cdot}572$ (m s^{-1}) where h is a heat transfer coefficient in cal cm^{-2} min^{-1} °C^{-1}; u is in cm s^{-1}.

[c] Estimates are based on projected needle and twig area; others are based on total surface area. $\sigma = 1{\cdot}75$, a typical value for upper levels of forest.

[d] From $r_V = r_H(\kappa/D)^{2/3} = r_M(v/D)^{2/3}$, where D, κ and v are taken as 0·24, 0·20 and 0·15 cm^2 s^{-1}, respectively, and adjusted from total needle surface area to projected area by multiplying by 0·389.

with others into a typical shoot, is approximately twice that of an isolated needle exposed to a similar wind speed.

Figure 7 shows the dependence on wind speed of the boundary-layer resistances for water vapour and momentum transfer from Sitka spruce shoots of two densities. The effect of shoot density is less at the higher wind speeds than at the lower ones, probably because there is less interference between the needles when the thickness of the boundary layer round each one is reduced. The lower resistance to momentum transfer than to water-vapour transfer is probably the result of bluff body effects which promote

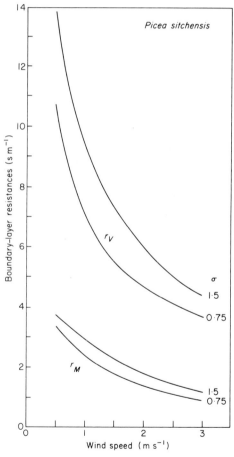

Fig. 7. The relationships between the boundary-layer resistances to the transfer of water vapour, r_V, and momentum, r_M, and wind speed for shoots of *Picea sitchensis* of two needle densities, σ (derived from Landsberg and Thom, 1971).

momentum exchange but not mass exchange. Making allowance for the difference between the diffusivity of water vapour in air ($\approx 0.24 \, \text{cm}^2 \, \text{s}^{-1}$) and the kinematic viscosity of air ($\approx 0.15 \, \text{cm}^2 \, \text{s}^{-1}$) leads to the expectation that

$$g_M = \left(\frac{0.15}{0.24}\right)^{2/3} \cdot g_V = 0.73 g_V. \tag{21}$$

In practice, however, it was found that

$$g_M = 3.27 g_V \quad (\sigma = 1.5, u = 2 \, \text{m s}^{-1}). \tag{22}$$

It is noteworthy that the discrepancy ($g_M > g_V$) is the reverse of that found in transfer from the canopy as a whole ($K_V > K_M$, see Section V.A).

C. Momentum Exchange within Canopies

1. *Momentum Balance*

Many of the wind profiles in forests described by Geiger (1965) are not monotonic, and this phenomenon (Fig. 13) is seen in most recently measured profiles (e.g. Allen, 1968; Oliver, 1971; Landsberg and James, 1971). The incidence of a more rapid air flow, or 'blow-through' below the crowns was ascribed by Geiger to penetration from the edge of the stand or from large breaks in the canopy. However, the finding that blow-through occurs also in near-ideal conditions of fetch, as at Thetford, leads to the conclusion that it is a general phenomenon, the result of a horizontal pressure gradient, which is sustained by persistent up-and-down draughts at varying points in the stand, as measured with vertical anemometers in wheat by Penman *et al.* (1970).

The occurrence of blow-through below the crowns precludes elaborate analysis of the exchanges of mass, heat and momentum within the lower levels of the canopy, using conventional, one-dimensional equations. It seems likely that as this phenomenon becomes more widely recognized in the canopies of shorter crops, within which the measurement of wind speed is more difficult, the use of one-dimensional models for analysis will also become suspect. Higher up within the canopy, the vertical transfer of momentum predominates and the effects of horizontal pressure gradients are not obvious since the shape of the wind profile is consistent with one-dimensional transfer. Horizontal divergence of momentum transfer may still occur in the upper canopy but may not be apparent because it is insufficient to reverse the wind speed profile. Thus a one-dimensional analysis of momentum transfer, such as that by Landsberg and Jarvis (1973), can do no more than elucidate canopy characteristics and provide orders of magnitude for exchange parameters.

Landsberg and Jarvis (1973) calculated a momentum balance for the upper layers of the Sitka spruce canopy at Fetteresso, down to the level at which du/dz approached zero ($z = 8$ m or $0.70h$). The shear stress $\tau(z)$ at successive levels, separated by $\Delta z = 0.25$ m was determined from

$$\tau(z) = \tau(h) - \sum_z^h F \tag{23}$$

and

$$F = \rho C_{M,s}(z) \cdot a(z) \cdot u^2(z) \cdot \Delta z$$

where $C_{M,s}(z)$ is a dimensionless drag coefficient, determined in the laboratory (see Section III.B), appropriate to the structure of the shoots of leaf area density $a(z)(\text{m}^2 \text{ m}^{-3})$, and at wind speed, u (m s^{-1}), at level z.

Comparison between $\sum_z^h F$ and $\tau(h)$ (Table VII) shows that the layer-by-layer estimate of momentum absorbed exceeded the shear stress at the surface of the canopy. Following Thom (1971), this discrepancy was interpreted as demonstrating interference between the shoots in the canopy, and a dimensionless canopy shelter factor, p_c, assumed to be the same at all levels, was calculated from

$$p_c = \sum F/\tau(h). \tag{24}$$

This shelter factor accounts for the fact that the individual shoots in a canopy interfere aerodynamically with one another. A consequence of Eqs (23) and (24) is that the effective drag coefficient of the shoots in the canopy, $C_{M,c}$, can be defined as:

$$C_{M,c} = C_{M,s}/p_c \tag{25}$$

so that

$$\sum F \cdot C_{M,c} = \tau(h) \cdot C_{M,s} \tag{26}$$

and

$$p_c = \frac{C_{M,s}}{C_{M,c}} = \frac{r_{M,c}}{r_{M,s}} \tag{27}$$

where, as previously, the subscript s refers to the property of needles grouped together into shoots, and the subscript c indicates the effective property of the shoots when in a canopy.

The canopy shelter factor, p_c, was found to increase with increasing wind speed (Table VII) and the following relationship with 'mean canopy wind speed', u_m (Thom 1971), found:

$$p_c = 1.33u_m^{0.22}. \qquad (28)$$

Thus at $u_m = 2 \text{ m s}^{-1}$, $p_c = 1.55$. A typical shoot shelter factor, p_s, was 2.0 ($\sigma = 1.5$, $u = 2 \text{ m s}^{-1}$). Thus the total shelter effect experienced by a needle in the canopy, $p_{s,c}$, at $u = 2 \text{ m s}^{-1}$, is:

$$p_{s,c} = \frac{C_{M,n}}{C_{M,c}} = \frac{r_{M,c}}{r_{M,n}} = p_s \times p_c = 3.1. \qquad (29)$$

This figure can be compared with estimates of $p_{s,c} = 3.5$ for beans (Thom, 1971) and 3.6 for Scots pine at Thetford (Stewart and Thom, 1973).

Landsberg and Thom (1971) found that the effective drag coefficient of needles grouped together in shoots depended on the shoot density:

$$C_{M,s} = 0.32\sigma^{-0.42}. \qquad (30)$$

Combining Eqs (27), (28) and (30) gives the wind speed and shoot density dependence of the shelter factor for the canopy as a whole:

$$C_{M,c} = C_{M,s}/p_c = 0.24\sigma^{-0.42}u^{-0.22}. \qquad (31)$$

The momentum exchange coefficient within the canopy, $K_M(z)$, was calculated from

$$K_M(z) = \frac{\tau(z)}{\rho \, du(z)/dz}. \qquad (32)$$

By allowing $C_{M,c}$ to vary with height as a function of σ and u (Eq. 31) in deriving $\tau(z)$ (Eq. 23), the striking result was obtained that $K_M(z)$ is approximately constant and equal to $K_M(h)$ in the upper part of the canopy. A similar result was obtained by Thom (1971) in a bean crop, by a similar analysis, and by Gillespie and King (1971) and Druilhet et al. (1972) in maize using entirely different techniques. This contrasts strongly with the earlier accepted view that K_M decreases exponentially with depth in the canopy and implies that the bulk fluxes of air from the upper parts of the canopy must be very large.

2. Turbulence

There are few measurements of the statistics of turbulence inside forest canopies although these are important determinants of the energy and momentum balance. If the pattern of air flow in the canopy cannot be simply related to that above it, we must assume that two-dimensional flow patterns must be considered to define in-canopy balances. McBean (1968) found indications that turbulence in an old pine stand ($h = 20 \text{ m}$) at the Petawawa forest was larger than in a much shorter and more dense stand of *Pinus contorta* at Marmot Creek ($h = 5$ m). The placing of instruments was a considerable problem and regions of preferred downdrafts were clearly identified, implying the existence of horizontal pressure gradients and regions of preferred updrafts. There were good correlations between measurements of wind speed in and above the 20 m high pine stand. When Allen (1968) measured power spectra in a plantation of *Larix leptolepis*, he observed little small-scale turbulence near the forest floor and most of the variation in horizontal air flow resulted from pressure waves associated with large-scale eddies, with a scale length approaching 100 m. Where the canopy was densest, the spectra showed increased relative contributions from small-scale eddies with length scales corresponding to the distances between the trees.

IV. HEAT AND WATER BALANCE

A. Partitioning of Energy by Canopies

1. *Available Energy*

Conservation of energy requires that under steady-state conditions the algebraic sum of the gains and losses of energy by unit volume of a stand is equal to zero. Whilst steady states are rare during the day, especially at times when the short-wave radiation flux increases and the temperature rises rapidly, rates of change averaged over an hour are generally small and the assumption reasonable.

Consider a volume of the stand of unit cross-section, extending from a planc below the soil surface at which there is negligible change in temperature and water content during the day, to a plane at some appropriate reference level above the canopy, for example, the level at which the fluxes of heat and water vapour are measured. The main terms in the energy balance are:

$$\mathbf{C} + \lambda\mathbf{E} + \mu\mathbf{F}_C + \mathbf{J} + \rho c_p \int_0^z u(z) \cdot \frac{\partial T}{\partial x}(z) \cdot dz + \rho\lambda \int_0^z u(z) \cdot \frac{\partial q}{\partial x}(z) \cdot dz = \mathbf{R}_n \quad (33)$$

where $\mathbf{C}$ is the upward flux of sensible heat by convection,
      $\mathbf{E}$ is rate of evaporation from soil and trees,
      λ is latent heat of vaporization of water,
      $\mathbf{F}_C$ is flux density of CO_2,
      μ is the specific energy equivalent of CO_2 fixation,
      $\mathbf{J}$ is the total storage of sensible and latent heat,
      c_p is specific heat of air at constant pressure,
      q is specific humidity.

Horizontal movements of energy in the soil have been neglected in Eq. (33). The horizontal transport or advection terms of sensible and latent heat transfer in the air (the last two terms on the left-hand side) are also usually ignored because of the added problems and costs of making measurements simultaneously in two dimensions. This procedure may be justifiable in the energy balance of short crops when the measurements are made close to the surface, but is questionable in the case of forest, where the fetch requirement is far greater. The simple one-dimensional model requires a large, flat, uniform forest, posing a serious problem because forests have rides and other gaps which disturb the concentration and wind profiles. According to Dyer (1965) and Bradley (1968), the fetch necessary for the equilibrium profile to become established is about $200h$ where h is the height of the stand. Over this distance the canopy should preferably be uniform and without interruption. If this fetch requirement is not satisfied, horizontal transfer (advection) may render a one-dimensional model inadequate. Solving the diffusion equation in two dimensions for Fourier components of an arbitrary profile, it can be shown (James, unpublished) that a profile reaches equilibrium with a 'distance-constant' (analogous to a time-constant) given by

$$x = \frac{ul^2}{\pi^2 K} \qquad (34)$$

where u is the average wind speed, K is the average transfer coefficient and l is the height of the profile. Taking the values of u and K given in Table VII, and l equal either to 4 m (as at Fetteresso) or 10 m, leads to estimates of x

Table VII

Momentum exchange properties of the Sitka spruce canopy at Fetteresso for six wind speed profiles[a]

	Wind speed $u(h)$(m s^{-1})					
	1	2	3	4	5	6
r_{aM} (s m^{-1})	12·0	6·00	3·96	2·98	2·38	1·98
g_{aM} (m s^{-1})	0·08	0·17	0·25	0·33	0·42	0·50
$K_M(h)$ (m^2 s^{-1})	0·16	0·33	0·45	0·60	0·75	0·91
$\tau(h)$ (N m^{-2})	0·10	0·40	0·91	1·61	2·52	3·63
F (N m^{-2})	0·09	0·46	1·15	2·20	3·66	5·64
p_c	0·93	1·16	1·26	1·26	1·45	1·55
d (m)[b]	9·7	9·4	9·3	9·3	9·3	9·3

[a] From Landsberg and Jarvis (1973).
[b] From Thom's (1971) centre of pressure method.

of about 11 and 70 m, respectively. A distance of $2x$ or $3x$ into a forest from the leading edge should be sufficient to establish equilibrium with negligible advection from the leading edge. However, the effect of a gap or road in the forest, within the required fetch, is difficult to estimate.

In the absence of any estimates or extensive theoretical treatment of the magnitude of advected energy in forest energy budgets, the horizontal transport terms in Eq. (33) will be neglected.

The energy available for sensible and latent heat exchange, $\mathbf{H}$ is then

$$\mathbf{R}_n - (\mathbf{F}_C + \mathbf{J}) = \mathbf{H} \tag{35}$$

where

$$\mathbf{H} = \lambda\mathbf{E} + \mathbf{C}. \tag{36}$$

The energy used in photosynthesis by a coniferous forest, $\mu\mathbf{F}_C$, may average about 3 % of the net radiation during the day (Denmead, 1969) and consequently is usually ignored, although it may reach much higher proportions in the early morning and evening (see Section V.A). The storage term, $\mathbf{J}$, can be appreciable (Stewart and Thom, 1973) but is sometimes neglected. The total flux of energy into storage as heat in a column of unit cross-section through a forest stand has the following main components:

change of sensible heat in the air: $\displaystyle\int_0^z (\rho . c_p . \Delta T_a)\, dz = \mathbf{J}_H$ (37)

change of latent heat in the air: $\displaystyle\int_0^z (\rho . \lambda . \Delta q)\, dz = \mathbf{J}_V$ (38)

change of sensible heat in the trees: $\displaystyle\int_0^z (m . c_w . \Delta T_w)\, dz = \mathbf{J}_{veg}$ (39)

change of sensible heat in the soil: $\displaystyle\int_0^z (\rho_s . c_s . \Delta T_s)\, dz' = \mathbf{G}$ (40)

ΔT is a change in temperature (°C) and Δq the change in specific humidity in one hour, m is mass of wood per unit canopy volume, c is specific heat and the subscripts a, w and s refer to air, wood and soil, respectively.

Representative values for forest are shown in Table VIII. As far as $\mathbf{J}_a$ and $\mathbf{J}_q$ are concerned, the height of the reference plane above the forest is important since that determines the height over which the integration is made.

The total flux of energy into or out of storage above ground, $(\mathbf{J}_a + \mathbf{J}_q + \mathbf{J}_w)$, is largest in the early morning and evening when the net radiation and air temperatures are changing most rapidly, with a tendency for positive

Table VIII

Maximum flux densities of heat into storage on clear summer days[a]

Species	Site ref. no.	$\int_0^z m \cdot dz$ (kg m^{-2})	z_R[b] (m)	J_a (W m^{-2})	J_q (W m^{-2})	J_w (W m^{-2})	G (W m^{-2})	J (W m^{-2})	Reference
Pinus radiata	1	—	—	—	—	—	—	80	Denmead (1969)
Pinus sylvestris	17	23	20·5	21	15	34	10	40	Stewart and Thom (1973)
Pinus taeda	18a	—	—	28	—	112	28	126	Lutzke (1966)
	7	—	—	—	—	—	—	40	Gay (1971b)
Pseudotsuga menziesii	14	—	7·8	—	—	—	—	40	Black and McNaughton (1971)
	10	—	—	—	—	—	—	96	Gay (1972)
	10	—	—	—	—	—	—	40	Gay and Stewart (1973)
Picea abies	13	59	20	18	—	24	66	100	Tajchman (1971, 1972a)
Picea sitchensis	13	—	—	12	—	170	30	200	Strauss (1971)
	19	29	13	12	10	25	3	35	James and Jarvis (unpublished)

[a] See Eqs (37) to (40) for symbols

[b] Reference height up to which the storage integration is made.

values of J_q to reinforce values of $(J_a + J_w)$ (Stewart and Thom, 1973). At these times, the flux of energy into storage in tall open stands is of the same order as the net radiation flux density and is thus a major component of the energy balance. In the middle of the day, however, air temperature and humidity change more slowly and the temperatures of the trunks and branches approach air temperature (Tajchman, 1972a), so that $(J_a + J_q + J_w) = 0$ one or two hours after midday (Stewart and Thom, 1973). In contrast the flux of heat into the soil, G, is maximum at about midday when the air temperature and net radiation flux on the forest floor are maximum (Baumgartner, 1965).

Under very dense canopies, as at Fetteresso, and on overcast days in more open canopies, the soil heat flux is less than $5\ W\ m^{-2}$ and may be neglected (Tajchman, 1972a; Table VIII). In more open canopies on sunny days, G is a significant fraction of R_n (Table VIII). In dense canopies, too, J_w is small because most radiation and energy exchange occurs in the upper parts of the canopy, and in young stands ($h \approx 10\ m$), J_a and J_q are also not large. For Sitka spruce at Fetteresso, the flux into storage amounts to less than 1 % of the net radiation in the middle of the day, but can be as large as 15 % at 0800, and larger storage fluxes are found in the Scots pine at Thetford. The daily (24 h) sum of storage fluxes, however, is only 1 to 4 % of the net radiation flux density (Tajchman, 1972a).

2. The Bowen Ratio

Assuming horizontal uniformity and equilibrium conditions the available energy, as defined by Eq. (35), is partitioned into the vertical flux densities of sensible and latent heat [Eq. (36)]. The distribution of available energy between these fluxes is sometimes expressed as a fraction of available energy, for example $\lambda E/H$, or more frequently expressed by the Bowen ratio:

$$\beta = \frac{C}{\lambda E}. \tag{41}$$

In the absence of advection, β can vary between $+\infty$ for a dry surface ($\lambda E = 0$) to -1 for an evaporating wet surface for which all the energy necessary for evaporation is supplied by convection, i.e. there is no net radiation gain so that $C = -\lambda E$.

The exchanges of sensible and latent heat are turbulent transfer processes and, under steady conditions in the absence of advection, may be written as:

$$C = -\rho c_p K_H (\partial T/\partial z + \Gamma) \tag{42}$$

$$\lambda E = -\rho \lambda K_V (\partial q/\partial z) \tag{43}$$

where K_H and K_V are the eddy diffusivities of heat and water vapour, respectively, and Γ is the dry adiabatic lapse rate. Assuming that the fluxes are constant with height above the canopy, so that the average gradients can be approximated by finite differences between two levels, and that $K_V = K_H$, the Bowen ratio becomes

$$\beta = \frac{c_p}{\lambda} \cdot \frac{T + \Gamma \, \Delta z}{\Delta q}. \tag{44}$$

The adiabatic lapse rate is included in Eqs (42) and (44) because over forests the temperature gradients are small and the vertical interval spanned by the measurements is large. Determination of the Bowen ratio, and hence of the fluxes of heat and water vapour, requires highly accurate measurements of the gradients of temperature and humidity above the canopy (estimation of C and λE by eddy correlation has hardly been tried over forest so far). Acceptable levels of accuracy are difficult to obtain because of the small size of the gradients (Table IX) which are much smaller than over field crops, because of the extreme aerodynamic roughness of the forest surface. Measuring the gradients close to the canopy surface where they are largest minimizes the fetch requirement (Eq. 34). As well as reducing the relative size of the measurement errors it introduces the need to sample horizontally because

Table IX

Examples of the size of the gradients over dry coniferous forest canopies during unstable daytime conditions

Species	Site ref. no	Height interval (m)	Condition[a]	grad T (°C m^{-1})	grade e (mbar m^{-1})	grad ϕ (mg m^{-4})
Pinus radiata	1	h to $(h + 2)$	s	0·05	0·1	1·0
Pinus taeda	3b	h to $(h + 2)$	s	<0·1	<0·2	<1·0
			o	0·05	0·1	0·5
Pinus sylvestris	17	h to $(h + 15)$	s	0·03	0·02	—
			o	0·03	0·02	—
Pinus sylvestris	20	$(h + 7)$ to $(h + 32)$	s	<0·03	<0·04	—
			o	0·01	0·02	—
Pinus sylvestris	18a	$(h + 4)$ to $(h + 7)$	s/o	0·02	0·03	—
Pseudotsuga menziesii	22	h to $(h + 1)$	s	0·3	0·3	—
Picea sitchensis[b]	19	h to $(h + 4)$	$S_t > 400$	0·22	0·08	0·90
			$S_t < 400$	0·08	0·05	0·80

[a] Symbols as in Table II.
[b] Median values.

of surface heterogeneity. To overcome these problems at Fetteresso, we have recently been measuring gradients differentially as differences in concentration between h and $(h + 4)$ m at four stations. At each station, four sampling points are distributed along a 5 m support boom at each of two levels. A typical diurnal course of the gradients is shown in Fig. 8. Black and

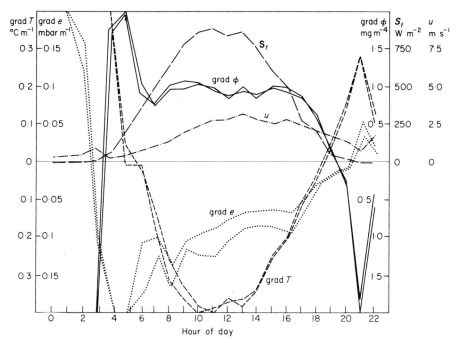

Fig. 8. The diurnal course of solar irradiance, S_t, wind speed, u, and duplicated gradients of temperature, grad T, vapour pressure, grad e, and carbon dioxide concentration, grad ϕ, on 28 June 1973, over a 4 m interval immediately above the canopy of *Picea sitchensis* at Fetteresso (Site ref. no. 19).

McNaughton (1971) have also measured gradients differentially close to the top of the canopy. However, the alternative of measuring *profiles* at greater heights above the canopy, as is usual over field crops, has been more widely applied over forest. Because of the extremely small gradients at greater heights (Table IX), adequate accuracy can be obtained only with enormous care and expense.

The size of the Bowen ratio over forest has an important bearing on afforestation and watershed management and it is therefore important to establish the range of values to be expected. Few data are yet available to

Table X

Ranges of Bowen ratios measured over coniferous forest canopies by the energy balance

Species	Site ref. no.	No. of days	Bowen ratios[a] canopy wet	canopy dry overcast	canopy dry sunny	Reference
Pinus radiata	1	3	—	—	0.1–0.8	Denmead (1969)
Pinus contorta	7	1	—	—	0.4–1.2	Gay (1971b)
Pinus taeda	3b	—	0.1–0.4	0.4–1.4	0.4–1.4	Sinclair et al. (unpublished)
Picea sp.	15	19	0.2–0.4	0.2–2.1	0.3–1.2	Storr et al. (1970)
Pinus resinosa/strobus	9a	5	—	0.45	0.6–2.0	Mukammal (1966)
Pinus sylvestris	18a	3	—	—	–0.1–0.8	Lutzke (1966)
	21a	4	—	0.2–0.7	0.3–3.5	Odin and Perttu (1966)
	12	1	—	—	0.5–1.3	Tajchman (1972b)
	17	8+	–0.3	0.5–2	1–4	Stewart and Thom (1973)
					4–10[c]	Gay and Stewart (1973)
						Stewart et al. (unpublished)
Pseudotsuga menziesii	20	7	—	—	–0.2–1.3	Berggren et al. (1971)
	10	3	—	0.3–0.4	0.3–1.8	Gay and Stewart (1973)
						Gay (1972)
	14	6	—	0.5–1.6	0.2–1.5	Black and McNaughton (1971)
						McNaughton and Black (1973)
Picea abies	13	4	—	–0.3–1.3	0.2–3.1	Tajchman (1967, 1972a)
		8	—	0–1.0	0.1–0.9	Strauss (1971)
Picea sitchensis[b]	19	35	–0.7–0.9	–0.7–1.9	0.5–3.0[d]	Jarvis et al. (unpublished)
					1.1–3.4[e]	

[a] Mean hourly estimates between 8.00 and 16.00 h in the summer.

[b] Top and bottom 10 % of range excluded; $n = 274$ h.

[c] Water stressed in the afternoon.

[d] $S_t > 400$ W m^{-2}; $r_i > 18.4$ s m^{-1}; i.e. relatively high δe.

[e] $S_t > 400$ W m^{-2}; $r_i < 18.4$ s m^{-1}; i.e. relatively low δe.

provide an assessment (Table X). For irrigated farm crops, daytime values of β range from about -0.5 to $+0.5$. Values less than -1 occur when advected heat is important and more positive values ($+1$ to 1.6) have been found for water-stressed crops (Penman *et al.*, 1967). Measurements over forest (Table X) fall into two groups. For most of the forest sites, irrespective of species, the daytime Bowen ratio of a dry canopy varies between 0.1 and 1.5, and when the canopy is wet with rain or dew, between -0.7 and $+0.4$. However, at two United Kingdom sites, Thetford (*Pinus sylvestris*) and Fetteresso (*Picea sitchensis*), much larger values of β have been found consistently, and similar larger values are found occasionally in other extensive records (e.g., Tajchman (1967), *Picea abies* at Ebersberger). At Thetford, β rises to a fairly steady value of between 1 and 4 soon after sunrise; in the afternoon, it may either decrease or, in conditions leading to stomatal closure, may climb to even higher values (Stewart and Thom, 1973). At Fetteresso, β may reach 2 to 3 in the middle of the day but declines again in the afternoon (Fig. 9). Why do the values of β for forest exceed those for field crops, in some cases by a very large margin, and why are the Bowen

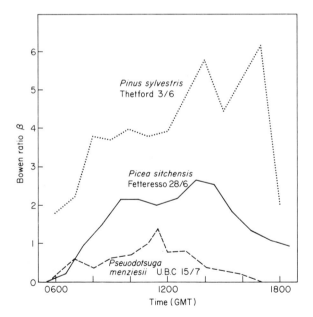

Fig. 9. Typical examples of the diurnal course of the Bowen ratio from three sites in the summer. Site details are given in the Appendix, Table XVI; r_c and r_i are shown in Fig. 11. *Pinus sylvestris* data from Stewart and Thom (1973); *Pseudotsuga menziesii* data from McNaughton and Black (1973).

ratios measured at the two U.K. sites generally larger than those reported from elsewhere?

Monteith (1965) generalized the Penman equation for evaporation from a short crop in the form (see Volume 1, pp. 88–90):

$$\lambda E = \frac{\Delta H + c_p \rho \, \delta e / r_a}{\Delta + \gamma (1 + r_c / r_a)} \tag{45}$$

where δe is the atmospheric water vapour saturation deficit (mbar), γ is the psychrometric constant (mbar/°C), r_a is the boundary-layer resistance of the canopy to mass transfer (s m^{-1}) (it is assumed that $r_a = r_{aV} = r_{aH}$), r_c is the canopy resistance to water vapour transfer (s m^{-1}), i.e., the resistance of all the leaves in the canopy summed in parallel and Δ is the rate of change of saturation vapour pressure with temperature. Other symbols are as before. Strictly, this equation treats the canopy as a single large, flat leaf, without taking into account the vertical and dissimilar distributions of sources and sinks of heat and water vapour. Nonetheless, it is more than adequate to answer the questions posed. By combining Eq. (45) with Eqs. (36) and (41), the Bowen ratio becomes

$$\beta = \frac{1 + \gamma (1 + r_c / r_a) \Delta}{1 + c_p \rho \, \delta e / \Delta H r_a} - 1. \tag{46}$$

Following Monteith (1965) and Stewart and Thom (1973), a 'climatological resistance' is defined as

$$r_i = \frac{\rho c_p \, \delta e}{\gamma H} \, (\text{s m}^{-1}). \tag{47}$$

Equation (46) can then be written as

$$\beta = \frac{\Delta / \gamma + 1 + r_c / r_a}{\Delta / \gamma + r_i / r_a} - 1. \tag{48}$$

It is apparent that β is approximately proportional to the canopy resistance, r_c, when r_c is large in relation to the boundary-layer resistance, r_a, (i.e. when stomata are closing); and similarly that β is approximately proportional to H and *inversely* proportional to δe when r_i is large in relation to the boundary-layer resistance, r_a.

For a sunny day, typical values of the variables in Eq. (46) are: H = 500 W m^{-2}; $\delta e = 5$ mbar (in an oceanic climate) or 20 mbar (in a continental climate) and $r_i = 20$ and and 80 s m^{-1}, respectively; $r_a = 5$ s m^{-1} (from Table VII, assuming $r_{aV} = r_{aM}$ or from Table VI taking $r_a = r/L$); $r_c = 50 - 100$ m^{-1} from Fig. 11 or from Table XII taking $r_c = r_s/L$) when the stomata are

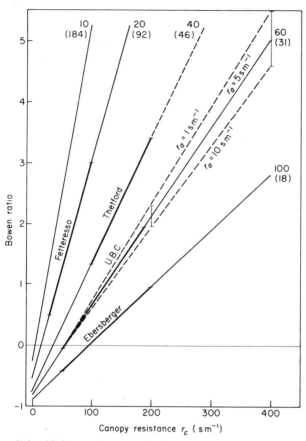

Fig. 10. The relationship between Bowen ratio, β, and canopy resistance, r_c, at different values of the climatological resistance, r_i, calculated from Eq. (48) for $H = 500$ W m^{-2} and $r_a = 5$ s m^{-1}. The figures by the curves are values of r_i (s m^{-1}); the figures in brackets are the corresponding ratios of $H/\delta e$ W m^{-2} mbar^{-1}). The heavy lines indicate appropriate ranges of β and r_c (see Table X and Fig. 11) for relevant values of r_i at sites in different climates.

open. The results of solving Eq. (48) for a range of values H, δe, r_c and r_a are shown in Fig. 10 (Δ/γ was taken as 1·65 throughout).

Figure 10 shows that on a sunny day with open stomata, Bowen ratios of 2 to 3 are to be expected in an oceanic climate where the vapour pressure deficit is only about 3 to 6 mbar, as at Fetteresso most of the summer. As the stomata close and r_c approaches 400 s m^{-1}, say, much higher Bowen ratios are expected, even in less humid oceanic climates ($\delta e = 10$ to 20 mbar) such as at Thetford. Conversely in continental climates, with generally higher air temperatures and vapour pressure deficits ($\delta e = 20$ to 30 mbar), such as at

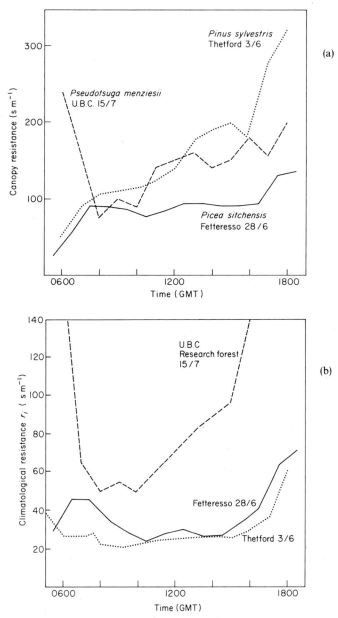

Fig. 11. Typical examples of (a) the diurnal course of the canopy resistance, r_c; and (b) the diurnal course of the corresponding climatological resistance, r_i, for the same three sites and summer days for which β is shown in Fig. 9.

Ebersberger, β would be expected to be less than unity when the stomata are open ($r_c \approx 50\,\text{s m}^{-1}$) and to rise above unity only in circumstances of virtually complete stomatal closure (r_c exceeding $300\,\text{s m}^{-1}$), or unusually high humidity. Changes in r_a as a result of changes in wind speed only slightly alter the positions of the curves in Fig. 10. A doubling of r_a to $10\,\text{s m}^{-1}$ reduces β as shown by the lower limit marked on the $r_i = 60$ curve; the upper limit shows the effect of reducing r_a to $1\,\text{s m}^{-1}$.

A wide range of values of β is therefore expected, depending particularly on the stomatal resistance but also on the general climatic and local weather conditions at the site. The very large values of β in the late afternoon at Thetford almost certainly result from stomatal closure increasing r_c, whereas the generally high values in the middle of the day at Fetteresso are largely the result of the prevailing small vapour pressure deficits: r_i is generally about 18·4 (i.e., δe increases with $\mathbf{R}_n$ by 1 mbar per $100\,\text{W m}^{-2}$). Bowen ratios listed in Table X for the other sites are generally smaller because the sites are nearly all much more continental, but also because measurements on clear warm summer days with large δe are most frequently reported. At any site, unusual values of β may occur when particular weather conditions produce unusual values of r_i, independent of changes in boundary-layer or stomatal resistances.

3. The Canopy Surface Resistance and Water Vapour Flux

Rearrangement of Eq. (48) enables the canopy surface resistance, r_c, to be derived from the Bowen ratio as follows:

$$r_c = \left(\frac{\Delta}{\gamma}\beta - 1\right)r_a + (\beta + 1)r_i. \tag{49}$$

Stewart and Thom (1973) give a more sophisticated treatment in which equality of r_{aV}, r_{aH} and r_{aM} is not assumed. However, taking typical values of $r_a = 5\,\text{s m}^{-1}$, $r_i = 40\,\text{s m}^{-1}$, and $\beta = 2$, it is readily shown that the ratio of the two terms on the right hand side of Eq. (49) is large, i.e.

$$[(\beta + 1)r_i]\left/\left[\left(\frac{\Delta}{\gamma}\beta - 1\right)r_a\right]\right. > 10.$$

Thus the estimate of r_c is insensitive to the exact value of r_a, especially at small values of β and large r_i/r_a. In some circumstances the term containing r_a can be omitted altogether giving $r_c \simeq (\beta + 1)r_i$ without introducing serious error. Figure 11a shows that r_c derived from Eq. (49) increases during the course of the day. The increase in the afternoon is particularly marked at Thetford, probably because of poor supply of water to the roots in the deep sandy soil.

The flux of latent heat, λE, is obtained from the Bowen ratio as $\lambda E = H/(1 + \beta)$. The amount of water evaporated from the canopy between 0800 1700 GMT at Fetteresso is shown in relation to the total net radiation flux in Fig. 12. When the canopy is wet after rain most of the energy from net

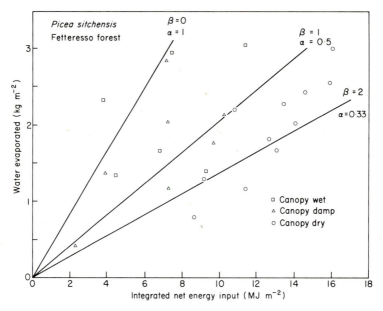

Fig. 12. The relationship between the quantity of water evaporated from the Sitka spruce stand at Fetteresso (Site ref. no. 19) and the input of net radiation between 08 00 and 17 00 GMT. for a number of days in summer 1971.

radiation is used for evaporation, but when it is dry, $\lambda E/R_n$ is only 0·25 to 0·5. If the surface of the canopy is completely covered with a film of water, the potential evaporation, λE_0, is obtained from Eq. (45) by setting $r_c = 0$. The relative rate of evaporation from the canopy in the same weather conditions, E/E_0, is then obtained as

$$\frac{E}{E_0} = \frac{\Delta + \gamma}{\Delta + \gamma(1 + r_c/r_a)}$$

and this fraction depends only on air temperature and the ratio r_c/r_a. Using Baumgartner's (1956) data for Scots pine, Monteith (1965) took r_c/r_a to be 15 and concluded that 'the transpiration rate is about one-fifth of the evaporation rate from leaves wetted by rain and exposed to the same weather'. For Sitka spruce at Fetteresso, the ratio r_c/r_a generally lies between

20 to 40, and for Scots pine at Thetford between 20 to 70, leading to the conclusion that the evaporation from dry foliage is only about 5% of that from wet foliage *exposed to the same weather* because of the large ratio. In contrast the ratio r_c/r_a is between one and two for field crops and E/E_0 is over 60% (Monteith, 1965).

B. Exchanges within Canopies

1. *Characteristics of Profiles*

Figure 13 shows vertical profiles through the canopy of wind speed, air temperature, humidity and carbon dioxide at Fetteresso. The shapes of these profiles are similar to those from a number of other sites (e.g. Thetford, Triangle, Canberra). None of the profiles is monotonic: the gradients

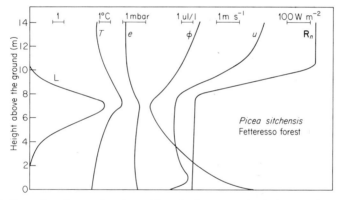

Fig. 13. Examples of mean hourly profiles of micrometeorólogical properties measured at Fetteresso between 12 00 and 13 00 GMT. on a sunny (but not completely clear) day in July 1970. $S_t = 605 \text{ W m}^{-2}$, $R_n = 524 \text{ W m}^{-2}$, $T_R = 11·80°C$, $e_R = 11·13 \text{ mbar}$, $\phi_R = 315·4 \text{ } \mu l/l$, $u_R = 2·9 \text{ m s}^{-1}$.

become zero at some point in the canopy and reverse in the trunk region below the crowns. This feature is expected for CO_2, since the soil is a source of CO_2; and for temperature, since the soil and tree trunks are sinks for sensible heat. However, the hollow in the humidity profile below the crowns indicates an apparent sink for humidity which does not exist. The hollow in the humidity profile must indicate divergence in the horizontal humidity flux, and, although not obvious in the profiles, in the horizontal fluxes of CO_2 and sensible heat. This mass flux divergence presumably has the same causes as the horizontal divergence in momentum transfer, indicated by the occurrence of blow-through (see Section II.C.1). Thus in the trunk region and lower levels of the canopy, the one-dimensional model is clearly inadequate. Higher up within the canopy, horizontal flux divergence is not apparent, as vertical transfer predominates and the shapes of the profiles are consistent with one-dimensional transfer. However, as discussed on Vol. 1, p. 96, horizontal divergence may render the conventional, one-dimensional analysis inaccurate.

For this reason, and also because the gradients in the canopy are very small, it is hazardous to determine the source or sink strengths for heat, water vapour or carbon dioxide within the canopy by differentiation of the flux profiles (see Vol. 1, p. 97). Although this seems to work well in farm crops, the evidence of blow-through in forest, where it occurs, invalidates the use of the one-dimensional flux model, and the small size of the gradients makes the second differentiation very inaccurate. Moreover, profiles drawn by hand through a few plotted points are very subjective, particularly when a point looks dubious and is not replicated. Notwithstanding these difficulties, Denmead (1964) calculated source-sink distributions of sensible and latent heat within a *Pinus radiata* canopy, using cumulative incomplete Beta functions to differentiate the vertical flux profiles of net radiation, sensible and latent heat as objectively as possible, and he obtained sensible looking profiles. However, we believe that alternative methods should be used to define the distribution and strengths of sources and sinks within coniferous forest canopies because of the uncertainties inherent in the micrometeorological approach.

2. *Exchanges of Heat and Water Vapour*

The distribution and strength of the sources of sensible heat within the canopy can be described from measurements of needle temperature using the relationship

$$C(z) = \rho c_p \frac{T_l(z) - T_a(z)}{r_{aH}(z)} \tag{50}$$

where $T_l(z)$ and $T_a(z)$ are needle (and branch) and ambient air temperatures, respectively, at height z in the canopy. Partitioning of the sensible heat flux in this way has rarely been attempted, perhaps because of its limited physiological interest, but also because the accurate measurement of needle temperatures is difficult (Perrier, 1971). Rutter (1967) found that needle temperatures in a wet Scots pine canopy were 0·3 to 1·0 °C below nearby air temperature, and that under normal conditions for transpiration, needle temperatures were 0 to 0·2 °C above air temperature. These small differences between needle and air temperatures result from the small boundary-layer resistance of needles (Table VI) and the consequent rapid exchange of sensible heat which leads to a very close coupling between needle and air temperature. Probable leaf-air temperature differences can be calculated by combining Eq. (50) with Eqs (36) and (45) for a range of values of net radiation, vapour pressure deficit, stomatal and boundary-layer resistances. Results of such calculations for a limited set of conditions are shown in Table XI. Under

Table XI
Calculated leaf-air temperature differences for needles
with a range of stomatal and boundary-layer resistances[a]

r_s (s m^{-1})	Leaf-air temperature difference (°C) r_{aH} (s m^{-1})			
	1	10	50	100
0	-6.83	-6.69	-6.07	-5.30
250	-0.03	-0.27	-0.76	-0.72
500	0.00	0.05	0.36	0.86
1000	0.00	0.23	1.11	2.14
2000	0.03	0.32	1.55	3.01
4000	0.04	0.36	1.80	3.53

[a] $R_n = 500$ W m^{-2}; $\delta e = 12$ mbar.

normal conditions of ventilation within the canopy ($0.5 < u > 2.0$ m s^{-1}; $10 < r_a < 50$ s m^{-1}; See Section III.B and C), needle temperatures of more than 2 °C above air temperatures are most unlikely even with closed stomata and strong radiation on the needles. Yamaoka (1958) provided experimental verification by elegant measurements with 20 thermocouples of the leaf-air temperature difference of needles of *Cryptomeria japonica*. He found that even with an irradiance of 700 W m^{-2}, the leaf air temperature difference exceeded 2.5 °C only at windspeeds of less than 1 m s^{-1} (air temperature 20 °C; relative humidity 60%). At a wind speed of 2 m s^{-1}, the leaf temperature was within ± 1 °C of air temperature at irradiances between 0 and 7 0 W m^{-2} (Fig. 14). This behaviour contrasts with the much higher temperature elevations (10 to 15 °C) of strongly irradiated broad leaves which result from their larger dimensions.

At Fetteresso, the water vapour source strength is estimated by extensive measurements of stomatal resistance or conductance of shoots within the canopy using a diffusion porometer (Beardsell *et al.*, 1972, W. R. Watts unpublished). The diffusion porometer measures the stomatal resistance or conductance of a needle or group of needles per unit (projected) area of needle. The rate of evaporation in a layer of the canopy of finite thickness Δz depends on the total stomatal and boundary-layer resistances of all the

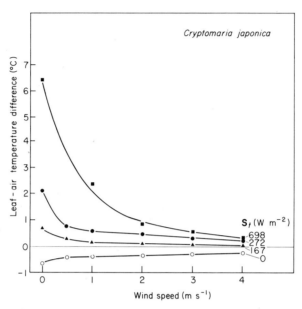

Fig. 14. The relationship between the leaf-air temperature difference of a shoot of *Crytomeria japonica* and wind speed at four irradiances. Air temperature 20°C, relative humidity 60% (from Yamaoka, 1958).

needles in that layer, $r_s(\Delta z)$ and $r_{a,v}(\Delta z)$, respectively, and is

$$\mathbf{E}(\Delta z) = \rho \frac{q_l(\Delta z) - q_a(\Delta z)}{r_s(\Delta z) + r_{a,v}(\Delta z)} \qquad (51)$$

where $q_l(\Delta z)$ and $q_a(\Delta z)$ are the mean specific humidities at the liquid-air interface in the needle and in the air, respectively, in the layer of the canopy. It is convenient to use resistances in the expression because the stomatal and boundary-layer resistances in series are additive. However, in what follows, it is more convenient to use conductances because evaporation is proportional to conductance and because the conductance of a number of leaves in a population can be added to give either the total or the mean conductance of the population. As the relevant resistances are in parallel, the simplest way to obtain the effective resistance of a population of n leaves is to take the reciprocal of the mean conductance, i.e., $\bar{r}_s = (\sum g_s/n)^{-1}$.

The layer of foliage of thickness Δz is made up of needles of several age classes which open their stomata to a varying extent in response to changes in the environment (Fig. 15; Section IV.C.2). The total conductance of the ith age class in a layer, $g_{s,i}$ is the sum of the conductances of all the needles

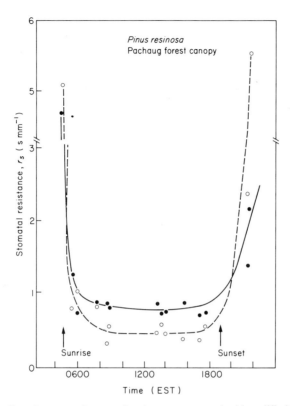

Fig. 15. The diurnal course of stomatal resistance, measured with a diffusion porometer, in needles of *Pinus resinosa* at Pachaug Forest (Site ref. no. 4) in July. ○ newly emerged needles, ● one-year-old needles (from Waggoner and Turner, 1971).

in that class, but since the conductance per unit area of a sample is measured, the total conductance is given by $g_{s,i} = \bar{g}_{s,i} \cdot L_i$, where $\bar{g}_{s,i}$ is the mean stomatal conductance per unit needle area and L_i is the leaf area index of needles in the ith age class. The total conductance of needles of different ages in a layer of the Sitka spruce canopy at Fetteresso is shown in Fig. 16a.

The total conductance of a layer (Fig. 16b) is the sum of the conductances of all needles of all age classes in that layer, namely,

$$g_s(\Delta z) = \sum (g_{s,i} \cdot L_i). \tag{52}$$

The total conductance of the canopy (Fig. 16b) is the sum of the conductances of all needles in all age classes and layers, namely,

$$g_s(0 - h) = \sum_{0}^{h} g_s(\Delta z) \tag{53}$$

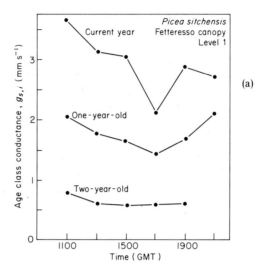

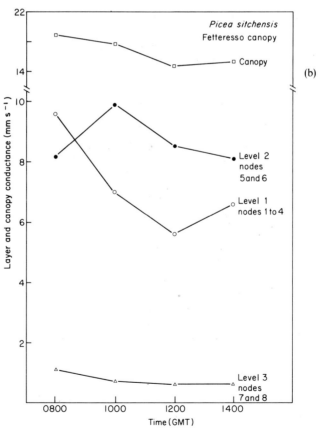

Since needle temperatures are closely coupled to air temperatures [this affects q_l in Eq. (51)], since humidity changes little through the canopy (affects q_n) and since r_{av} is small in relation to r_s, the stomatal conductances of the age classes and layers of foliage shown in Fig. 16 are a good approximation to the distribution of evaporation within the canopy. The stomatal conductance varies with time of day, with height in the canopy and with the weather.

C. Exchange Characteristics of Shoots

1. Boundary-Layer and Cuticular Resistances

The boundary-layer resistance of needles and shoots to heat and water vapour transfer has been considered in Section III.C together with the boundary-layer resistance to momentum transfer. Because of its small size, resulting from the small width of needles, it is relatively unimportant in relation to r_s in influencing transfer processes.

The cuticular resistance to water vapour (or CO_2) transfer of conifer needles has rarely been determined because of difficulties in ensuring complete stomatal closure (Jarvis, 1971). Whiteman and Koller (1967) give a figure of $3\,600\,\text{s}\,\text{m}^{-1}$ for *Pinus halepensis* based on transpiration measurements in the dark. Even larger values are expected because of the thick cuticle (Lange and Schultz, 1966) and because of the large quantities of waxes on the surface of the needles. These waxes are present as an amorphous covering layer which merges into the cuticle, and from which outgrowths of tubes and rods usually appear (Leyton and Juniper, 1963; Leyton and Armitage, 1968; Johnson and Jeffree, 1970; Jeffree *et al.*, 1971; Hanover and Reicosky, 1971; Lehela *et al.*, 1972; Campbell, 1972). Preliminary estimates of cuticular resistance of droughted Sitka spruce needles in the dark lie in the range $9\,000$ to $15\,000\,\text{s}\,\text{m}^{-1}$ (C. L. Beadle, unpublished). Consequently, the cuticle may be regarded as impervious to water vapour and CO_2.

2. Stomatal Resistances

Minimal stomatal resistances of a number of conifers are listed in Table XII. In general, they are somewhat larger than minimal values for deciduous

---

Fig. 16. (a) The diurnal course of the total conductance of needles of different ages, $g_{s,i}$ ($= \bar{g}_{s,i} \cdot L_i$), in the upper part (nodes 1 to 4) of the Sitka spruce canopy at Fetteresso on a sunny day in July. (At 13·00h $S_t = 770\,\text{W}\,\text{m}^{-2}$, $\delta e = 11·4\,\text{mbar}$, $T_a = 20·2°\text{C}$, (b) The diurnal course of the total conductance of the layers, $g_s(\Delta z)$, into which the canopy is divided, and of the conductance of the entire canopy ($\sum_0^h g_s(\Delta z)$). (At 13·00h $S_t = 780\,\text{W}\,\text{m}^{-2}$, $\delta e = 7·1\,\text{mbar}$, $T_a = 17·2°\text{C}$.) (a and b from W. R. Watts, unpublished.)

Table XII
Minimal stomatal resistances to water vapour transfer of coniferous needles

Species	Needle age	Minimal r_s (s m^{-1})	Method	Reference
Pinus halepensis	—	420	1	Whiteman and Koller (1967)
	—	1 000	2	Kaufmann (1973)
Pinus strobus	—	1 200	3	Gates (1966)
Pinus resinosa	—	780	3	Gates (1966)
Pinus contorta	Current year	290	1	Dykstra (1974)
Pinus resinosa	Current year	310	2	Waggoner and Turner (1971)
	1 year	430	2	Waggoner and Turner (1971)
	2 year	860	2	Waggoner and Turner (1971)
	3 year	890	2	Waggoner and Turner (1971)
	4 year	1 100	2	Waggoner and Turner (1971)
Pinus ponderosa	Current year	950	2	Running (1974)
Pinus sylvestris	—	*ca.* 200	4	Rutter (1967)
	—	400	2	Robins (*In* Stewart and Thom, 1973)
Tsuga heterophylla	Current year	450	2	Running (1973)
Sequoia sempervirens	—	1 500	2	Kaufmann (1973)
Thuja occidentalis	—	2 700	3	Gates (1966)
Abies grandis	Current year	240	2	Running (1974)
Picea mariana	—	1 700	3	Gates (1966)
Picea Breweriana	Current year	140	2	Running (1974)
Picea engelmannii	—	2 400	2	Kaufmann (1973)
	—	400	2	Kaufmann (unpublished)
Picea sitchensis	Current year	120	1	Ludlow and Jarvis (1971), Neilson *et al.* (1972)
	1 year	480	1	Ludlow and Jarvis (1971), Neilson *et al.* (1972)
	2 year	650	1	Ludlow and Jarvis (1971), Neilson *et al.* (1972)
	3 year	1 400	1	Ludlow and Jarvis (1971), Neilson *et al.* (1972)

Table XII—*continued*

Species	Needle age	Minimal r_s (s m^{-1})	Method	Reference
	Current year	140	2	W. R. Watts (unpublished)
	1 year	220	2	W. R. Watts (unpublished)
	2 year	330	2	W. R. Watts (unpublished)
	Current year	290	2	Running (1974)
Pseudotsuga menziesii	Current year	120	2	Running (1974)
	Current year	300	1	Leverenz (1974)

Method: 1 assimilation chamber; 2 diffusion porometer; 3 leaf energy balance; 4 loss in weight.

trees, farm crops and herbaceous plants (Holmgren *et al.*, 1965; Whiteman and Koller, 1967; Turner 1974). Some of the values are so large that it is questionable whether they are the lowest obtainable but the smaller minimal resistances of *Picea sitchensis* and *Pinus resinosa* are not exceptional.

The stomatal resistance, estimated from the water vapour flux in assimilation chambers or diffusion porometers, includes components additional to the resistance of the stomatal pore itself (Jarvis, 1971). Any hydraulic or diffusion resistance or salt accumulation which lowers the vapour pressure at the liquid-air interface in the leaf is included (Jarvis and Slatyer, 1970): there is no relevant evidence for conifers. However, the stomatal pores of many conifers are located at the bottom of an indentation in the leaf epidermis so that there is a distinct ante-chamber above the pore shaped like a cup or inverted cone. This chamber is filled with wax tubes in a haphazard arrangement, rather like a bundle of pencils tossed into a waste paper basket (Jeffree *et al.*, 1971; Hanover and Reicosky, 1971; Lehela *et al.*, 1972; Campbell, 1972). In Sitka spruce, Jeffree *et al.* estimated from geometrical considerations that approximately one third of the stomatal resistance was accounted for by the ante-chamber, one third by the wax tubes in the ante-chamber, and one third by the pore and intercellular spaces in the leaf. The wax tubes begin to grow out of the cuticle lining the ante-chamber at a very early stage in leaf development, before the stomatal pore itself is differentiated.

The stomatal resistance of conifers declines in response to increasing irradiance in the usual way (Fig. 17) (Ludlow and Jarvis, 1971, Waggoner and Turner, 1971). The resistance is small over a wide range of leaf temperatures, but increases considerably below 5 °C and above 20°C (Neilson and

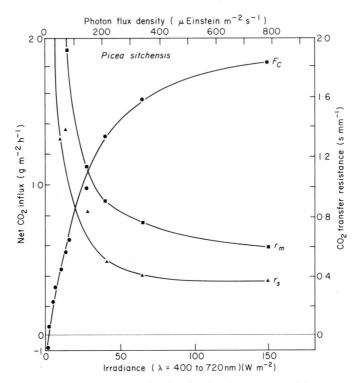

Fig. 17. The relationships amongst the flux density of radiant energy and the net CO_2 influx, F_C, stomatal, r_s, and mesophyll, r_m, resistances to CO_2 transfer ($\phi_a = 315 \mu l/l$, $T_l = 20.0°C$, $\delta e = 6$ mbar) (from Ludlow and Jarvis, 1971).

Jarvis, 1975). There is no response of stomatal resistance to CO_2 concentration in Sitka spruce between 100 and 600 $\mu l/l$ (Ludlow and Jarvis, 1971; Neilson and Jarvis, 1975). There is little or no response to oxygen at normal CO_2 concentrations (Ludlow and Jarvis, 1971; Cornic and Jarvis, 1972). Increase in stomatal resistance in response to increasing vapour pressure deficits may begin at a threshold of about 10 mbar (Ludlow and Jarvis, 1971; Whiteman and Koller, 1964) but in the canopy, stomatal closure of Sitka spruce is more nearly proportional to vapour pressure deficit (Fig. 18). Both Sitka spruce and *Pinus resinosa* (Waggoner and Turner, 1971) show no change in stomatal resistance down to a shoot-water potential of -15 bar. Lopushinsky (1969) found stomatal closure in *Pinus ponderosa* and *Pinus contorta* in the range -14 to -17 bar and in *Picea engelmannii* at a shoot-water potential of -16 bar. Little closure occurred in *Pseudotsuga menziesii* at -19 bar or in *Abies grandis* at -25 bar. Caldwell (1970) found no reduc-

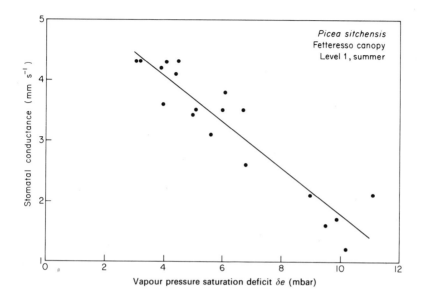

Fig. 18. The relationship between stomatal conductance of current year shoots in the upper part of the Sitka spruce canopy at Fetteresso and vapour pressure saturation deficit, δe, during the summer months; each point is the mean conductance of six shoots and 12 measurements of vapour pressure deficit made in the same hour ($r^2 = 0.86$) (from W. R. Watts, unpublished).

tion in transpiration by *Pinus cembra* at wind speeds of up to $15 \, \mathrm{m \, s^{-1}}$. Reduction in soil temperature to $2 \, ^\circ \mathrm{C}$ has little or no effect on stomatal resistance of Sitka spruce over a wide range of leaf temperatures and vapour pressure deficits (Turner and Jarvis, 1975) as expected from the lack of response of transpiration rate in Scots pine (e.g. Linder, 1972). However, stomata of *Pinus radiata* are much more sensitive to soil temperature (Babalola *et al.*, 1968). Stomatal resistance declines during the first few weeks of needle maturation after bud break (Ludlow and Jarvis, 1971) and then increases as the needles age (Table XII).

V. CARBON DIOXIDE

A. Exchange by Canopies

The flux of carbon dioxide in the canopy is perhaps the most biologically significant of the exchanges originating at the leaves. During photosynthesis in daytime, the vertical flux of CO_2, $\mathbf{F}_C$, into the canopy from the bulk air above is given formally by

$$\mathbf{F}_C = -K_C \frac{\partial \phi}{\partial z}. \tag{54}$$

The gradient of CO_2 concentration, $\partial \phi/\partial z$, is usually measured over the same height interval as $\partial T/\partial z$ and $\partial q/\partial z$ [Eqs (42) and (43)], and the eddy diffusivity for CO_2 transfer, K_C, is obtained by solving Eqs (35), (42), (43) and (54) for the eddy diffusivity of energy K_E, and assuming that $K_E = K_H = K_V = K_C$. Thus

$$K_E = \frac{\mathbf{H}}{-\rho c_p(\partial T/\partial z + \Gamma) - \rho \lambda (\partial q/\partial z) + \mu(\partial \phi/\partial z)}. \tag{55}$$

Over forest, K_E is generally an order of magnitude larger than over field crops because of the greater aerodynamic roughness of the surface [Eqs (9) and (16), Table V] and lies in the range 0·4 to 2 $m^2\ s^{-1}$.

Several alternative methods for deriving $\mathbf{F}_C$ from the fluxes of heat, water vapour and momentum are summarized by Denmead and McIlroy (1971) Methods based on the flux of momentum or shearing stress, use wind profiles to derive the eddy diffusivity for momentum, K_M, from Eq. (16) or a related equation. The assumption is then made that $K_M = K_C$ in Eq. (54). In deriving K_M from wind profiles, empirically-based corrections to the wind profile taken from the literature, e.g. Dyer and Hicks, 1970; Webb, 1970), are often made for departures from neutral stability (see Vol. 1, p. 81). The validity of the assumption $K_M = K_E$ and the applicability of such stability corrections, determined over grass and *short* crops, are, however, open to question over forest.

Over Scots pine at Thetford, Thom *et al.* (1975) found that the ratio K_E/K_M (K_M uncorrected for instability) increased from 2 to 10 as Ri decreased from +0·01 to −0·4. Furthermore, the measured ratio of K_E/K_M was consistently 2·5 times larger than that expected from application of the Dyer and Hicks (1970) correction for instability. Such pronounced instability is uncommon at Fetteresso, but in the range $-0·002 > Ri > -0·03$ median values of K_E/K_M for all weather conditions increased from 1·8 to 2·9. These results from two separate studies suggest that the eddy diffusivity over forest is significantly larger than that at the same height above much smoother surfaces. Thom *et al.* (1975) suggest that the enhanced eddy diffusivity is the result of the pressure of wake diffusion in the region below $d + 25\ z_0$, perhaps aided by free convective thermals originating within the canopy, where (and when) Ri is relatively large (and negative).

Whether or not such an explanation is correct, the studies at both sites indicate that the assumption $K_E = K_M$ is not applicable over forest, and hence that it is inadvisable to calculate mass fluxes, such as the CO_2 flux, using eddy diffusivities for momentum in normal daytime conditions of instability. However, the energy balance approach is subject to large errors when net radiation is very small, and the momentum approach may then be the better choice, provided that the relationship between K_E and K_M is known.

The flux of CO_2 derived from Eq. (54) is subject to a number of errors. The gradient of CO_2 itself over a coniferous forest canopy is small (Figs 8 and 13, Table IX) and hence a measurement resolution of the order of $0 \cdot 1 \, \mu l/l$ is necessary. Secondly, the derivation of K_E depends upon the accurate measurement of temperature and humidity gradients which are themselves very small over coniferous forest (Table IX). Thus appreciable scatter in the derived CO_2 fluxes is to be expected.

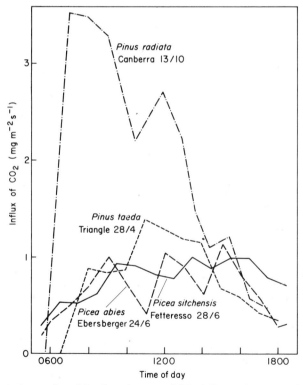

Fig. 19. Typical examples of the diurnal course of CO_2 influx to the canopy from the bulk air above at four sites in the early summer. *Pinus radiata* data from Denmead, 1969; *Pinus taeda* data from Sinclair *et al.*, unpublished; *Picea abies* data from Puller and Baumgartner, unpublished.

Diurnal variation in the downward flux of CO_2 into four forest canopies on sunny days is shown in Fig. 19. Examination of a large number of hourly mean CO_2 fluxes measured at Fetteresso over the past three years shows an asymptotic increase with short-wave irradiance and a decline at higher vapour pressure deficits (Fig. 20). For typical conditions, e.g. $S_t = 300$ W m^{-2},

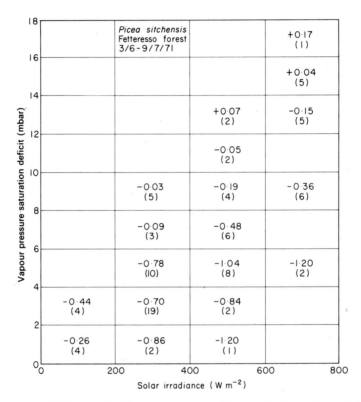

Fig. 20. Influx of CO_2 into the Sitka spruce canopy at Fetteresso in June and early July 1971 in relation to solar irradiance and vapour pressure saturation deficit. The figures in the rectangles are the mean hourly CO_2 fluxes (mg m^{-2} s^{-1}) and the number of observations in brackets.

$\delta e = 3$ mbar, the carbon dioxide flux is about 0·8 mg m^{-2} s^{-1}. An increase in δe to about 14 mbar, a rare event, is sufficient to reduce F_C to zero, probably because of stomatal closure (see Fig. 18). This decline is approximately exponential and a provisional expression for carbon dioxide flux to the canopy is

$$\mathbf{F}_C = \frac{aS_t}{b + S_t} \cdot e^{-c\delta e} \qquad (56)$$

where a, b and c are empirical constants, which vary with season. Typical values of the constants for 21 to 29 June 1973 are: $a = 2.88$, $b = 454$, $c = 0.069$. At Fetteresso, the average short-wave irradiance in the photosynthetically useful waveband from 380 to 720 nm is approximately 53% (Landsberg *et al.*, 1973). A short-wave irradiance (whole spectrum) of 100 W m^{-2} corresponds to a photon flux of 230 μEinstein m^{-2} s^{-1} (unpublished 1973 data).

In addition to the CO_2 influx to the canopy from the bulk air above the forest, CO_2 respired in the soil also reaches the canopy by turbulent transfer where it is assimilated (see the shape of the CO_2 profile in Fig. 13). This vertical flux of CO_2 cannot be measured quantitatively using Eqs (52) and (53) because the weak net radiation and the blow-through below the tree crowns prevent the accurate estimation of K_C there. Consequently alternative methods of measuring CO_2 efflux from the soil, usually involving some kind of enclosure and an air flow system, must be used. Little informa- is available for coniferous forest. Witkamp and Frank (1969) found rates of CO_2 efflux from the soil in a *Pinus echinata* forest to vary between 0.022 and 0.027 mg m^{-2} s^{-1} depending on soil temperature. Chiba and Tsutsumi (1967) found that the evolution of CO_2 from forest soils increased exponentially with temperature. At 20 °C the following rates were obtained in mg m^{-2} s^{-1}: *Picea abies* 0.05, *Cryptomeria japonica* 0.07, *Chamaecyparis obtusa* 0.035, *Pinus densiflora* 0.04 to 0.05. In general these rates are lower than are found in deciduous woodland, crops and grassland.

Night-time efflux of CO_2 from the canopy as a result of respiration by the soil, wood and leaves at Fetteresso lies between zero and 0.5 mg m^{-2} s^{-1} with a median value of 0.15 mg m^{-2} s^{-1} at typical summer air temperatures of 5 to 15 °C.

About 10^4 J of radiant energy are fixed by the plant in the reduction of one gram of CO_2 to sugars [this is the value of μ in Eqs (33) and (55)]. Thus a mass flux of CO_2 of 1 mg m^{-2} s^{-1} or 3.6 g m^{-2} h^{-1} is equivalent to a flux of energy of 10 W m^{-2}. In the middle of the day, this flux amounts to only 2% of the net radiation flux density. However, in the early morning when F_C is already large and R_n is lagging behind but increasing fast, $\mu F_C / R_n$ can reach 15% at Fetteresso, and Denmead (1969) found that it was about 12% over *Pinus radiata* at 0800 (Fig. 19).

B. Exchanges within Canopies

For the reasons advanced earlier (Section IV.B.2) in connection with the source-sink distribution of heat and water vapour within the canopy, the CO_2 sink distribution within coniferous forest canopies cannot be accurately derived from the second differentiation of the CO_2 concentration profile (Fig. 13) and we are not aware of any published attempts to apply this approach in coniferous forest. However, two alternative techniques—feeding $^{14}CO_2$ for a short period and environmentally controlled assimilation chambers—are in wide use to describe the distribution of photosynthetic activity in space and time and in relation to weather.

1. Uptake of $^{14}CO_2$

The technique of feeding leaves with air containing a mixture of $^{14}CO_2$ and a carrier for a short exposure of 10 s to 1 min was developed on farm crops (e.g. Turner and Incoll, 1971) and its application to conifers has followed the successful use of a diffusion porometer with needles. The method, successfully used at Fetteresso by R. E. Neilson, allows frequent, rapid and extensive measurement of the photosynthetic rate within the canopy (Fig. 21). The rates of CO_2 influx at any one level show considerable variability because of the variations in the area distribution of irradiance (Section II.B.2). The influx of CO_2 is influenced not only by the photon flux during the period in which the isotope is fed (30 s), but also by the photon flux experienced by the shoot immediately before the measurement. Because the photon flux within the lower levels of the canopy is well below saturation, measured rates of CO_2 uptake all fall on approximately the same curve relating rate of CO_2 uptake to photon flux density (cf. Fig. 23), although with some scatter resulting from the rapidly changing photon flux. Thus these results show that photosynthesis within the canopy is strongly dependent upon the local radiation environment.

2. Carbon Dioxide Uptake in Assimilation Chambers

Assimilation chambers have the advantage over the $^{14}CO_2$ method that they allow continuous, intensive measurements of CO_2 influx with some possibility of experimental manipulation of the shoot environment. They have the disadvantage that the conditions in the chamber cannot be identical with those outside, although this may not be of great consequence (Jarvis, 1970).

When available, the capability to manipulate the shoot environment is a feature of great importance since it enables the investigation of CO_2 flux in relation to conditions under-represented in the normal range of weather conditions, such as the blank or poorly filled squares in Fig. 20. Hypotheses about the responses of photosynthesis to canopy environment can then be tested more thoroughly, but this technique has hardly been tried with conifers.

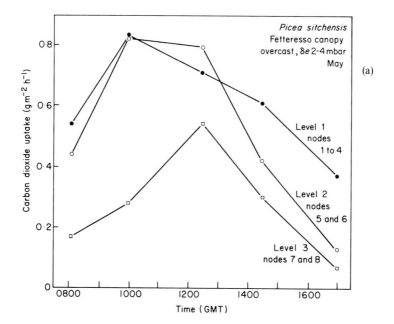

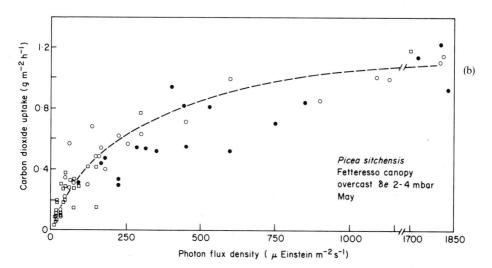

Fig. 21. (a) The diurnal course of CO_2 uptake of the previous years shoots measured with $^{14}CO_2$ feeding technique in the three levels of the Sitka spruce canopy at Fetteresso on an overcast day in mid-May. Each point is the mean of six determinations. (b) All individual measurements of CO_2 uptake on the same day plotted as a function of the photon flux density at the time of measurement [symbols as in (a)] (a and b from R. E. Neilson, unpublished).

Because of the expense of controlled environment assimilation chambers and the associated problems of maintenance, they are unsuitable for extensive measurements of photosynthesis in the canopy. However they are eminently suitable for describing the response of selected shoots to changing environmental conditions, and have been widely used for this purpose in coniferous forest. Early work of this kind is summarized by Polster (1950), Kramer and Kozlowski (1960), Kozlowski and Keller (1966), Lyr *et al* (1967) and Tranquillini (1968). Some recent studies of CO_2 exchange in canopies in relation to weather are listed in Table XIII. Neuwirth (1968) and Helms (1970)

TABLE XIII
Recent studies of CO_2 exchange in canopies using assimilation chambers

Pinus sylvestris	Ungerson and Scherdin (1965, 1968)
	Künstle and Mitscherlich (1970)
	Künstle (1971)
Pinus aristata	Mooney *et al.* (1966)
	Schulze *et al.* (1967)
Pinus cembra	Transquillini (1957, 1959)
	Pisek and Winkler (1958)
Pinus ponderosa	Helms (1970)
Pinus taeda	Higginbotham and Strain (1972)
Pseudotsuga menziesii	Helms (1965)
	Künstle and Mitscherlich (1970)
	Woodman (1971)
Picea abies	Pisek and Tranquillini (1954)
	Pisek and Winkler (1958)
	Ungersen and Scherdin (1965)
	Neuwirth (1968)
	W. Koch, O. L. Lange and E. D. Schulze (unpublished)

describe preliminary observations on the influence of crown structure and position in the canopy on photosynthetic rate measured with assimilation chambers in *Pinus sylvestris* and *Pinus ponderosa*, respectively (see Fig. 22).

C. Photosynthetic Characteristics of Shoots

The CO_2 exchange characteristics of coniferous shoots were comprehensively reviewed by Kozlowski and Keller (1966), Larcher (1969), Ferrell (1970) and Walker *et al.* (1972). Some properties are listed in Tables XIV and XV. Maximum CO_2 fluxes are small compared with the CO_2 fluxes in deciduous and herbaceous plants (Larcher, 1969; Šesták *et al.*, 1971) but the reasons for this cannot be considered here.

Needles in different parts of a canopy are exposed to large differences in irradiance which is the dominant conditioning environmental influence. The needles at the top of the canopy are physiologically and anatomically

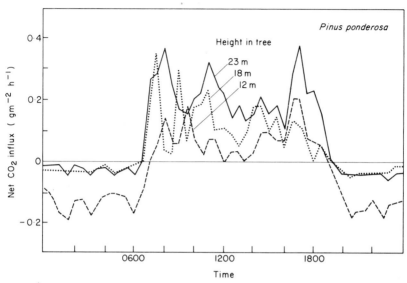

Fig. 22. The diurnal course of CO_2 uptake of current year shoots measured with assimilation chambers at three levels in a stand of *Pinus ponderosa* on a sunny day (from Helms, 1970).

conditioned by the high irradiances they experience, in comparison with needles lower down which are shade-adapted. The maximum rate of photosynthesis is larger in shoots which have developed at the higher levels in the canopy (Fig. 23), and both minimal stomatal and mesophyll resistance increase with increasing depth. The dark respiration rates and the compensation irradiances become progressively smaller with increasing depth but the photochemical efficiency is similar at all levels. Because of the low

Table XIV

Carbon dioxide exchange characteristics of coniferous shoots[a]

Characteristic	No. of species	Median	Range
Maximum net CO_2 influx			
—projected area basis (g m^{-2} h^{-1})	11	0·7	0·4–1·8
—dry weight basis (mg^{-1} g^{-1} h^{-1})	19	7	1·5–2·0
Maximum dark CO_2 efflux			
—dry weight basis (mg g^{-1} h^{-1})	19	0·7	0·2–1·7
Optimum temperature (°C)	16	20	12–23
Minimum compensation temperature (°C)	10	−5	−3–−8
Maximum compensation temperature (°C)	12	42	34–46

[a] From Larcher (1969).

Table XV

Mesophyll resistance and other CO_2 exchange characteristics of current year needles

Species	n	$F_{C,\,max}$ (g m^{-2} h^{-1})	r_s (s m^{-1})	r_m (s m^{-1})	Compensation irradiance (W m^{-2})	Dark respiration (g m^{-2} h^{-1})	Einstein per mole CO_2	Source
Picea sitchensis	12	1·48	270	840	3·1	0·16	17	a
Picea sitchensis	6	1·74	430	500	5·6	0·11	13	b
Pinus halepensis	—	0·48	670	1 700	—	0·05	—	c
Abies balsamea	6	1·70	350	890	—	—	—	d

[a] From Ludlow and Jarvis (1971); glasshouse grown.
[b] From Ludlow and Jarvis (1971); detached forest shoots.
[c] From Whiteman and Koller (1964, 1967).
[d] From Little and Loach (1973).

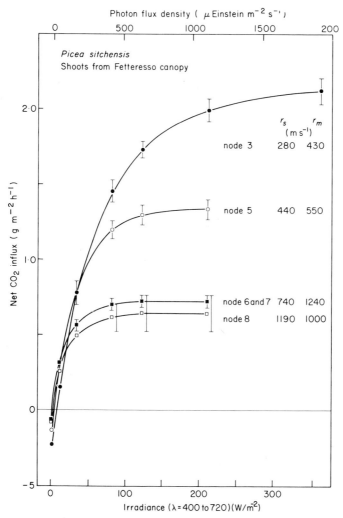

Fig. 23. The effect of developmental position in the canopy on the capacity to photosynthesise and respire of shoots of Sitka spruce from Fetteresso. The figures by the curves are minimal values of r_s and r_m, respectively, in s m^{-1}. The bars are two standard errors of five replicates (from R. E. Neilson, unpublished).

irradiances within the canopy, photosynthesis at depth is not restricted by the limitation to the radiation-saturated rate (cf. Fig. 21b), but is largely limited by the photochemical efficiency of the needles at the lower levels. Thus increases in pigment content and absorptivity in the lower levels would be expected to lead to increases in photosynthesis there.

APPENDIX

Table XVI
Location and features of the main sites providing data used in this preview

Site ref. no.	Name of		Country	Position	Species	Year planted	Mean (median) tree height (m)	Tree density (no./ha)	Leaf area index[a]
	Forest/Site	Nearest town/city							
1	—	Canberra	Australia	35° 18′ S, 149° 08′ E	*Pinus radiata*	1964	7·5	700	—
2a	Beit Oren	Tel Aviv	Israel	32° 44′ N, 35° 00′ E	*Pinus halepensis*	—	6·5	—	—
2b	Mount Carmel	Tel Aviv	Israel	32° 50′ N, 35° 00′ E	*Pinus halepensis*	—	6·0	—	—
3a	Dute	Durham	U.S.A.	36° 00′ N, 79° 00′ W	*Pinus taeda*	1933	23·4	750	—
3b	Triangle	Chapel Hill	U.S.A.	36° 00′ N, 79° 00′ W	*Pinus taeda*	1957	12·5	1 700	2·6
3c	Fox Park	Laramie	U.S.A.	41° 04′ N, 106° 06′ W	*Pinus contorta*	1900	10·0	1 700	2·8
3d	Tateno	Tsuchiura	Japan	36° 06′ N, 140° 06′ E	*Pinus densiflora*	*ca.*1964	4·5	2 300	0·8
3e	Omyojin	Morioka	Japan	39° 42′ N, 140° 54′ E	*Pinus densiflora*	*ca.*1934	23·0	750	1·7
4	Pachaug	Voluntown	U.S.A.	41° 32′ N, 71° 51′ W	*Pinus resinosa*	1931	15	2 500	2·6
5	—	Ithaca	U.S.A.	42° 41′ N, 76° 35′ W	*Larix leptolepis*	—	10·4	1 100	3·6
6	C. Lathrop Pack	Warrensburg	U.S.A.	43° 30′ N, 73° 50′ W	*Pinus resinosa*	1931	11·8	2 400	—

No.	Location	Place	Country	Coordinates	Species	Year			
7	Crane Prairie	Bend	U.S.A.	43° 50' N, 121° 45' W	Pinus contorta	1928	7.0	—	—
8	Saddleback	Deerfield	U.S.A.	44° 57' N, 70° 31' W	Pinus strobus	s.s.[b]	6.0	—	3.1
9a	Petawawa	Chalk River	Canada	46° 03' N, 77° 22' W	Pinus resinosa/strobus	—	22.0	600	—
9b	Petawawa	Chalk River	Canada	46° 03' N, 77° 22' W	Picea glauca	—	—	—	—
10	Cedar River	Seattle	U.S.A.	47° 20' N, 122° 30' W	Pseudotsuga menziesii	1935	28.0	570	—
11	Marcell Expt.	Dulwth	U.S.A.	47° 31' N, 93° 28' W	Picea mariana	—	11.9	—	—
12	—	Hartheim	B.R.D.	47° 55' N, 07° 35' W	Pinus sylvestris	—	2.7	12 000	—
13	Ebersberger	Munich	B.R.D.	48° 10' N, 11° 50' E	Picea abies	1895	27.5	800	8.4
14	U.B.C.	Haney	Canada	49° 10' N, 122° 48' W	Pseudotsuga menziesii	1959	7.8	1 700	—
15	Marmot Creek	Calgary	Canada	50° 57' N, 115° 10' W	Picea sp.	—	26	—	—
16	Bangley	Oxford	U.K.	51° 45' N, 01° 15' W	Picea abies	—	15	—	—
17	Thetford	Cambridge	U.K.	52° 25' N, 00° 39' E	Pinus sylvestris	1930	15.5	800	4.3
18a	Eberswalde	Berlin	D.D.R.	52° 50' N, 13° 50' E	Pinus sylvestris	1865	21	—	—
18b	Eberswalde	Berlin	D.D.R.	52° 50' N, 13° 50' E	Pinus sylvestris	1939	12.6	—	—
19	Fetteresso	Aberdeen	U.K.	56° 58' N, 02° 24' W	Picea sitchensis	1948	11.5	4 100	9.8
20	Velen	Mariestad	Sweden	58° 48' N, 14° 15' E	Picea abies	1915	22.0	1 400	—
21a	Dundret	Gällivare	Sweden	67° 08' N, 20° 30' E	Pinus sylvestris	1962	1.5	7 000	—
21b	Linolompolo	Gällivare	Sweden	67° 15' N, 20° 20' E	Pinus sylvestris	s.s.[b]	12	1 000	—

[a] On a projected area basis.
[b] Self-sown.

REFERENCES

Allen, L. H. Jnr. (1968). *J. appl. Meteorol.* **7**, 73–78.
Anderson, M. C. (1966). *J. appl. Ecol.* **3**, 41–54.
Atzet, T. and Waring, R. H. (1970). *Can. J. Bot.* **48**, 2163–2167.
Babalola, O., Boersma, L. and Youngberg, C. T. (1968). *Pl. Physiol.* **43**, 512–521.
Barry, R. G. and Chambers, R. E. (1966). *Quart. J. R. met. Soc.* **92**, 543–548.
Baumgartner, A. (1956). *Ber. dt. Wetterdienst* **5**.
Baumgartner, A. (1965). *In* 'Methodology of Plant Eco-Physiology' (F. E. Eckardt, ed.) pp. 495–512. UNESCO, Paris.
Beardsell, M. F., Jarvis, P. G. and Davidson, B. (1972). *J. appl. Ecol.* **9**, 677–690.
Belt, G. H. (1969). *Bull. Am. met. Soc.* **50**, 469.
Bergen, J. D. (1971). *For. Sci.* **17**, 314–322.
Berggren, R., Bringfelt, B., Hårsmar, P-O., Sandgren, W. and Ljungqvist, S. (1972). Report on IHD-project No. 19. Swedish Meteorol. and Hydrol. Institute, Stockholm.
Berglund, E. R. and Mace, A. C. (1972). *J. appl. Met.* **11**, 806–812.
Birkebak, R. and Birkebak, R. (1964). *Ecology* **45**, 646–649.
Black, T. A. and McNaughton, K. G. (1971). *Boundary-layer Meteorol.* **2**, 246–254.
Bradley, E. F. (1968). *Quart J. R. Met. Soc.* **94**, 361–379.
Bradley, E. F. (1971/72). *Agric. Met.* **9**, 183–190.
Brown, G. W. (1973). *Agric. Met.* **11**, 115–112.
Caldwell, M. M. (1970). *Pl. Physiol.* **46**, 236–239.
Campbell, R. (1972). *Ann. Bot.* **36** 307–314.
Chiba, K. and Tsutsumi, T. (1967). *Bull. Kyoto Univ. Forestry* **39**, 91–99.
Clark, J. B. and Lister, G. R. (1975a) *Pl. Physiol.* **55**, 401–406.
Clark, J. B. and Lister, G. R. (1975b). *Pl. Physiol.* **55**, 407–415.
Cornic, G. and Jarvis, P. G. (1972). *Photosynthetica* **6**, 225–239.
Deacon, E. L. and Swinbank, W. C. (1958). *In* 'Climatology and Microclimatology' pp. 38–41. UNESCO, Paris.
Denmead, O. T. (1964). *J. Appl. Met.* **3**, 383–389.
Denmead, O. T. (1969). *Agric. Met.* **6**, 357–372.
Denmead, O. T. and McIlroy, I. C. (1971). *In* 'Plant Photosynthetic Production. Manual of Methods' (Z. Šesták, J. Čatsky and P. G. Jarvis, eds), pp. 467–516. Dr. W. Junk n.v., The Hague.
Dirmhirn, I. (1964). 'Das Strahlungsfeld in Lebensraum.' Frankfurt.
Druilhet, A., Perrier, A., Fontan, J. and Laurent, J. L. (1972). *Boundary-Layer Meteorol.* **2**, 173–187.
Dyer, A. J. (1965). *Quart J. R. met, Soc.* **91**, 151–157.
Dyer, A. J. and Hicks, B. B. (1970). *Quart. J. R. met. Soc.* **96**, 715–721.
Dykstra, G. F. (1974). *Can. J. For. Res.* **4**, 201–206.
Egan, W. G. (1970). *For. Sci.* **16**, 79–94.
Federer, C. A. (1968). *J. appl. Met.* **7**, 789–795.
Federer, C. A. and Tanner, C. B. (1966). *Ecology* **47**, 555–560.
Ferrell, W. K. (1970). Paper presented at First North American Forest Biology Workshop, East Lancing, Michigan.
Freyman, S. (1968). *Can. J. Pl. Sci.* **48**, 326–328.
Fritschen, L. J. (1967). *Agric. Met.* **4**, 55–62.
Gates, D. M. (1966). *Quart. Rev. Biol.* **41**, 353–365.
Gates, D. M. and Benedict, C. M. (1963). *Am. J. Bot.* **50**, 563–573.

Gates, D. M. and Tantraporn, W. (1952). *Science* **115**, 613–616.
Gates, D. M., Keegan, H. J., Schleter, J. C. and Weidner, V. R. (1965a). *Appl. Optics* **4**, 11–20.
Gates, D. M., Tibbals, E. C. and Kreith, F. (1965b). *Am. J. Bot.* **52**, 66–71.
Gay, L. W. (1971a). *Arch. met. geoph. Bioklimat.* B **19**, 1–14.
Gay, L. W. (1971b). *Proc. Oregon State Univ.* **7**, 11–23.
Gay, L. W. (1972). *In* 'Proceedings—Research on Coniferous Forest Ecosystems—A Symposium' (J. F. Franklin, L. J. Dempster and R. H. Waring, eds), pp. 243–253. Portland, Oregon.
Gay, L. W. and Knoerr, K. R. (1970). *Arch. Met. Geophysk. Bioklimt.* B **18**, 187–196.
Gay, L. W. and Stewart, J. B. (1973). Presented to Swecon Seminar.
Gay, L. W., Knoerr, K. R. and Braaten, M. O. (1971). *Agric. Met.* **8**, 39–50.
Geiger, R. (1965). 'The Climate near the Ground'. Cambridge, Mass.
Gillespie, T. J. and King, K. M. (1971). *Agric. Met.* **8**, 59–68.
Hanover, J. W. and Reicosky, D. A. (1971). *Am. J. Bot.* **58**, 671–687.
Helms, J. A. (1965). *Ecology* **46**, 698–708.
Helms, J. A. (1970). *Photosynthetica* **4**, 243–253.
Higginbotham, K. O. and Strain, B. R. (1972). Triangle Research Site, Eastern Deciduous Forest Biome Memo Report 72–88.
Holmgren, P., Jarvis, P. G. and Jarvis, M. S. (1965). *Physiologia Pl.* **18**, 557–573.
Idso, S. B., Baker, D. G. and Blad, B. L. (1969). *Quart. J. R. Met. Soc.* **95**, 244–257.
Impens, I. and Lemeur, R. (1969). *Arch. Met. Geophysk. Bioklimt.* B **17**, 403–412.
Jarvis, P. G. (1970). *In* 'Prediction and Measurement of Photosynthetic Productivity' (I. Šetlík, ed.), pp. 353–367, Centre for Agricultural Publishing and Documentation, Wageningen.
Jarvis, P. G. (1971). *In* 'Plant Photosynthetic Production, Manual of Methods' (Z. Šesták, J. Čatsky and P. G. Jarvis, eds), pp. 566–631. W. Junk, The Hague.
Jarvis, P. G. and Slatyer, R. O. (1970). *Planta* **90**, 303–322.
Jeffree, C. E., Johnson, R. P. C. and Jarvis, P. G. (1971). *Planta* **98**, 1–10.
Johnson, R. P. C. and Jeffree, C. E. (1970). *Planta* **95**, 179–182.
Kaufmann, M. R. (1973). *Am. Soc. Pl. Physiol. Abstracts. Pl. Physiol.* Suppl. **51**, 8.
Kinerson, R. S. (1973). *J. appl. Ecol.* **10**, 657–660.
Kondo, J. (1971). Wind profile and solar radiation in and above the forests. Report on the Water Resources in the Basin of Omyojin, Iwate. Iwate Univ. pp. 15–30 (in Japanese).
Kozlowski, T. T. and Keller, T. (1966). *Bot. Rev.* **32**, 1–382.
Kramer, P. J. and Kozlowski, T. T. (1960). 'Physiology of Trees'. McGraw Hill, New York.
Künstle, E. (1971). *Allg. Forst-u. Jagdztg.* **142**, 105–108.
Künstle, E. and Mitscherlich, G. (1970). *Allg. Forst-u. Jagdztg.* **141**, 89–94.
Landsberg, J. J. and James, G. B. (1971). *J. appl. Ecol.* **8**, 729–741.
Landsberg, J. J. and Jarvis, P. G. (1973). *J. appl. Ecol.* **10**, 645–655.
Landsberg, J. J. and Ludlow, M. M. (1970). *J. appl. Ecol.* **7**, 187–192.
Landsberg, J. J. and Thom, A. S. (1971). *Quart. J. R. met. Soc.* **97**, 565–570.
Landsberg, J. J., Jarvis, P. G. and Slater, M. B. (1973). *In* 'Plant Response to Climatic Factors' Uppsala Symposium 1970, pp. 411–418. UNESCO, Paris.
Lange, O. L. and Schulze, E-D. (1966). *Forstwiss. Centralbl.* **85**, 1–64.
Larcher, W. (1969). *Photosynthetica* **3**, 167–198.
Lehela, A., Day, R. J. and Koran, Z. (1972). *For. Chron.* **48**, 32–34.

Leonard, R. E. (1968). Presented at Forest and Fire Meteorology Conference, Salt Lake City.

Leonard, R. E. and Eschner, A. R. (1968). *Water Resources Res.* **4**, 931–935.

Leonard, R. E. and Federer, C. A. (1973). *J. appl. Met.* **12**, 302–307.

Leverenz, J. (1974). M.Sc. Thesis, University of Washington, Seattle, U.S.A.

Leyton, L. and Armitage, I. P. (1968). *New Phytol.* **67**, 31–38.

Leyton, L. and Juniper, B. E. (1963). *Nature, Lond.* **198**, 770–771.

Linacre, E. T. (1968). *Agric. Met.* **5**, 49–63.

Linder, S. (1971). *Physiologia Pl.* **25**, 58–63.

Linder, S. (1972). Dissertation, Umeå University, Sweden.

Little, C. H. A. and Loach, K. (1973). *Can. J. Bot.* **51**, 751–758.

Lopuschinsky, W. (1969). *Bot. Gaz.* **130**, 258–263.

Ludlow, M. M. and Jarvis, P. G. (1971). *J. appl. Ecol.* **8**, 925–953.

Lutzke, R. (1966). *Arch. forstwesen* **15**, 995–1015.

Lutzke, R. (1969). *Arch. forstwesen* **18**, 921–927.

Lyr, H., Polster, H. and Fiedler, H. J. (1967). (Gehölz-Physiologie.) Gustav Fischer, Jena.

Martin, H. C. (1971). *J. appl. Met.* **10**, 1132–1137.

McBean, G. A. (1968). *J. appl. Met.* **7**, 410–416.

McNaughton, K. G. and Black, T. A. (1973). *Water Resources Res.* **9**, 1579–1590.

Monteith, J. L. (1963). *In* 'Environmental Control of Plant Growth' (L. T. Evans, ed), pp. 95–112. Academic Press, New York.

Monteith, J. L. (1965). *Symp. Soc. exp. Biol.* **19**, 205–234.

Monteith, J. L. (1969). *In* 'Physiological Aspects of Crop Yield' (J. D. Eastin, F. A. Haskins, C. Y. Sullivan and C. H. M. Van Bavel, eds), pp. 89–111. American Society of Agronomy and Crop Science Society of America, Washington.

Monteith, J. L. and Szeicz, G. (1961). *Quart. J. R. met. Soc.* **87**, 159–170.

Mooney, H. A., West, M. and Brayton, R. (1966). *Bot. Baz.* **127**, 105–113.

Mukammal, E. I. (1966). Unpublished report, Canadian Meteorological Branch and Canadian Department of Forestry.

Mukammal, E. I. (1971). *Arch. Met. Geophysk. Bioklimt.* B **19**, 29–52.

Neilson, R. E. and Jarvis, P. G. (1975). *J. appl. Ecol.* **12**, 879–892.

Neilson, R. E., Ludlow, M. M. and Jarvis, P. G. (1972). *J. appl. Ecol.* **9**, 721–745.

Neuwirth, G. (1968). *In* 'Klimaresistenz und Stoffproduktion' 1st Internationale Baumphysiologen Symposium, pp. 179–188 Berlin.

Norman, J. M. (1971). Ph.D. Thesis, Univ. Wisconsin, Univ. Microfilms, Ann Arbor, Michigan.

Norman, J. M. and Jarvis, P. G. (1974). *J. appl. Ecol.* **11**, 375–398.

Norman, J. M. and Jarvis, P. G. (1975). *J. appl. Ecol.* **12**, 839–879.

Odin, H. and Perttu, K. (1966). Dept. Reafforestation, *Res. Notes* 9. Royal College of Forestry, Stockholm.

Oliver, H. R. (1971). *Quart. J. R. met. Soc.* **97**, 548–553.

Penman, H. L., Angus, D. E. and Van Bavel, C. H. M. (1967). *Agronomy* **11**, 483–505.

Penman, H. L., Long, I. F. and French, B. K. (1970). Rep. Rothamsted exp. Stn. for 1969, Pt. 1, p. 37.

Perrier, A. (1971). *In* 'Plant Photosynthetic Production. Manual of Methods' (Z. Šesták, J. Čatský, and P. G. Jarvis, eds), pp. 632–671. (Dr. W. Junk, n.v. The Hague.

Perttu, K. (1970). *Studia Forest. Suec.* **72**.

Pisek, A. and Tranquillini, W. (1954). *Flora* **141**, 237–270.

Pisek, A. and Winkler, E. (1958). *Planta* **51**, 518–543.

Polster, H. (1950). 'Die Physiologischen Grundlagen der Stofferzeugung im Walde'. Bayerischer Landwirtschaftsverlag, Munich.

Reifsnyder, W. E. (1967). *In* 'International Review of Forestry Research' (J. A. Romberger and P. Mikola, eds), Vol. 2, pp. 127–179. Academic Press, New York.

Reifsnyder, W. E. and Lull, H. W. (1965). U.S.D.A. Forest Science Technical Bulletin No. 1344.

Reifsnyder, W. E., Furnival, G. M. and Horowitz, J. L. (1971). *Agric. Met.* **9**, 21–38.

Running, S. W. (1974). M.Sc. Thesis, Oregon State University, Corvallis, U.S.A.

Rutter, A. J. (1967). *In* 'Forestry Hydrology' (W. Sopper and H. W. Lull, eds), pp. 403–416, Pergamon Press, Oxford.

Rutter, A. J. (1968). *In* 'Water Deficits and Plant Growth' (T. T. Kozlowski, ed.), pp. 23–84. Academic Press, New York and London.

Schulze, E. D., Mooney, H. A. and Dunn, E. L. (1967). *Ecology* **48**, 1044–1047.

Šesták, Z., Jarvis, P. G., and Čatský, J. (1971). *In* 'Plant Photosynthetic Production. Manual of Methods' (Z. Šesták, J. Čatský and P. G. Jarvis, eds), pp. 1–48. Dr. W. Junk, n.v., The Hague.

Sinclair, T. R. and Knoerr, K. R. (1972). Triangle Research Site, Eastern Deciduous Forest Biome Memo, 72–75.

Stanhill, G. (1970). *Solar Energy* **13**, 59–66.

Stanhill, G., Hofstede, G. J. and Kalma, J. D. (1966). *Quart. J. R. met. Soc.* **92**, 128–140.

Stanhill, G., Fuchs, M. and Oguntoyinbo, J. (1971). *Arch. Met. Geophysyk. Bioklimt.* B **19**, 113–132.

Stewart, J. B. (1971). *Quart. J. R. met. Soc.* **97**, 561–564.

Stewart, J. B. and Thom, A. S. (1973). *Quart. J. R. met. Soc.* **99**, 154–170.

Storr, D., Tomalin, J., Cork, H. F. and Munn, R. E. (1970). *Water Resources Res.* **6**, 705–716.

Strauss, R. (1971). *Wiss. Mitt. met. Inst., München* Nr. 22.

Szeicz, G., Endrodi, G. and Tajchman, S. (1969). *Water Resources Res.* **5**, 380–394.

Tajchman, S. (1967). *Wiss. Mitt. met. Inst., München* Nr. 12.

Tajchman, S. (1971). *Water Resources Res.* **7**, 511–523.

Tajchman, S. (1972a). *J. appl. Ecol.* **9**, 359–375.

Tajchman, S. (1972b). *Allg. Forst-u. Jagdztg.* **143**, 35–38.

Thom, A. S. (1971). *Quart. J. R. met. Soc.* **97**, 414–428.

Thom, A. S., Stewart, J. B., Oliver, H. R., and Gash, J. H. C. (1975). *Quart. J. R. met. Soc.* **101**, 93–105.

Tibbals, E. C., Carr, E. K., Gates, D. M., and Kreith, F. (1964). *Am. J. Bot.* **51**, 529–538.

Tranquillini, W. (1957). *Planta* **49**, 612–661.

Tranquillini, W. (1959). *Planta* **54**, 107–129.

Tranquillini, W. (1968). *In* 'Klimaresistenz und Stoffproduktion' 1st Internationale Baumphysiologen Symposium, pp. 153–169, Berlin.

Turner, N. C. (1974). *In* 'Mechanisms of Regulation of Plant Growth' Bulletin No. 12, Royal Soc. N.Z. 423–432.

Turner, N. C. and Incoll, L. D. (1971). *J. appl. Ecol.* **8**, 581–591.

Turner, N. C. and Jarvis, P. G. (1975). *J. appl. Ecol.* **12**, 561–576.

Ungerson, J. and Scherdin, G. (1965). *Planta* **67**, 136–167.

Ungerson, J. and Scherdin, G. (1968). *Flora* **157**, 381–434.

Vézina, P. E. (1964). *Agric. Met.* **1**, 54–65.

Vézina, P. E. and Péch, G. Y. (1964). *For. Sci.* **10**, 443–451.

Waggoner, P. E. and Turner, N. C. (1971). *Bull. Conn. Agric. exp. Station* 726.
Walker, R. B., Scott, D. R. M., Salo, D. J. and Reed, K. L. (1972). *In* 'Research on Forest Ecosystems' (J. F. Franklin, L. J. Dempster and R. H. Waring, eds), pp. 211–226. Pacific Northwest Forest and Range Experimental Station, Portland, Oregon.
Webb, E. K. (1970). *Quart. J. R. met. Soc.* **96**, 67–90.
Whiteman, P. C. and Koller, D. (1964). *Israel J. Bot.* **13**, 166–176.
Whiteman, P. C. and Koller, D. (1967). *J. appl. Ecol.* **4**, 363–377.
Witkamp, M. and Frank, M. L. (1969). *Pedobiologia* **9**, 358–365.
Woodman, J. N. (1971). *Photosynthetica* **5**, 50–54.
Woolley, J. T. (1971). *Pl. Physiol.* **47**, 656–662.
Yamaoka, Y. (1958). *Trans. Am. geophys. Union* **39**, 266–272.

8. Deciduous Forests

JU. L. RAUNER

Institute of Geography, Academy of Sciences of the U.S.S.R., Moscow

This chapter presents the essential characteristics of the micrometeorological regime of deciduous forests. The examples to be cited refer to natural and cultivated forests in the zone of broad-leaved forests (the Moscow region) and in the central forest-steppe of the Russian plain (the Kursk region).

The micrometeorological regime has been defined by complex heat-balance measurements analysed for the following purposes: (1) definition of the upward and downward fluxes of radiation including net radiation above the crown and within the canopy at different levels; (2) definition of the vertical profiles of temperature, air humidity and wind velocity above the crown and within the canopy as well as of the temperature of the surface and upper layers of the soil; (3) definition of the water content in the root zone of the soil and of surface evaporation; (4) specification of structural parameters of the forest stand and of the vertical profile of the leaf and non-leaf surfaces of the forest and grass canopy.

A detailed account of all techniques of measurement and calculation and a full description of the deciduous forests where the measurements were made was presented in a book by Rauner (1972) and the principal results summarized here are taken from this source.

I. ARCHITECTURE OF LEAF CANOPY

The geometrical structure peculiar to a plant canopy determines its inter-action with fluxes of energy. This structure, in turn, can be evaluated

241

Table I

Vertical distribution of the area index for leaves (L) and other surfaces (L') for deciduous forest stands (15 to 20 years); h_1 and h_2 are heights for top and bottom of crown

	Type of forest stand									
	Oak (Quercus robur)		Maple (Acer platanoides)		Aspen (Populus tremula) + undergrowth		Linden (Tillia cordata) + undergrowth		Birch (Betula verrucosa)	
$1 - z/h_1$	L	L'	L	L'	L	L'	L	L'	L	L'
0·05	0·40	0·02	0·60	0·00	0·08	0·01	0·30	0·01	0·80	0·05
0·10	1·00	0·05	1·40	0·01	0·30	0·02	1·45	0·06	1·65	0·13
0·20	1·60	0·80	1·40	0·02	1·35	0·04	1·40	0·08	1·95	0·17
0·30	1·00	0·08	0·80	0·03	1·70	0·05	0·70	0·07	1·30	0·14
0·40	0·50	0·06	0·60	0·03	1·00	0·03	0·23	0·05	0·60	0·10
0·50	0·10	0·06	0·20	0·02	0·30	0·02	0·18	0·03	0·00	0·05
0·60	0·00	0·05	0·02	0·02	0·20	0·02	0·01	0·01	0·00	0·02
0·70	0·00	0·04	0·00	0·02	0·50	0·02	0·01	0·01	0·00	0·02
0·80	0·00	0·03	0·00	0·01	1·40	0·03	0·20	0·00	0·00	0·03
0·90	0·00	0·01	0·00	0·00	0·30	0·03	0·30	0·00	0·00	0·03
$\Sigma =$	4·60	0·48	5·02	0·16	7·13	0·27	4·78	0·32	5·30	0·74
	$L + L' = 5\cdot08$		5·18		7·40		5·10		6·04	
h_1(m)	6·50		7·00		10·50		11·50		7·60	
h_2(m)	3·00		4·00		5·50		5·50		3·50	

quantitatively from the characteristics of the leaf and non-leaf surfaces and from their relationships. Parameters defining structure are of special importance for micrometeorological studies in the forest. In the first place, they can be used to estimate the solar and turbulent heat exchange for a given type of forest. In the second place, they are vital for establishing general relationships between physical characteristics of the environment and the architecture of forest canopies.

Table I shows values for the typical vertical profiles of the relative leaf and non-leaf surface L and L' as well as the total area index L_t of young deciduous forests of the Central Black-Soil zone of the Russian plain (the Kursk region). The layer with the first maximum of biomass is in the upper part of the crown. The second maximum lies in the lower part of the canopy and depends upon the evolution of the lower plant layers of the forest stand (brushwood). The highest values of the leaf area index appear in a complex forest stand (e.g. aspen plus undergrowth).

Figures 1 and 2 show the curves of the vertical distribution of the leaf and non-leaf surface areas a_L and a_S, for a deciduous forest stand with a complex structure (Fig. 1) and a complex composition (Fig. 2). The second forest stand shows a close similarity in distribution curves, and a coincidence in the maximum values of a_L and a_S in different lead species as well as in their absolute values (the highest values are found in oak).

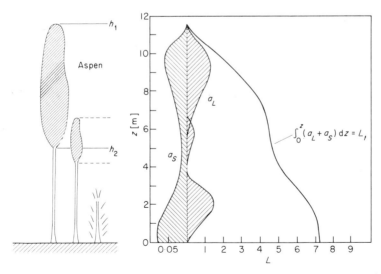

Fig. 1. Profile of leaf area density a_L and non-leaf area density a_S in a complex deciduous forest composed of a main layer aspen (*Populus tremula*) and subsidiary layers (regrowth and bushes of hazel and other species); about 25 years old.

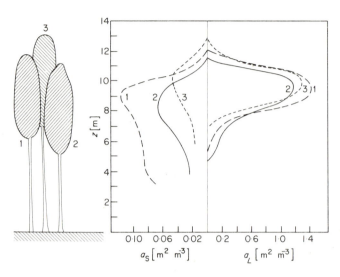

Fig. 2. Profile of the leaf area density a_L and non-leaf area density a_S in a complex deciduous forest with the main layer composed of several deciduous species (linden, *Tillia cordata*; oak, *Quercus robus*; aspen, *Populus tremula*). Stand age about 25 years. (1) oak; (2) linden; (3) aspen.

If the crown-cover in a deciduous forest is not very dense, grass can develop to form a substantial canopy. Figure 3 shows that the vertical profile of the leaf surface in such cases becomes very evident whilst the leaf area index is of the same order as in the forest stand (oak).

Figure 4 shows the changing value of L_t and of other plant quantities in an oak stand with adequate soil water. The leaf area index reaches a maximum exceeding 5 at the age of 30 to 50 years and thereafter remains practically unchanged to 70 or 80 years. From 90 to 100 years of age up to 220 years the value of L declines from 4·6 to 3·6. At the expense of the decreasing relative area of the leaf surface the non-leaf surface increases and at the end of the life span in oak reaches respectively $0·60 + 0·70$ and $0·40 + 0·45$. The accompanying forest storeys, as well as the grass canopy, are known to begin their intensive growth during this period (see for example, Sukachev and Dylis, 1964). Hence the total photosynthesizing surface of the deciduous forest community does not change much with age. The ways in which biometric indexes changing with age are typical for other leaf species as well; they may differ only in the absolute values of L_t, a_L and a_S and in their relationships.

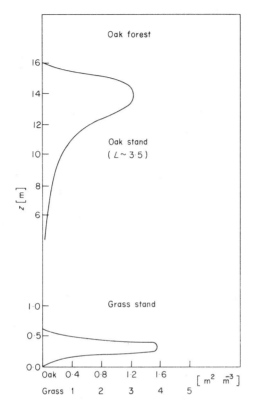

Fig. 3. Profile of leaf area density of oak and grass canopy in an oak stand, 40 years old. (The records were obtained by V. D. Utekhin, see Grin, Rauner, Utekhin, 1970).

II. RADIATION CHARACTERISTICS OF DECIDUOUS FORESTS

The distribution of radiation in forests depends mainly on the optical properties of individual plant elements and of the canopy as a whole. Table II shows the dependence on solar elevation of the albedo of some deciduous forests. The table shows this dependence to be most marked at lower elevations (from 10 to 30°). The dependence is rather weak at large elevations and in winter forests (covered with snow).

Figure 5 shows annual changes in the albedo of the leaf canopy which reveal the dynamics of vegetation phases in forest stands. Under snow, the albedo reaches a maximum but after melting and before foliage appears,

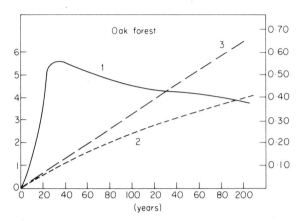

Fig. 4. Changes of leaf area index with age and the relationships between leaf and non-leaf densities in oak stands of the forest-steppe zone.

1. $L = \int_0^h (a_L + a_S)\,dz$ (left-hand axis).

2. $a_S/(a_L + a_S)$ (right-hand axis).

3. a_S/a_L (right-hand axis).

minimal values are observed. With the evolution of the leaf canopy the albedo increases somewhat and then remains unchanged, on average, up to the time of leaf yellowing and leaf fall when it slightly decreases. The difference in the albedo during February and March, presented in Curves 1 and 2 of Fig. 5, are typical for the albedo of forest stands in the western and eastern parts of the European belt of deciduous forests. This difference is

Table II

Change of albedo with solar elevation in some types of natural vegetation in bright weather

Type of forest	10°	20°	30°	β 40°	50°	60°
Oak with full foliage (summer)	0·30	0·19	0·17	0·160	0·155	0·15
Oak without foliage (spring)	0·25	0·18	0·15	0·135	0·120	0·12
Birch-aspen forest without foliage (end of the winter, snow cover)	0·31	0·30	0·28	0·26	—	—

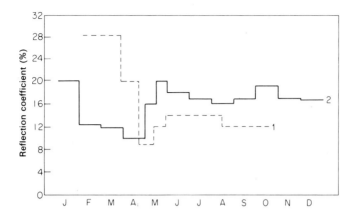

Fig. 5. Annual course of albedo (%) in deciduous forests: (1) a birch-aspen forest in the Moscow region (height 15 to 17 m, age 40 to 50 years); (2) an oak forest in Belgium (height 9 to 10 m) according to Grulois (1968).

due to the effects of the local snow cover which is very stable during winter in eastern Europe.

Figures available for the emissivities of leaves range between 0·93 and 0·97, the most reliable values fall between 0·94 and 0·96.

A theory of radiation transfer has been developed to describe the transmission of the average flux of solar radiation within a homogeneous leaf canopy. This theory describes the extinction of upward and downward fluxes of short-wave radiation as a function of the cumulative characteristics of the stand (see Ch. 2). Recently the theory has been expanded to the heterogeneous plant canopies (Ross, 1972).

In forests an exponential function has proved to be a good approximation to the downward flux of short-wave radiation S_t and the net radiative flux $\mathbf{R}_n$ (Rauner, 1968, 1972):

$$S_t(L) = S_t(0) \exp\left(-\mathcal{K}_t L\right)^{m_t}$$
$$\mathbf{R}_n(L) = \mathbf{R}_n(0) \exp\left(-\mathcal{K}_n L\right)^{m_n} \tag{1}$$

where $\mathcal{K}_t$; $\mathcal{K}_n$ and m_t, m_n are constants $(0 < \mathcal{K} < 1, \ m \geqq 1)$. The values $\mathcal{K}_t$ and $\mathcal{K}_n$ differ slightly and $\mathcal{K}_t < \mathcal{K}_n$ for the downward fluxes of PAR, $\mathcal{K}$ has a maximum value in the range 0·6 to 1·0. The parameter $\mathcal{K}$ shows a marked dependence on solar elevation, with the maximum values at low elevation between 15 and 20°.

Tables III and IV present generalized relative fluxes of radiation as functions of the cumulative leaf area index and of solar elevation as well as a daytime total for simple and complex forest stands. Such radiation

Table III

Relative fluxes of total and net radiation as a function of leaf area index (oak with full foliage and without foliage)

| $L' = \int_0^z (a_L + a_s)\,dz$ | Total radiation $S_t(L)/S_t(0)$ | | | | | Net radiation $R_n(L)/R_n(0)$ | | | | Without foliage | | |
| | solar elevation β | | | | | solar elevation β | | | | solar elevation $>45°$ | | |
	$>50°$	$50\text{-}40°$	$40\text{-}30°$	$<30°$	Total for day time	$<45°$	$45\text{-}30°$	$<30°$	Total for day time	$L = \int_0^z a_s\,dz$	$S_t(L)/S_t(0)$	$R_n(L)/R_n(0)$
0·0	1·00	1·00	1·00	1·00	1·00	1·00	1·00	1·00	1·00	0·0	1·00	1·00
1·0	0·90	0·85	0·78	0·67	0·87	0·78	0·73	0·58	0·78	0·1	0·94	0·93
1·5	0·85	0·77	0·65	0·53	0·78	0·64	0·64	0·43	0·62	0·2	0·87	0·85
2·0	0·76	0·65	0·53	0·44	0·69	0·53	0·52	0·33	0·53	0·3	0·77	0·72
2·5	0·68	0·50	0·49	0·33	0·58	0·37	0·38	0·27	0·45	0·4	0·69	0·58
3·0	0·63	0·41	0·38	0·27	0·48	0·31	0·29	0·21	0·35	0·5	0·64	0·50
3·5	0·53	0·32	0·27	0·23	0·37	0·23	0·23	0·19	0·28			
4·0	0·43	0·25	0·20	0·18	0·28	0·18	0·18	0·16	0·20			
4·5	0·35	0·17	0·15	0·14	0·21	0·14	0·14	0·14	0·14			
5·0	0·28	0·14	0·12	0·11	0·15	0·12	0·12	0·12	0·11			
5·5	0·14	0·10	0·09	0·09	0·10	0·10	0·10	0·10	0·08			

Table IV

Relative fluxes of total radiation, net radiation and short-wave back radiation as a function of leaf area index (aspen + undergrowth; age ~20 years)

$L_t = \int_0^z (a_L + a_S)\,dz$		$\dfrac{S_t(L)/S_t(0)}{45\text{-}30°}$	$\dfrac{R_n(L)/R_n(0)}{>30° \quad <30°}$		$\dfrac{S_r(L)/S_r(0)}{>30° \quad <30°}$	
Aspen	0·0	1·00	1·00	1·00	1·00	1·00
	1·0	0·75	0·78	0·68	0·73	0·62
	1·5	0·67	0·68	0·57	0·64	0·50
	2·0	0·58	0·57	0·50	0·56	0·43
	2·5	0·52	0·50	0·42	0·50	0·33
	3·0	0·43	0·43	0·33	0·44	0·26
	3·5	0·35	0·34	0·26	0·37	0·22
	4·0	0·22	0·19	0·15	0·30	0·18
Undergrowth	4·5	0·11	0·13	0·10	0·24	0·15
	5·0	0·10	0·10	0·09	0·21	0·13
	5·5	0·09	0·09	0·09	0·18	0·12
	6·0	0·07	0·08	0·08	0·15	0·11
	6·5	0·06	0·07	0·08	0·13	0·10
	7·0	0·06	0·06	0·08	0·13	0·09

characteristics in their generalized form can be used in ecological studies, for instance, for evaluating the fluxes within a leaf canopy when its leaf index is known, or, vice versa, for finding the value of L when the radiation flux in the given forest stand is already measured.

Measurements of upward fluxes of radiation within stands can be well described by the exponential law:

$$\frac{S_r(L)}{S_r(0)} = \exp\left(-\mathcal{K}_t L\right)^m \tag{2}$$

where $m \simeq 1$ and K_t is 0·25 and 0·27 for oak and aspen respectively.

The relative absorbed solar flux within a canopy is expressed by:

$$\alpha_c = \left[1 - \frac{S(L)_t}{S(0)_t}\right] - \rho(h)\left[1 - \frac{S(L)_r}{S(0)_r}\right] \tag{3}$$

where $\rho(h)$ is the reflection coefficient at level h. Table V presents results of the calculations with Eq. (3). The table shows that the radiation absorbed by an aspen canopy is somewhat smaller than that absorbed by a meadow grass stand. It seems likely that the differences are due to the inter-crown gaps particularly in the upper half of the crown-cover because in the lower half when $L > 5$, the absorption function for both types of vegetation is

Table V
Relative absorbed total radiation as a function of the leaf area index

Type of vegetation	Leaf area index						
	1	2	3	4	5	6	7
Aspen	0·20	0·35	0·52	0·65	0·76	0·82	0·86
Open grass meadow	0·28	0·48	0·61	0·72	0·79	0·82	0·85

almost identical. Unfortunately available measurements of infra-red radiation are very scarce. Figure 6 shows some results of measurements taken separately for upward and downward infra-red radiation fluxes within a linden stand. It is interesting to observe a maximum upward infra-red flux and a balance of fluxes near the layer with the maximum area density in the upper part of the crown space.

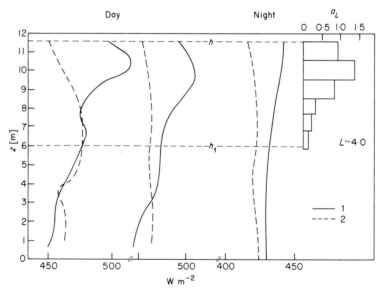

Fig. 6. Vertical profiles of long-wave radiation within the leaf canopy of a linden stand, 25 years old, during the day and at night: ——————— upward flux; —————— downward flux.

Table VI shows typical features of the changes with age of the relative values S_t and R_n (oak) for full foliage and for no foliage ($\beta > 40°$). Important differences are found not only in the absolute values but in the nature of

Table VI

Change with age of relative fluxes under forest canopy (oak forest)

	Age (years)													
	10	20	30	40	50	60	70	80	90	100	120	140	160	180
Total radiation														
Full foliage	0·70	0·18	0·12	0·12	0·13	0·15	0·17	0·18	0·19	0·21	0·23	0·23	0·25	0·27
Without foliage	0·80	0·70	0·63	0·55	0·50	0·43	0·40	0·38	0·35	0·33	0·32	0·30	0·30	—
Net radiation														
Full foliage	0·58	0·13	0·08	0·08	0·10	0·11	0·13	0·14	0·15	0·15	0·17	0·18	0·19	0·20
Without foliage	0·80	0·65	0·50	0·44	0·40	0·35	0·30	0·28	0·27	0·26	0·26	0·25	0·25	—

changes occurring with age. In the case of full foliage, the minimum radiation penetrating below the oak canopy is observed at the age of 30 to 50 years (12% to 13% for S_t and 8% to 10% for R_n). This penetration increases to 23% to 27% (S_t) and 18% to 20% (R_n) at the age of 140 to 180 years. Contrary to this change, when there is no foliage present, the penetration of both solar and net radiation decreases from 70% to 80% in young stands to 25% to 30% in stands over 150 years old.

The effects of the wind regime upon the vegetation canopy are associated with random space-and-time fluctuations in radiation fluxes. The movement of canopy elements can therefore reveal the statistical structure of the wind flow. This phenomenon has been well illustrated by American micro-meteorologists led by E. R. Lemon (Desjardins and Sinclair, 1969). According to these scientists, the normalized spectral densities for a horizontal wind component and the light flux within the plant canopy are characterized by similar distributions.

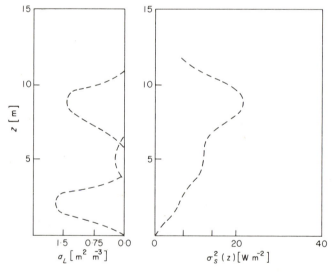

Fig. 7. The dispersion of total radiation within a leaf canopy (aspen and undergrowth) with $\beta > 40°$. The conditions outside the canopy are quasi-stationary.

Figure 7 shows a typical curve for the dispersion distribution of the downward flux σ_s^2 within a leaf canopy under quasi-stationary external conditions. The maximum dispersion coincides with the layer which has the maximum surface area $a_L + a_S$. In the lower layer the value of σ_s^2 decreases sharply. The penetration of radiation into a plant canopy can be considered as a

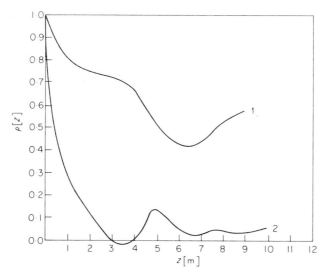

Fig. 8. Normalized auto-correlation function of (z) for (1) the downward flux of total radiation and (2) net radiation. The conditions outside the canopy are quasi-stationary and $\beta > 40°$.

stochastic process. Figure 8 presents normalized autocorrelation functions for the flux S_t and net radiation R_n. The decrease of $\rho(z)$ is approximately exponential i.e. $\rho(z) \simeq \exp(-\gamma z)$ with greatly different values of γ for S_t and R_n. The spectral distribution of space fluctuations (wave numbers) for the flux S_t is more narrowly defined, with a typical preponderance of small wave numbers. On the other hand, the spectral distribution for the net radiation is likely to approach the so-called 'white noise', i.e. uncorrelated values. The difference can be easily explained physically since the value R_n is determined by the balance of upward and downward fluxes, in other words, it depends upon a combination of factors acting above and below the given level z.

III. AERODYNAMIC CHARACTERISTICS OF LEAF CANOPIES

In developed forest vegetation, the values of the zero plane displacement and of the roughness parameter depend strongly on the density and height of the stand. In uniform stands, there is good evidence for a dependence between the normalized values z_0/h and d/h, on the one hand, and the average wind velocity above the forest, on the other (Fig. 9).

Figure 9 shows the value d/h decreases to almost 0·5 whereas the value of z_0/h rises to 0·25 with increasing wind speeds. This behaviour can be explained

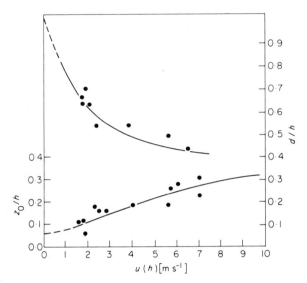

Fig. 9. Normalized values of the roughness parameter z_0 and zero plane displacement d as a function of the average wind velocity $u(h)$ near the stand top.

physically as follows. As the average wind increases at the top of the stand, the level of the zero plane gradually drops down to a layer which presents the maximum value of the leaf area density. As the velocity of the wind increases, there is an increase of roughness and in the value of z_0.

Figure 10 illustrates the vertical profile of the eddy diffusivity K above a deciduous forest and within a leaf canopy down to the soil surface as well as the relation between this diffusivity and the position of zero plane displacement.

The entire leaf canopy can be divided into two parts. In the upper part K changes sharply but in the lower part it changes much less. Similar results were also found for other types of natural forests. The displacement layer in a dense forest stand may be considered as a hypothetical 'hard wall' below which the condition of $dK/dz =$ constant is approximately true, but K in this layer is, on average, one order of magnitude lower than at the top of the stand.

The decrease of wind flow within the crown for different types of deciduous forests depends on the cumulative foliage area and may be described by an exponential function (see Fig. 11), i.e. $u(L) = u(0) \exp(-nL)$ where n lies between 0·25 and 0·35. This relationship is typical for values of wind velocity $v(0)$ between 1 and 3 m s^{-1}. With a higher wind velocity, the value of n begins to decrease and when the velocity reaches 5 m s^{-1}, n decreases

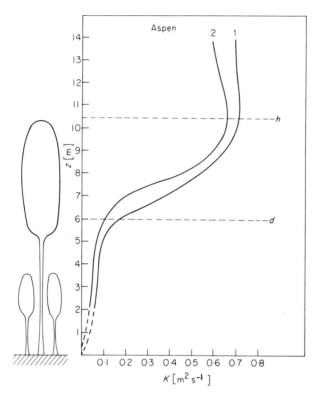

Fig. 10. Vertical profile of eddy-diffusivity K within a leaf canopy (aspen) at different solar elevations: (1) $\beta > 40°$; (2) $\beta < 40°$.

by a factor of 2 (Rauner and Ananiev, 1971). Experimental measurements in deciduous forests agree well with theoretical predictions by Perrier (1967).

Table VII contains some determinations of momentum fluxes near the top of a stand and within the canopy for a number of wind speed ranges. When analysed, these records show that the decrease of shearing stress τ within the canopy also follows an exponential law, the extinction parameter being a function of the wind average velocity above the canopy.

The following approximate formula can be used to determine the drag coefficient of a leaf canopy for a horizontal flux (see Vol. 1, p. 67)

$$C_{aM} \approx \frac{\tau(h)}{\rho[\int_0^h (a_L + a_S) \bar{u}^2 \, dz]} \tag{4}$$

based on the assumptions that C_{aM} const and $\tau(0) = 0$ (Uchijima and Wright, 1964).

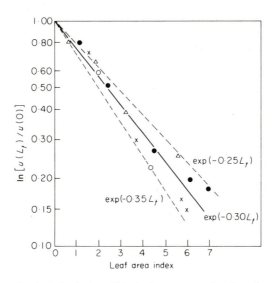

Fig. 11. Normalized wind velocity within leaf canopy as a function of cumulative total surface area $L = \int_0^z (a_L + a_S)\,dz$ for different types of deciduous forests: ($\times$) maple; ($\bullet$) aspen; ($\bigcirc$) linden; ($\triangle$) birch.

Table VII

Vertical distribution of momentum fluxes (N m^{-2}) in forest stands

z(m)	Wind velocity (m s^{-1})				
	1·0–2·0	2·1–3·0	3·1–4·5	4·6–6·0	6·1–9·0
Birch					
10·0	0·11	0·39	0·94	2·97	6·61
9·0	0·12	0·36	0·92	2·90	6·50
7·0	0·04	0·36	0·40	1·57	3·68
6·0	0·01	0·06	0·14	0·78	1·84
5·0	0·005	0·03	0·07	0·52	1·29
4·0	0·003	0·02	0·03	0·34	0·98
Aspen					
13·7	0·20	0·83	2·10		
10·0	0·09	0·87	1·04		
9·0	0·06	0·27	0·79		
8·0	0·04	0·16	0·53		
7·0	0·02	0·11	0·35		
6·0	0·01	0·08	0·23		

Table VIII gives values of τ for the leaf canopies of birch, aspen and pine (all about 15 to 20 years old). The coefficient for deciduous species is smaller than for pine by a factor of 2 to 3.

Table VIII
Drag coefficient as function of wind speed

Type of forest stand	$u(h)$, m s^{-1}				
	1–2	2–3	3–4	4·0–6·0	6·0
Birch	0·023	0·019	0·016	0·016	0·027
Aspen	0·015	0·018	0·020	—	—
Pine	0·038	0·046	0·051	0·057	0·064

IV. HEAT BALANCE OF LEAF CANOPIES

Components of the heat balance of forests in leaf are presented in Table IX. Evaporation plays a dominant role particularly in the second part of summer. During this period forests with a small diameter, less than a kilometre or two, surrounded by agricultural crops, create an 'oasis' effect which reduces the value of the sensible heat flux to zero or even to stable negative quantities. In other words, the surrounding countryside supplies the forest with heat. For larger forest areas of 3 to 5 km, say, this effect is rather weak (see Table IX, last line). Figure 12 illustrates features of the annual change of turbulent and latent heat fluxes for a vast birch-aspen forest area in the Moscow region. During cold seasons with snow cover, as well as in spring after snow melting and before foliation, the most important term in the heat balance is convective heat exchange with the atmosphere. During this period, land surfaces covered by deciduous forests play a very significant part in the thermal transformation of air masses. As leaves appear during spring, the relationship between λE and C changes radically but from June to August, the ratio $\lambda E/C$ remains almost constant. During leaf yellowing and especially after defoliation (the second half of September and October) $\lambda E/C$ decreases again. The figures in Table IX present a typical picture of the annual changes of λE and C for deciduous forests. Differences in other climatic regions may be chiefly attributed to shifts in phenophase.

Figure 13 shows the relationship between the normalized total evaporation and leaf transpiration of forests (oak) and soil moisture in the root zone of

Table IX
Components of heat balance for some types of deciduous forests

Types of the forest stand		Integrated flux (MJ m^{-2})				
		R_n	λE	C	G	E mm
Natural mixed deciduous forest, (oak, aspen, maple, 19–12 years)	July	419	406	9	4	162
	August	398	402	−4	0	160
	September (first half)	126	105	25	−4	42
Natural oak forest (40–60 years)	May	369	260	88	21	103
	June	335	276	50	9	110
	July	419	410	0	9	163
	August	318	318	−4	4	127
	September (first half)	117	100	17	0	40
Planted maple forest (10–12 years)	June	440	385	42	13	153
	July	452	436	12	4	174
	August	293	289	0	4	115
	September	205	184	21	0	73
	October (first half)	63	42	25	4	17
Natural linden-oak forest (20–25 years)	July–August (second half– first half)	432	324	104	4	128

the soil (2 m deep) (Rauner, 1966). The general form of the curve is in good agreement with well-known data for other types of vegetation (for example, see Budagovsky, 1964; Slatyer, 1967 and others). At the same time, it is important to note that 'critical' values of soil moisture for forests show that the actual evaporation rate of the system remains rapid even when the available water is small, about 5 to 6 % by volume for oak.

Deciduous forests seem able to use soil water very effectively. The structure of their heat balance depends less on soil water than herbaceous vegetation or crops under other comparable conditions.

Figure 14 illustrates typical vertical fluxes of latent and sensible heat within a leaf canopy during the day when the root zone of the soil contains adequate water. It is characteristic that the sign of the convective heat flux changes in the upper layer of the leaf canopy. Below the level where the surface area density is a maximum, the turbulent heat flux is directed towards

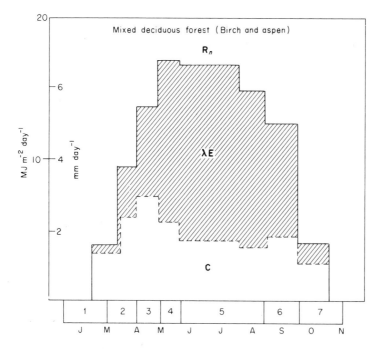

Fig. 12. Annual course of eddy fluxes of sensible and latent heat in deciduous forests near Moscow (birch, *Betula verrucosa* and aspen, *Populus tremula*; height 15 to 17 m, 40 to 50 years old): (1) snow cover before the beginning of snow melting; (2) snow melting; (3) after snow melting before foliation; (4) foliation; (5) full foliage; (6) yellowing and leaf fall; (7) after complete defoliation.

the soil surface or the surface of the vegetation mass. In other words, an additional source of heat is available for transpiration. The transpiration rate reaches a maximum in the upper half of the crown whereas evaporation from the soil surface constitutes only a very small fraction of total water loss. The calculated components of heat balance under a leaf canopy also show that for daytime totals, the heat flux from warmer parts of the crown and the conductive flux into the soil are almost in balance (i.e. $-\mathbf{C} \approx \mathbf{G}$). Hence, energy for evaporation from the soil and for transpiration of the herbaceous canopy are approximately equal to the radiant energy available below the leaf canopy. It is possible, therefore, to use the following approximate formula when calculating $\lambda \mathbf{E}_t$, the daytime totals of latent heat of transpiration for a leaf canopy (Rauner, 1972):

$$\lambda \mathbf{E}_t \approx \lambda \mathbf{E} - (\lambda \mathbf{E}_i + \mathbf{R}_n) \tag{5}$$

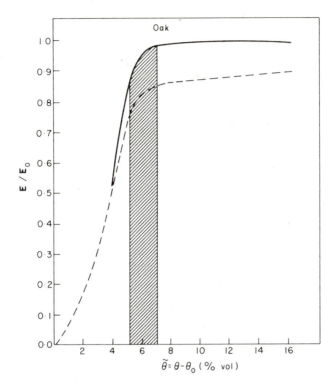

Fig. 13. Normalized total evaporation of oak forest (40 to 50 years old) as a function of average soil moisture of a 2-m root zone of soil (volume %). E_0 is the potential transpiration calculated from the heat balance equation for a wet surface. θ soil-water content minus θ_0 the water content at wilting point, is the so-called 'wilting coefficient' ($\theta_0 \approx 1.4$ of maximum hygroscopicity). The soil is a black soil under the forest (Chernozem). ———— total evaporation, – – – – – – transpiration.

where λE is latent heat of total evaporation, λE_i is latent heat of evaporation of intercepted precipitation, R_n is net radiation measured below the crown.

Using the estimates of the vertical flux of latent heat and of the distribution of transpiring surfaces (Fig. 14), it is possible to draw up tables of normalized fluxes of water vapour as a function of the leaf area index (Table X). Like similar tables for radiation fluxes (see above) these tables can be applied to ecological and physiological studies of the water regime and transpiration of forest communities.

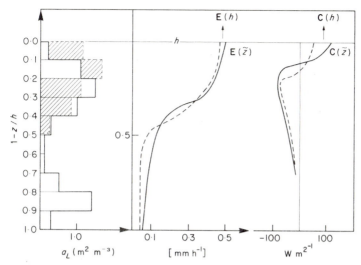

Fig. 14. Vertical profiles of fluxes of sensible and latent heat (1100 to 1300) in two types of deciduous forest stands and distribution of transpiring surface: ————— aspen; ─ ─ ─ ─ ─ maple.

Table X
Relative evaporation rate as a function of leaf area index
(oak and aspen + undergrowth)

$L = \int_0^z a_L \, dz$	Oak				Aspen + undergrowth		
	Solar elevation β			Daytime total	Solar elevation β		Daytime total
	$> 45°$	$45–30°$	$< 30°$		$> 40°$	$< 30°$	
0·0	1·00	1·00	1·00	1·00	1·00	1·00	1·00
0·5	0·96	0·87	0·50	0·87	0·95	0·87	0·92
1·0	0·86	0·73	0·32	0·72	0·90	0·75	0·88
1·5	0·64	0·57	0·26	0·57	0·85	0·65	0·83
2·0	0·48	0·46	0·22	0·48	0·80	0·57	0·78
2·5	0·38	0·38	0·19	0·39	0·74	0·50	0·72
3·0	0·32	0·30	0·17	0·32	0·65	0·43	0·64
3·5	0·26	0·24	0·16	0·27	0·55	0·38	0·53
4·0	0·23	0·18	0·14	0·21	0·41	0·32	0·38
4·5	0·18	0·12	0·13	0·15	0·24	0·27	0·23
5·0	0·10	0·09	0·10	0·08	0·17	0·22	0·15
5·5					0·13	0·18	0·13
6·0					0·11	0·15	0·11
6·5					0·10	0·11	0·09
7·0					0·09	0·09	0·08

V. PHOTOSYNTHESIS AND TRANSPIRATION EFFICIENCY
OF DECIDUOUS FORESTS

The energy efficiency of the primary production process of any type of vegetation is expressed by its coefficient of photosynthetic efficiency $\varepsilon_s = \mu \, \delta W / S^*$; by its coefficient of transpiration efficiency $\varepsilon_e = \mu \, \delta W / \lambda E_t$; and by the ratio $\varepsilon_s / \varepsilon_e = \lambda E_t / S^*$ (μ is the energy equivalent of photosynthesis usually regarded as 16.7 kJ g^{-1} for forest vegetation, and δW is the increase of total dry weight g m^{-2} for a period in which the income of PAR is S^* J/m^2). In respect of heat and water use, the primary organic products develop most effectively when ε_s and ε_e tend to a maximum value and $\varepsilon_s / \varepsilon_e$ to an optimum.

Table XI presents energy characteristics of the production process and the photosynthesis and transpiration efficiency in an oak forest at different ages, based on measurements and calculations. The largest solar radiation absorption, as well as the most intense relative transpiration are found between the ages of 20 and 40 years. During this period, as shown above, the forest stand develops its most extensive photosynthesizing and transpiring apparatus and the storage of dry matter above the ground is greatest. After this optimal period the values of the absorbed PAR and transpiration begin gradually dropping down and by the end of the life span (above 150 years), they account for respectively 75% and 55% of the incoming PAR and of total water loss by evaporation.

The highest coefficient of photosynthetic efficiency ε_s (above 2%) is found in oak forests of 25 to 35 years of age. Thereafter it decreases considerably and remains almost constant from 60 to 120 years. Because of the prevailing processes of respiration and senescence within the stands, this coefficient sharply declines. The coefficient of transpiration efficiency reaches a maximum during the period of maximum storage, and behaves like ε_s as the stand ages. The ratio $\varepsilon_s / \varepsilon_e$ never reaches unity, i.e. the energy efficiency of photosynthesis is always less than that of transpiration. This implies that with the same quantity of vegetation the latent heat of transpiration is used more economically by the leaf canopy than the comparable absorption of solar energy in the PAR region.

It can be also shown that the period when the annual storage of dry matter reaches a maximum (on average 6 to 8 m^3 ha^{-1} year^{-1}) is characterized by optimal values of L between 4 and 6. The ratio of the total non-assimilating surface to the leaf surface which is a measure of losses by respiration does not reach 0.5. The critical value of L for stand at which the loss of biomass begins to exceed its accumulation does not exceed 2 to 2.5; the ratio of total assimilating surface to leaf surface increases to 1.5 to 2.0.

Table XI

Relative absorption of PAR, relative transpiration $\mathbf{E}_t/\mathbf{E}_0$ and efficiency of phytosynthesis and transpiration as function of age in an oak forest

Characteristics	Age (years)														
	10	20	30	40	50	60	70	80	90	100	120	140	160	180	200
Relative absorbed PAR	0·35	0·91	0·93	0·93	0·92	0·91	0·91	0·90	0·90	0·88	0·86	0·85	0·83	0·80	0·75
Relative transpiration															
$\mathbf{E}_t/\mathbf{E}_0$	0·43	0·80	0·90	0·90	0·85	0·80	0·76	0·73	0·71	0·69	0·68	0·67	0·63	0·60	0·55
ε_s (%)	1·20	1·50	1·70	1·40	1·20	1·00	0·90	0·80	0·80	0·80	0·70	0·70	0·60	0·50	0·40
ε_e (%)	1·40	1·70	2·10	1·80	1·40	1·20	1·15	1·10	1·10	1·10	1·00	0·95	0·80	0·70	0·60
$\varepsilon_s/\varepsilon_e$	0·86	0·88	0·81	0·78	0·85	0·83	0·78	0·73	0·73	0·73	0·70	0·73	0·75	0·70	0·67

REFERENCES

Budagovsky, A. I. (1964). 'Evaporation of the Soil Moisture', p. 242, Nauka, Moscow.

Desjardins, R. L. and Sinclair, T. R. (1969). *In* 'Annual Progress Report', Microclimate Investigations. pp. 21–29, U.S. Dep. of Agriculture, Bradfield Hall, Ithaca, New York.

Grin, A. M., Rauner, Ju. L., Utekhin, V. D. (1970). Proceedings of the U.S.S.R. Acad. of Sciences, Geographycal Series, No. 4, Moscow, pp. 10–23.

Grulois, J. (1968). *Bull. Soc. R. Bot. Belg.* **101**, 141–153.

Oicawa, T. and Saeki, T. (1972). *In* 'Photosynthesis and Utilization of Solar Energy", IBPL, Level III experiments, Tokyo, pp. 107–116.

Perrier, A. (1967). *Meteorologie*, 1–4, 527–550.

Rauner, Ju. L. (1966). Proceedings of the U.S.S.R. Acad. of Sciences, Geographical Series, No. 3, pp. 17–29, Nauka, Moscow.

Rauner, Ju. L. (1968). *In* 'Actinometry and Atmospheric Optics', pp. 335–342, Valgus, Tallin.

Rauner, Ju. L. (1972). 'Heat Balance of the Plant Cover', p. 209. Hydrometeorological Press, Leningrad.

Rauner, Ju. L., Ananiev, I. P. (1971). Proceedings of the U.S.S.R. Acad. of Sciences, Geophysical Series, No. 2, pp. 70–79, Nauka, Moscow.

Russ, Ju. (1972). *In* 'Solar Radiation and Productivity of Plant Stand', pp. 122–147, Tartu.

Sukachev, V., and Dylis, N. (eds.) (1964). 'Fundamentals of Forest Biogeocoenology', p. 573, Nauka, Moscow.

Slatyer, R. O. (1967). 'Plant-water Relationships', Academic Press, London and New York.

Uchijima, Z., Wright, J. (1964). *Bull. Nat. Inst. Agric. Sci. Tokyo*, Ser. A, No. 11.

9. Carbon Dioxide Exchange and Turbulence in a Costa Rican Tropical Rain Forest

L. H. ALLEN, JR. and E. R. LEMON

USDA Microclimate Investigations, Bradfield Hall, Cornell University, Ithaca, New York, U.S.A.

I. DESCRIPTION OF THE FOREST

In November 1967, micrometeorological measurements were made in a regrowth tropical rain forest near Turrialba, Costa Rica (Lemon *et al.*, 1970; Allen *et al.*, 1972). The objectives were to describe the micro-climate of this forest system and to obtain flux densities of carbon dioxide, water vapour, and sensible heat to or from the forest. Concurrently, Stephens and Waggoner (1970) used leaf chamber techniques to determine the relation between net photosynthesis and light for leaves of several of the species prevalent in the forest.

The site was 2·4 km WSW from the headquarters of the Instituto Interamericano de Ciencias Agricolas de la Organización de los Estados Ameri-
canos (OEA)), Turrialba, Costa Rica, and was called Bosque de Florencia.
The forest site lay at an elevation of 700 m and sloped towards the north-east.
The main trees were *Goethalsia meiantha* (Donn. Smith) Burret, which
formed a top canopy between 30 and 40 m above the forest floor. Foliage
was sparse between 10 and 30 m, but more abundant between ground level
and 10 m. The leaf area index (L) of the top canopy (30 to 40 m) was about
2 and that of the bottom canopy was about 1. *Cecropia peltata* L., a fast
growing pioneer species, persisted in the middle story of Bosque de Florencia,
and especially around the borders. The crowns were umbrella shaped. The
distribution of leaf elevation angles tended to be predominantly horizontal.
The horizontal dispersion of leaves tended to be regular, rather than random
or clumped. Average annual rainfall was 2488 mm. March averaged 69 mm
and June 290 mm. Stephens and Waggoner (1970), Lemon *et al.* (1970),
and Allen *et al.* (1972) describe the site and vegetation more completely.

II. MICROMETEOROLOGICAL MEASUREMENTS

Two micrometeorological methods were attempted to determine fluxes:
an energy balance and a momentum balance (Lemon, 1967). Vertical profiles
of net radiation, temperature, water vapour, CO_2, and heat storage in the
biomass are needed for the energy balance method. Vertical profiles of wind
and CO_2 are required for the momentum balance method (see Vol. 1, Ch. 3).
    In the Bosque de Florencia, two rope and pulley systems were rigged to
a 40-m *Goethalsia* tree to support the sensors and sampling systems (Fig. 1).
The mast tree was located in the forest about 200 to 250 m downwind from
the leading edge which faced the north easterly prevailing wind. The trailing
edge of the forest was about 50 m downwind from the mast tree.
    Measurements were made and air samples collected during sixteen 15- or
20-minute 'runs' on November 14 and 15, 1967. The air samples were collected
simultaneously at 10 heights and stored in laminated polyethylene-aluminum-
mylar bags for subsequent analysis. Carbon dioxide concentration was
measured with an infra-red gas analyser, and water vapour concentration
was measured by absorption in magnesium perchlorate absorption tubes.
Air temperature was measured at the same 10 heights with unaspirated but
radiation-shielded, copper-constantan thermocouples.
    Wind speed was measured with five heated-thermocouple anemometers.
Meter readings were recorded simultaneously every 5 s with a manual
time-lapse movie camera. Net radiation and photosynthetically active

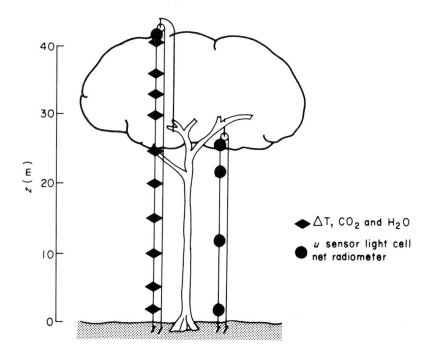

Fig. 1. Schematic diagram showing the rigging in the mast tree used to support the sensors and the air-sampling system. ΔT represents thermocouples for measuring air temperature, and u represents the heated thermocouple anemometers for measuring horizontal wind speed.

radiation (PAR) were also measured at several heights in the forest as illustrated in Fig. 1. Allen *et al.* (1972) described the forest rigging, instrumentation, and measurements in much more detail.

III. LEAF CHAMBER MEASUREMENTS

Stephens and Waggoner (1970) described in detail the procedure used to determine net CO_2 assimilation as a function of irradiance for detached leaves of several of the plant species in the top story, middle story, and lower story of the forest. Photosynthesis and dark respiration rates were determined from the concentration change in CO_2 between inlet and outlet gas of the net assimilation chamber at known flow rates. The concentration difference was kept generally between 10 and 20 ppm with flow rate ranging from 1 to 7·5 litres/min. A fan in the base of the chamber was used to maintain air turbulence. Three high-intensity incandescent lamps submerged in water supplied light and a series of fibreglass screens were used to decrease radiation from a maximum of 840 Wm^{-2} (as measured by an Eppley pyrheliometer) through several steps of lower radiation flux density. The lamp emission and water absorption spectra indicated that this radiation was about two-thirds as effective for photosynthesis as sunlight. The leaf chamber temperature was controlled by circulating water between double walls of the upper and lower surfaces. Fresh leaves with petiole cut under water and maintained under water were used in the chamber.

Stephens and Waggoner (1970) reported CO_2 exchange characteristics of other species as well as *Goethalsia* and *Cecropia*. *Cordia alliodora* (Ruiz and Pavon) Cham. and *Virola koschnyi* were tested in the leaf chamber as representative of the middle story. For the bottom layer, *Croton glabellus* L. and *Protium glabrum* (Rose) Engler were examined as representatives of plants 1 to 5 m tall, and *Heliconia latispatha* Benth. and an unidentified Melastomaceae were chosen as plants below 2 m.

IV. CARBON DIOXIDE FLUX MODEL

Figure 2 shows measured profiles of temperature, wind speed, CO_2 concentration, and water vapour concentration. From inspection, there is much uncertainty in the temperature profiles and especially the water vapour profiles. Since the energy balance method of determining fluxes relies on profile gradients of these entities, no attempt was made to use this method. However, the wind speed profiles looked smooth, so the momentum balance method was attempted.

No profile records were collected in the surface boundary layer above the forest. The wind profiles of Fig. 2 were extrapolated above the canopy by use of known profile characteristics (see Kung and Lettau, 1961; Kung, 1961; Lemon *et al.*, 1970 for further explanation). Water vapour and temperature profiles were extrapolated above the forest using records from a climatological station maintained by the Instituto Interamericano de Ciencias Agricolas de la OEA at Turrialba (data supplied by H. Trojer, OEA). Absolute CO_2 concentrations were estimated from CO_2 flux densities at the top of the forest and turbulent eddy diffusivities were derived from the extrapolated wind profiles. The equations illustrating the methods will be shown later.

Figure 3 shows the idealized CO_2 flux model devised by Lemon *et al.* (1970) to explain CO_2 flows into and out of this forest system. The forest was treated as a four-layer system based on specific features of the CO_2 and wind profiles and the distribution of leaf area with height.

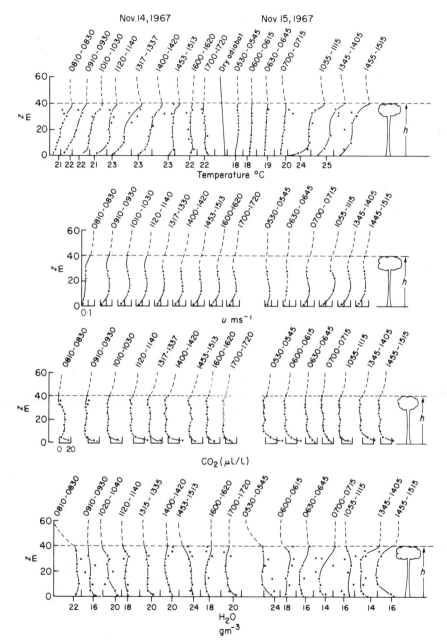

Fig. 2. (Top) Temperature profiles; (upper middle) wind speed (u) profiles; (lower middle) carbon dioxide concentration (CO_2) profiles; and (bottom) water-vapour concentration (H_2O) profiles in the Bosque de Florencia during 15- or 20-min runs, November 14 and 15, 1967.

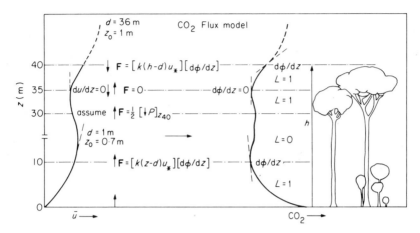

Fig. 3. Idealized wind speed, $\bar{u}$, and carbon dioxide, CO_2 profiles used in the four-layer CO_2 flux model of the Costa Rican Forest. See text for fuller explanation.

Layer 1. The bottom 10 metres included all leaves of the lower canopy with $L = 1$.

Layer 2. The next layer had few leaves and extended from 10 m to 30 m. Figure 3 shows that wind speed and CO_2 profiles exhibit small local maxima in this region. Epiphytes were plentiful in the upper part of this layer.

Layer 3. This layer began at the base of the upper canopy and extended to the height of a local minima in the wind speed and CO_2 profiles at about 35 m. This 30- to 35-m layer had $L \simeq 1$.

Layer 4. The top half of the upper canopy extended from 35 to 40 m and had $L = 1$. Thus the upper canopy, consisting of layers 3 and 4, had a total leaf area index of 2.

The heavy arrows in Fig. 3 indicate the direction of CO_2 flow at the boundaries of the layers during daylight. At ground level, respired CO_2 was always diffusing upward. As the forest floor was wet and diurnal temperature changes were small, this flow was assumed nearly constant day and night.

There was a continuous flow of CO_2 through the top boundary of the bottom layer, day and night. During darkness, the considerable upward flow through this boundary represents CO_2 coming from the soil and litter at ground level plus a respiratory contribution from the leaves in the lower canopy.

The horizontal arrow in the next layer indicates that CO_2 was being supplied horizontally from outside the forest, creating a layer of higher concentration between the top and bottom canopies. Respiration of the tree trunks probably did not contribute much. We cannot explain the small but persistent local CO_2 minimum in the middle of this layer. The one-dimensional vertical flux model does not apply in this region, but there is little green vegetation or biological activity either.

Near the base of layer 3, the CO_2 concentration decreased, indicating an upward flow down a mean concentration gradient caused by photo-synthesis in the lower half of the upper canopy. Midway in the upper canopy, however, a minimum in the CO_2 profile occurs where the mean gradient goes to zero. Evidently, an upward flux of CO_2 just balances a downward flux of CO_2, creating zero net flux at this boundary. This balance is indicated by the up and down arrows at the boundary.

At the top layer of the forest, CO_2 diffuses downwards from the atmosphere above the forest during the day as indicated by decreasing mean CO_2 content with decreasing height. The direction of flow is indicated by the downward arrow at the top of the forest.

A. Micrometeorological Computations

Assuming no flux divergence, the vertical flux density of CO_2 through a horizontal plane z distance above the ground is defined by:

$$\mathbf{F}(z) = K_C(d\phi/dz) \tag{1}$$

where $\mathbf{F}(z)$ is the vertical flux density at z level, K_C is the eddy 'diffusivity' for CO_2, and $d\phi/dz$ is the mean CO_2 concentration gradient. The CO_2 gradient, $d\phi/dz$, is the slope of the CO_2 profile at specified levels of z (Fig. 3). An equation similar to Eq. (1) will be used later to compute the fluxes through the upper boundary of the top canopy at $z = 40$ m and the upper boundary of the lower canopy at $z = 10$ m.

In Eq. (1), K_C is evaluated by assuming that matter, e.g. CO_2 in a turbulent air stream, is transferred in a manner similar to momentum:

$$\tau(z) = \rho K_M(du/dz) \tag{2}$$

where $\tau(z)$ is the 'shearing stress' or the vertical flux density of momentum through a horizontal plane z distance above the ground, ρ is the density of the air, K_M is the eddy 'diffusivity' for momentum, and du/dz is the mean wind speed gradient. K_M can be evaluated from the wind speed profiles and substituted for K_C in Eq. (1). The assumption that K_M equals K_C is reasonable in fully turbulent flow when there are no strong temperature gradients. The relation between wind speed, $u(z)$, and height, z, can be expressed by the logarithmic profile equation (see Vol. 1, p. 63):

$$u(z) = (u_*/k) \ln [(z - d)/z_0] \tag{3}$$

where $u_* = \sqrt{\tau/\rho}$ is the friction velocity, $k = 0.41$ is von Karman's constant, d is zero plane displacement and z_0 is the roughness length. Manipulation of Eqs (1) to (3) then gives:

$$\mathbf{F}(z) = k(z - d)u_*(\mathrm{d}\phi/\mathrm{d}z). \tag{4}$$

Equation (4) is presented in Fig. 3 at the two levels from which the CO_2 vertical flux calculations are made. (At the top of the canopy, $z = h = 40$ m.)

To evaluate d and z_0, two assumptions have to be made about the momentum balance; (a) at the top of the upper canopy all momentum transferred downward is completely extracted by the plant surfaces between 40 and 35 m so that momentum flux is zero at 35 m; and (b) at the top of the bottom canopy all momentum transferred downward is extracted by the plant and soil surfaces between 10 and 0 m so that momentum flux is zero at 0 m.

Since wind was not measured *above* the trees, known profile characteristics above other forests were used but information is scarce. While values for z_0 are of the order of 1 m in most forests, the ratio of d/h varies widely. Lemon *et al.* (1970) made duplicate calculations with $d = 12.2$ m ($d/h = 0.3$) and $d = 36$ m ($d/h = 0.9$). The photosynthesis rates computed by Eq. (4) using a value of $d = 12.2$ m were 1.8 times as large as the rates were using a value of $d = 36$ m.

It was assumed that the logarithmic wind profile shape immediately below $z = 10$ m represents the vertical momentum transfer of the wind across this plane. In Fig. 3, an extension of this profile above $z = 10$ m is indicated by a dashed curve. An analysis of this curve yielded the values $z_0 = 0.7$ m and $d = 1$ m for the lower canopy.

B. Results of Micrometeorological Computations

The CO_2 flux density calculations reported by Lemon *et al.* (1970) for CO_2 flow into the top canopy and out of the bottom canopy are too small by a factor of 10 due to a decimal point error in the CO_2 gradients. The CO_2 flux densities for each of the sixteen runs were recomputed and are presented in Table I. Wind speed profile parameters of $d = 36$ m and $z_0 = 1$ m were used in these new computations. Diurnal CO_2 budgets for November 14 and 15 (Table II) were based on the summed daylight hour fluxes of Table I and on inputs from Stephens and Waggoner (1970). These diurnal budgets are shown in Table II.

The components of the diurnal CO_2 budget of the Bosque de Florencia are explained as follows.

Table I

Summary of CO_2 flux densities across the 40-m level ($\mathbf{F}_{40}$) and the 10-m level ($\mathbf{F}_{10}$) in the Bosque de Florencia, Turrialba, Costa Rica, November 14 and 15, 1967. Total CO_2 uptake rate in the top canopy was estimated by $P_{40} \times 1.5$. (See text for more information)

Total	40 m					10 m			
	$u(h)$ m s^{-1}	$K(h)$ m^2 s^{-1}	$\partial\phi/\partial z$ g m^{-4}	$\mathbf{F}_{40}$ g m^{-2} h^{-1}	$\mathbf{F}_{40} \times 1.5$ g m^{-2} h^{-1}	$u(10)$ m s^{-1}	$K(10)$ m^2 s^{-1}	$\partial\phi/\partial z$ g m^{-4}	$\mathbf{F}_{10}$ g m^{-2} h^{-1}
(Nov. 14)									
0810	0.67	0.306	0.0006	0.66	0.99	0.53	0.270	−0.0008	−0.78
0910	0.68	0.310	0.0013	1.45	2.18	0.54	0.274	−0.0004	−0.40
1010	0.78	0.356	0.0021	2.69	4.04	0.62	0.314	−0.0004	−0.45
1120	0.85	0.388	0.0016	2.24	3.36	0.68	0.343	−0.0008	−0.99
1317	0.83	0.378	0.0018	2.45	3.68	0.66	0.334	−0.0006	−0.72
1400	0.81	0.369	0.0024	3.19	4.78	0.65	0.326	−0.0008	−0.94
1453	0.79	0.360	0.0026	3.37	5.06	0.63	0.318	−0.0003	−0.34
1600	0.76	0.347	0.0020	2.50	3.75	0.61	0.306	−0.0004	−0.45
1700	0.72	0.328	0.0022	2.60	3.90	0.58	0.290	−0.0004	−0.42
(Nov. 15)									
0530	0.14	0.064	0.0018	0.41	0.62	0.11	0.0564	−0.0012	−0.24
0600	—	—	0.0012	—	—	—	—	−0.0010	—
0630	0.35	0.160	0.0012	0.69	1.04	0.28	0.1410	−0.0012	−0.61
0700	0.63	0.287	0.0006	0.62	0.93	0.50	0.2540	−0.0010	−0.91
1055	1.10	0.502	0.0018	3.25	4.88	0.88	0.4430	−0.0004	−0.64
1345	0.94	0.429	0.0022	3.40	5.10	5.10	0.3790	−0.0012	−1.64
1445	0.83	0.378	0.0018	2.45	3.68	0.66	0.3340	−0.0010	−1.20

Table II

Diurnal CO_2 Budget, Bosque de Florencia, Turrialba, Costa Rica, based on the momentum budget and on net CO_2 assimilation chamber studies of Stephens and Waggoner (1970)

	Nov. 14	Nov. 15	Average
PAR, 400 to 700 nm (MJ m^{-2} day^{-1})	10·9	10·5	10·7
Gross production[a]	39.0	44.2	41·6˙
Photosynthesis[a]			
upper canopy (day)	33·9	39·1	36·5
lower canopy (day)	2·8	2·8	2·8
	36·7	41·9	39·3
Respiration[a]			
upper canopy (night)	1·92	1·92	1·92
lower canopy (night)	0·35	0·35	0·35
ground (day and night)	20·0	20·0	20·0
	22·3	22·3	22·3
Net production[a]	14·4	19·6	17·0
Daytime efficiency[b]	3·78%	4·49%	4·13%
24-hour efficiency[b]	1·48%	2·10%	1·79%

[a] Unit is g CO_2 per m^2 land area per day.
[b] Based on 11·3 kJ per g CO_2.

A. Photosynthesis in upper canopy (day)	Downward flux at 40 m during daylight multiplied by 1·5 to account for the second unit of leaf area index.
B. Photosynthesis in lower canopy (day)	Laboratory value (Stephens and Waggoner, 1970) of about 0·28 g m^{-2} h^{-1} × 10 h of effective sunlight.
C. Respiration in upper canopy (night)	Laboratory value (Stephens and Waggoner, 1970) of 0·068 g m^{-2} h^{-1} × 10 h or darkness × 2(L).
D. Respiration in lower canopy (night)	Laboratory value (Stephens and Waggoner, 1970) of 0·025 g m^{-2} h^{-1} × 14 h of effective darkness.
E. Non-leaf respiration at the ground (day and night)	Average daytime upward flux density at 10 m plus lower canopy photosynthesis rate × 10 h, plus night time upward flux density × 14 h.

Figure 3 reported in Allen *et al.* (1972) showed that the lower half of the top canopy (Layer 3) intercepted about 30 % as much PAR as did the upper half (Layer 4). However, because of light saturation of the *Goethalsia* leaf at low illumination (Stephens and Waggoner, 1970), the leaves of Layer 3 should assimilate CO_2 faster than only 30 % of the rate of Layer 4. We estimated the assimilation rate of Layer 3 to be 50% of that of Layer 4. (Simulations presented later indicated that Layer 3 should assimilate at least 50 % or more of the rate of Layer 4.) Hence the photosynthesis rates computed for the upper half of the top canopy (Layer 4) were multiplied by 1·5 to estimate the photosynthesis rates of the whole top canopy (Layer 3 and Layer 4).

Stephens and Waggoner (1970) reported that the maximum net CO_2 assimilation rate of *Goethalsia* leaves was about $0.65 \text{ g m}^{-2} \text{ h}^{-1}$. Even if we assume that the exposure of all the leaves of the top canopy allowed photosynthesis at this rate, the CO_2 uptake rate would be only 1·3 g per m² ground area per hour for the top canopy. The maximum rates computed by the aerodynamic approach were almost four times larger (Table 1). The rate of CO_2 liberation by litter and soil (including root) respiration appears to be exceptionally large also. In studies of the Forét de l'Anguédédou in the Ivory Coast, Müller and Nielsen (1965) concluded that about 8 % of gross production was lost by litter fall. Including root respiration raised the loss of gross production to about 15 %. Kira *et al.* (1967), Kira and Shidei (1967), and Kira (1969) presented data from a Khao Chong, Cambodia, rain forest showing that about 19 % of gross production would be lost by litter fall plus death of standing trees. With root respiration added, the estimated ground level losses would be about 23 %. The computations in Table II indicate that about 48 % of gross photosynthesis (net day-time CO_2 uptake plus estimated day-time respiration) would be returned by ground-level respiration.

If the day-time photosynthesis rate is assumed constant throughout the season, about 143 tonnes (metric) of CO_2 per hectare would be taken up annually. With estimated respiration added, the gross annual production would be equivalent to 152 t ha^{-1} of CO_2. Assuming a factor of 0·61 for the appropriate molecular weight ratio, the gross annual production of dry matter would be 92 t ha^{-1}. Müller and Nielsen (1965) estimated a gross annual production of 52.5 t ha^{-1} of dry matter in an Ivory Coast forest and Kira *et al.* (1967) estimated 123 t ha^{-1} in a Thailand rain forest.

In the diurnal CO_2 budget, both the top canopy photosynthesis and the release of CO_2 at ground appear to be large compared to canopy respiration. The momentum balance method appears to be giving flux densities that are too large.

C. Turbulence and its Effects on Carbon Dioxide Exchange

Turbulence was less in the tall (40 m) Costa Rican forest than in a short (10·4 m) Japanese larch plantation near Ithaca, New York (Allen, 1968; Allen *et al.*, 1972). Figure 4 shows wind speed records at five heights in the

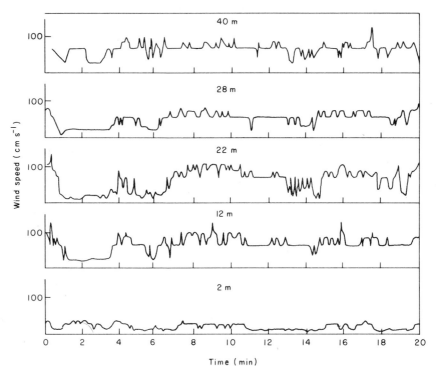

Fig. 4. Wind speed records at five heights in the Bosque de Florencia, 0910 to 0930, November 14, 1967. Data were taken at 5-sec intervals by photographing the meters of heated-thermocouple anemometers. The record shows some periods of very steady air flow. Data plotted at 5-sec intervals.

Bosque de Florencia, and Fig. 5 shows wind speed records in the larch plantation. The Costa Rican records show some periods of very steady air flow. Visual observations of the meters recording output from the heated-thermocouple anemometer probes at the time that the wind speeds were being measured showed long periods (up to about 1 min) of very steady wind. This low-turbulence air flow was prevalent even at the highest anemometer (40 m). One tentative conclusion is that turbulent transport is not as effective as the theory behind Eqs (1) to (4) requires. One result of less turbulent

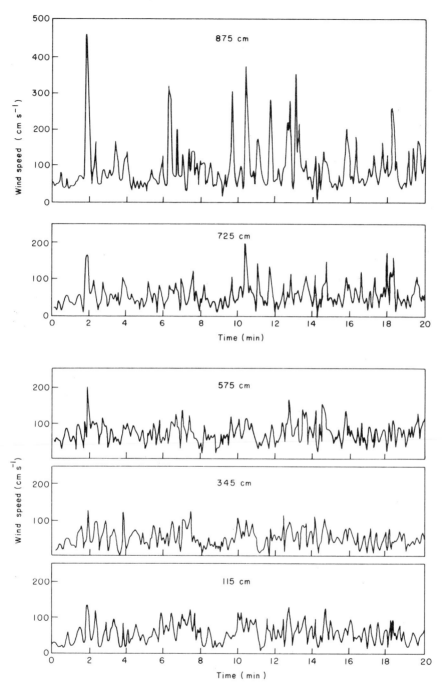

Fig. 5. Wind speed records in a Japanese larch plantation near Ithaca, New York, on October 30, 1964. The first (left) half of the records is from 1517 to 1527 EST, and the second (right) half is from 1545 to 1555 EST. The records show very turbulent air flow, with gustiness increasing with height. Data plotted at 4·8-sec intervals.

transport would be the build up of larger mean gradients of CO_2 in the forest system than is commonly found in vegetation communities with more turbulence. For instance, the average difference in CO_2 concentration between 40 and 35 m was $-4.5\ \mu l/l$, and the average gradient from 10 to 5 m was $+5.5\ \mu l/l$. Gradients in the Japanese larch plantation (unpublished) were about 1 to $2\ \mu l/l$ over the same vertical distances. The wind speed in the top canopy of the spire crowns of the Japanese larch was faster than in the top canopy of the umbrella crowns of the *Goethalsia*, so the results are not strictly comparable. However, the speeds in the trunk space (Fig. 2) were similar to the speeds in the Japanese larch trunk space (Allen, 1968).

Several factors may have contributed to the low level of turbulence in the Costa Rican forest. The umbrella-shaped crowns may have reduced the penetration of eddies. Wide spacing of turbulence-generating elements would have allowed more decay of turbulence. The temperature inversion throughout the height of the forest (Fig. 2) may have inhibited vertical exchange, especially when the large distance (40 m) is considered. The small fetch-to-height ratio (5:1), especially with a downhill slope facing the on-coming wind, may have created the mid-canopy blow-through or jet, shown in Fig. 2, which would suppress turbulence like the contracting section of a wind tunnel. Furthermore, the mid-canopy blow-through implies divergence and a concomitant net upward movement of mass and momentum (Kutzbach, 1961; Stearns and Lettau, 1963; and Shinn, 1971). This latter effect would also reduce the downward penetration of gusts since a gust front would itself enhance the mid-canopy jet.

Because the air sampling system in the Costa Rican forest drew point samples in a vertical line, the samples lacked horizontal space averaging. Since the gradient at the top of the forest was measured through the centre of an umbrella-shaped crown where vertical air motion would be restricted, it is likely that the CO_2 gradient obtained at the top of the forest was larger than would be expected for the whole forest system.

The turbulence shown in the wind records and the comparison of CO_2 regimes of the two forest systems suggests that the use of aerodynamic formulae may not be successful in computing fluxes in this Costa Rican forest. This is especially a problem since the upwind fetch-to-height ratio was only 5:1 and since information above the top of the forest had to be obtained by extrapolation.

V. SIMULATION OF CARBON DIOXIDE EXCHANGE

Monsi and Saeki (1953), de Wit (1965), Monteith (1965), Duncan *et al.* (1967), Stewart and Lemon (1969), Waggoner (1969), Duncan and Barfield (1970),

Stewart (1970), Uchimima and Inoue (1970), Allen *et al.* (1971), Duncan (1971), Lemon *et al.* (1971) and Sinclair (1971) have all used leaf photosynthesis models to predict photosynthesis in vegetation communities. Hozumi *et al.* (1969b) used the Monsi and Saeki (1953) approach to predict photosynthesis of a Cambodian forest which was described in detail by Hozumi *et al.* (1969a). So after experiencing uncertainties with the aerodynamic techniques in the Costa Rican study, a modelling approach was then tried.

A. Description of the SPAM Model

A soil-plant-atmosphere (SPAM) model developed by Stewart and Lemon (1969) and Stewart (1970) was tested thoroughly against measurements from an experimental cornfield (maize) reported by Lemon *et al.* (1971). SPAM was modified for simulating photosynthesis in the Costa Rican forest.

SPAM is basically a one-dimensional computer simulation model of vertical fluxes of micrometeorological entities in a plant community. The fluxes predicted include radiant energy, latent heat, sensible heat, and CO_2. Steady-state is assumed along the downwind direction so that there is no horizontal flux divergence. Two boundaries are defined for the model: the soil surface, and an imaginary plane at a reference height above the vegetation but still in the aerodynamic boundary layer. The vegetation is divided into layers (usually 15), each with equal leaf area index. Vegetation architecture (e.g., leaf angle distribution) in each layer must be specified.

Conditions at the soil (lower boundary) which must be specified are soil heat flux density, surface soil moisture potential, and rate of CO_2 evolution from the soil. The top boundary reference height for this 40-m forest was chosen to be 60 m. Several climate input conditions must be specified for this height, including: wind speed, air temperature, air vapour pressure, CO_2 concentration, short-wave radiation (photosynthetically active radiation, PAR, and near infra-red radiation, NIR), diffuse components of short-wave radiation, and net radiation. Latitude, longitude, and time must also be specified so that solar elevation angle can be computed.

Plant input factors which must be specified are: net photosynthesis rate and stomatal diffusion resistance as functions of irradiance, leaf boundary-layer resistance, and leaf respiration rate. The inputs are computed from equations or in sub-routines requiring more fundamental input parameters. The rectangular hyperbola describing the relation of stomatal diffusion resistance to irradiance (Stewart and Lemon, 1969; Stewart, 1970; Shawcroft, 1970; 1971) is:

$$r_s = r_s^* + \frac{\beta}{I + \beta(r_s' - r_s^*)} \tag{5}$$

where,

r_s = stomatal diffusion resistance;

r_s^* = minimum asymptotic resistance in bright light;

I = irradiance;

β = parameter which relates the approach of the rectangular hyperbolic to the origin. In our practical application, it relates the rate of decrease of stomatal resistance with light near zero illumination;

r_s' = cuticular resistance, the maximum stomatal diffusion resistance in zero illumination.

r_s^*, β, and r_s' are the fundamental input parameters for rate of opening of stomata with light. The boundary-layer resistance equation used (Stewart and Lemon, 1969; Stewart, 1970) was:

$$r = 0.604 \, (w/u)^{1/2}, \tag{6}$$

where,

w = leaf width ($= 5$ cm),

u = local wind speed (cm s^{-1})

The quantities w and 0.604 are the basic input parameters, with wind speed in the canopy computed from an exponential equation (Uchijima and Wright, 1964) (see Vol. 1, p. 73). Leaf respiration rate was computed from a measured rate at a measured reference temperature of 23°C (Stephens and Waggoner, 1970), assuming a Q_{10} value of 2 for respiration (see Waggoner (1969) for the Arrhenius type equation used).

Photosynthesis is computed from a sub-routine, so the inputs are much more complex. For the earlier SPAM simulations they included mesophyll resistance, carboxylation resistance, maximum rate of photosynthesis at high light intensities, and the slope of the photosynthesis/light curve in weak light.

1. Three Stages of SPAM

a. Light regime. The first stage of SPAM computes the light regime in the vegetation canopy. The model assumes a random horizontal distribution of leaf elements. The first sub-routine computes the solar elevation angle based on latitude, longitude, time of year, and time of day, based on the method of Robertson and Russelo (1968). The next sub-routine computes the penetration of direct-beam solar radiation and the penetration of diffuse sky radiation in nine annular sky-band widths of 10° from the horizon to the zenith. In the model, the sky is assumed to have equal brightness, but any brightness distribution as a function of elevation angle could be used. The brightness of the solar aureole is not considered separately; it is included with the direct solar beam. The method for computing penetration is basically that of de Wit (1965) and Duncan *et al.* (1967). The vegetation geometry (i.e. the angular distribution of leaf elements), as well as the source geometry, is important in predicting penetration of solar and sky radiation.

The next sub-routine computes the redistribution of radiation throughout the canopy of vegetation based on the optical properties of leaves and soil. Inputs of leaf transmissivity, leaf reflectivity, and soil reflectivity for both PAR and NIR radiation are required. The outputs of this sub-routine are direct-beam radiation, sky radiation, total downward diffuse radiation (includes direct-beam and sky radiation that has been intercepted by leaves and then scattered downward, as well as uninterrupted sky radiation), total downward radiation, and upward diffuse (reflected) radiation. The light model of S. E. Jensen (unpublished) was modified for use in this part of SPAM.

The last sub-routine of the radiation model divides the PAR load into 20 radiation load classes for each of the 15 layers of leaves, based on the maximum possible radiation load on a leaf exposed normally to the solar beam. Then the leaf elements of each layer are separated into 18 azimuth angle classes and 9 elevation angle classes. The nine leaf elevation angle classes are read as inputs. The amount of leaf in each azimuth angle class is assumed equal, but this information could be specified as well as the leaf elevation angle distribution (Loomis and Williams, 1969).

The fraction of leaf material in each radiation load class is computed, based on both the exposure of leaf elements to direct-beam radiation in the 18×9 angle classes and the diffuse components of transmitted and reflected radiation in the canopy. This information is stored for later use.

Stomatal diffusion resistance is computed as a function of radiation load in each of the 20 radiation load classes and 15 layers in this sub-routine. This information is also stored for later use.

b. Momentum budget and energy budget. The second stage of the SPAM model computes the momentum budget (wind) and energy budget of the vegetation community. The wind speed and eddy diffusivity above the canopy system is computed from the log-exponential wind profile of Swinbank (1964). This wind profile is coupled to an exponential decrease of wind in the canopy (Uchijima and Wright, 1964).

Next, the long-wave radiation energy balance between leaf layers was computed. Leaf-to-air transfer of heat and water vapour were computed based on air flow and radiation balance. Then the energy balance at the soil surface was solved, based on theory by Owen and Thompson (1963), and finally the air concentrations of water vapour and heat were computed by matrix algebra (see Vol. 1, p. 210). This energy balance portion was in an interactive loop which repeated itself up to five times. At least one transfer back to the wind and eddy diffusivity computations was made. After that, the interactions continued until the process converged to a suitable final approximation of heat and water vapour flux out of or into the system.

   c. Net photosynthesis. Leaf respiration rates were computed from leaf temperatures. Then a sub-routine computed net photosynthesis of leaves of each layer based on radiation exposure, stomatal diffusion resistance, boundary-layer resistance, background CO_2 concentration, and leaf respiration rate. Other factors in the photosynthesis response curve were read in as fixed input parameters so that only these variables predicted in the earlier stages remained.

2. *Output Summary*

Finally, the summary output of the model was printed, including the height distribution of wind, eddy diffusivity, boundary-layer resistance, net radiation components, and leaf temperature, as well as concentrations, sources, and flux densities of CO_2, water vapour, and air temperature.

B. Modifications of SPAM

Several modifications and additions to SPAM were used in the Costa Rican forest simulations. First, diffuse downward-moving radiation and diffuse upward-moving radiation in the canopy due solely to sky sources were separated from radiation due solely to scattered direct-beam radiation. This separation enabled us to predict the proportion of photosynthesis in the canopy supported by sky sources alone. Since the photosynthesis rate reported by Stephens and Waggoner (1970) for the dominant upper story tree, *Goethalsia*, reached saturation in relatively weak light, our hypothesis was that diffuse sky radiation would be very important in the total photo-synthesis of the system (de Wit, 1965).

   Another modification improved the method of computing stomatal diffusion resistance. Partial conductances for elements in each radiation load class were computed, summed, and reciprocated to give the effective stomatal diffusion resistance for each class. The method in the original program contained a small error.

One major modification was to remove the Swinbank (1964) log-exponential wind speed profile from the program and use a neutral wind profile for computing wind speed and eddy diffusivity above the surface of the forest. This change can be justified on four counts. First, previous experience showed that computation of the Monin-Obukhov length in the log-linear profile sometimes diverged, particularly under conditions where heat advection may occur, including early morning and late afternoon hours. Secondly, the forest presents a relatively rough surface and the departure of the neutral profile approximation under unstable temperature profile conditions is not as serious over rough surfaces as over smooth surfaces. Third, wind speed above the canopy was not measured and had to be extrapolated from measurements at the top of the canopy, i.e. the wind was not known precisely anyhow. Simulations by Waggoner (1969), Uchijima and Inoue (1970) and Sinclair et al. (1972) with steady-state models showed that eddy diffusivity could vary over a wide range without affecting CO_2 fluxes from canopies materially. Over the range of eddy diffusivities at the canopy top, i.e. $0\cdot2$ to $2\,m^2\,s^{-1}$, Waggoner's (1969) simulation model predicted a 10-fold variation in CO_2 depletion but less than 2% change in CO_2 flux density. Varying wind at the canopy top from $0\cdot22$ to $12\cdot25\,m\,s^{-1}$ produced a $7\cdot5$-fold change in boundary-layer resistance but only a 1% change in predicted photosynthesis. Larger changes in CO_2 flux density would be expected by changing these parameters simultaneously. Also, it is not clear how temperature effects on respiration were handled. However, with all interactions considered, it is not likely that large changes in CO_2 fluxes would have been predicted.

For eddy diffusivities over the range of $0\cdot025$ to $0\cdot3\,m^2\,s^{-1}$ Uchijima and Inoue (1970) found significant dependence of simulated net photosynthesis only at smaller diffusivities. Their simulations substantiated Waggoner's conclusions at the larger diffusivities. Hence, we concluded that an exact knowledge of wind speed is not necessary to obtain good simulations of photosynthesis by plant communities. However, profile shapes are strongly influenced by eddy diffusivity (Waggoner, 1969; Uchijima and Inoue, 1970; Allen et al. 1971; Allen, 1973).

The original SPAM model computed leaf temperatures for each of the leaf radiation load classes of the 15 layers. This procedure was simplified to compute only the average leaf temperature of each layer. The simplification was tested using records for a corn canopy on August 18, 1968. Photosynthesis was only $2\cdot4\%$ less, and the program was executed twice as fast. Also, in the original version of SPAM, bulk air temperature, vapour pressure, and CO_2 concentration were computed by standard matrix inversion (Stewart and Lemon, 1969; Stewart, 1970). This part of the program was changed to utilize a tridiagonal matrix (Carnahan *et al.*, 1969), since all coefficients in the matrix to be solved were zero except for the subdiagonal, diagonal, and superdiagonal values. This method also required much less central processor unit time and somewhat less core region on the IBM 360/65 system.†

The most significant changes and additions were made in the method used for computing photosynthesis. In the original SPAM model, a special sub-routine was used which involved detailed information on stomatal and boundary-layer resistances, as well as internal mesophyll and carboxylation resistances (Stewart and Lemon, 1969; Stewart, 1970). C. T. de Wit (personal communication, 1972) suggested using a best-fit hyperbolic approximation to light response curves. The photosynthesis sub-model of Stewart and Lemon (1969) and Stewart (1970) depends heavily on accurate knowledge of stomatal diffusion resistance. Since stomatal diffusion resistances were not known for the *Goethalsia* leaves, de Wit's suggestion was used here to fit the leaf chamber light response curves.

$$P = \frac{\alpha I P^*}{\alpha I + P^*} \tag{7}$$

where,

$$P = \text{gross photosynthesis rate,}$$

$$I = \text{irradiance (PAR),}$$

$$\alpha = \lim_{I \to 0} \left(\frac{\Delta P}{\Delta I} \right), \text{ or initial slope,}$$

$$P^* = \text{asymptotic } P \text{ at large } I.$$

We simulated gross photosynthesis and dark respiration separately, and obtained net photosynthesis by difference. Our operational definition of gross photosynthesis includes any internal recycling of CO_2 due to photo-respiration.

In the original SPAM model, photosynthesis predictions were made after the leaves were separated into 20 radiation load classes. In the Costa Rica forest simulations, photosynthesis predictions were made before separation into the 20 classes on the basis of the predicted radiation loads on each of the 18 azimuth angle classes × 9 leaf elevation angle classes. Then these 162 individual photosynthesis calculations were separated into the 20 radiation load classes. The values for each class were summed taking account of the relative amount of leaf area involved in each of the 162 computations.

---

† Mention of these or other proprietary products is for the convenience of the reader only and does not imply endorsement or preferential treatment by the U.S. Department of Agriculture.

In some cases, this refinement may be insignificant for total canopy predictions but it can be significant for weak radiation classes on the steep portion of the photosynthesis/light curves.

Components of photosynthesis were computed and printed in much more detail in these simulations than in previous SPAM simulations. The computations made for each of the 18 azimuth angle classes × 9 leaf elevation angle classes were summarized in two printouts. The first printout listed many components of photosynthesis *per unit leaf area* in each radiation load class of each of the 15 layers.

The second printout listed the components of photosynthesis *weighted* by the computed leaf area in each class. The values printed out were:

Gross photosynthesis
Respiration
Skylight photosynthesis
Skylight/Gross photosynthesis ratio
Diffuse light photosynthesis
Diffuse/Gross photosynthesis ratio
Gross—Skylight photosynthesis
Net—Skylight photosynthesis
Gross—Diffuse Photosynthesis
Net—Diffuse Photosynthesis
Net photosynthesis
Skylight/Net photosynthesis ratio
Diffuse/Net photosynthesis ratio
Absorbed PAR
Efficiency = Net/PAR

Skylight photosynthesis is defined as the rate of photosynthesis estimated for leaf elements in the total absence of direct-beam radiation. Diffuse light photosynthesis is the rate of photosynthesis which is predicted from diffuse radiation in the canopy. It includes direct-beam radiation scattered by leaves, but not the direct-beam radiation itself. Gross photosynthesis is the photosynthesis rate predicted before *dark* respiration is subtracted. Photorespiration is not considered as a separate input; it is handled implicitly in the hyperbolic relation between photosynthesis and PAR. Net photosynthesis is simply predicted gross photosynthesis minus predicted dark respiration. Efficiency of net photosynthesis is the ratio of the energy equivalent of photosynthesis to the predicted absorbed PAR.

The sum of each of the photosynthesis components (and the average of each of the ratios) for each *layer* was also computed. Finally, many of the above components were summed for each radiation load class across each of the 15 layers of the system. The following ratios and components of photo-synthesis for each radiation load class for the whole system were printed:

  Gross photosynthesis
  Respiration
  Skylight photosynthesis
  Net photosynthesis
  Skylight/Net photosynthesis ratio
  Diffuse light photosynthesis
  Diffuse/Net photosynthesis ratio
  Absorbed PAR
  Efficiency = Net/Absorbed PAR

Also, the totals and ratios for the whole canopy system were printed.

Following the detailed breakdown of photosynthesis into leaf radiation exposure classes, net photosynthesis is summed layer by layer. Finally, all components of the microclimate, and sources and fluxes are printed out as in the original SPAM model.

C. Inputs for Carbon Dioxide Exchange Simulations

Light response curves of leaves are the first requirements for simulation of photosynthesis of the Costa Rican forest. The second requirement is a description of PAR source geometry and source intensity. The third require-ment is a description of the vegetation geometry, which affects the PAR load on plants. In the revised SPAM model, these are the factors which have a direct input. Other climatic and plant (e.g. stomatal resistance) factors which affect temperature also affect net photosynthesis indirectly.

1. *Climate and Soil Input Parameters*

Table III lists climate input parameters incorporated in the simulation of net photosynthesis by the Costa Rican forest for a one-day period (November 14, 1967). Since the field data were obtained at irregular intervals (Table I), simulations were run at intervals of one hour throughout the day, beginning at 0712 apparent solar time and going to 1612 apparent solar time. (The runs were made on a clocktime basis and reported as apparent solar time after corrections for longitude (86·6°) and ephemeris of the sun (+ 15 min)). The global solar radiation, air temperature, and vapour pressure were supplied by Dr H. Trojer from the records of the climatological station at the Instituto Interamericano de Ciencias Agricolas de la OEA, Turrialba, Costa Rica. Sky brightness distribution was assumed uniform. Net radiation was taken as 70 % of total solar radiation. This first approximation was obtained by

Table III

Climate input parameters to the SPAM simulations of photosynthesis of the Bosque de Florencia. Temperature, vapour pressure, and global solar radiation were based on November 14, 1967, data from the climate station located at the Instituto Interamericano de Ciencias Agricolas de la OEA, Turrialba, Costa Rica

Apparent solar time	Air[b] Wind[a] speed (m s^{-1})	Air[b] Tempera-ture (°C)	Air[b] Vapour[c] pressure (mbar)	Total solar radiation (W m^{-2}) 10% Diffuse	30% Diffuse	60% Diffuse	Total PAR (400 to 700 nm) (W m^{-2}) 11% Diffuse	33% Diffuse	66% Diffuse	Diffuse (Sky) PAR (400 to 700 nm) (W m^{-2}) 11% Diffuse	33% Diffuse	66% Diffuse
0712	3·00	19·7	17·7	175	140	95	85	70	40	35	35	30
0812	3·60	22·9	17·6	420	340	230	200	160	105	35	55	70
0912	3·96	23·2	18·4	630	515	340	300	245	160	35	85	105
1012	4·24	21·2	18·6	810	665	440	390	315	210	40	105	140
1112	4·40	25·0	21·5	840	685	455	405	330	215	40	110	140
1212	4·52	25·1	21·5	980	795	530	470	385	250	50	125	168
1312	4·48	24·5	21·9	925	755	505	440	365	235	50	120	154
1412	4·40	23·0	23·4	805	655	440	385	315	210	40	105	140
1512	4·24	21·2	23·0	455	370	245	215	175	120	35	55	75
1612	3·92	20·4	23·0	210	175	110	95	85	55	35	35	35

[a] Extrapolated to 60 m.

[b] Decreased by 0·2 °C to account for adiabatic lapse rate and decrease in temperature from 40 m up to the reference height at 60 m.

[c] Decreased by 0·1 to account for profile gradient decrease from 40 m up to the reference height at 60 m. Explanation in text.

comparing records with climate station total solar radiation. Soil heat flux, a minor component of the energy budget, was estimated at 1 % of net radiation. Wind speed at the reference height (60 m) was computed beforehand using the 40-m anemometer values with $d = 36$ m and $z_0 = 2$ m.

As the soil and litter surface of the Costa Rican forest was always damp, we estimated the surface moisture tension at only -5 bars.

2. *Plant Canopy and Leaf Scale Inputs*

Leaf area index was approximately 3·3. For the upper canopy, $L = 2$ and for the lower canopy $L = 1$ (see Fig. 3, Lemon *et al.*, 1970, and Stephens and Waggoner, 1970). A value of $L = 0·3$ was estimated for the mid-canopy since it was not devoid of green vegetation. The distribution of leaf elevation angles was not measured, but a planophile distribution was used. The relative amount of leaf elements in the leaf angle classes were:

0°–10°	0·20
10°–20°	0·17
20°–30°	0·15
30°–40°	0·13
40°–50°	0·11
50°–60°	0·09
60°–70°	0·07
70°–80°	0·05
80°–90°	0·03

The shape of this cumulative distribution of leaf elevation angles is similar to that described by de Wit (1965) for an idealized planophile distribution, but it is not as extreme as de Wit's example. Indeed, it resembles an 'upside down' spherical distribution (Nichiporovich, 1961).

Field observations and photographs showed that the leaves of the tree crowns were not clumped or randomly distributed but were arranged in a *regular* dispersion (i.e. leaves tended to be uniformly distributed in horizontal space). Based on Nilson's (1971) treatment of gap frequency in plant stands, we estimated a dispersion factor of 1·2 to simulate radiation penetration.

Measurements of stomatal diffusion resistance were not available. We estimated minimum stomatal diffusion resistance at high illumination (r_s^*) to be 2·5 s cm^{-1}, consistent with minimum values reported by Woods and Turner (1971) and Gee and Federer (1972) for temperate broad-leaf trees.

Dark respiration rates at a reference temperature of 23°C are summarized in Table IV (Stephens and Waggoner, 1970). Equation [7] described the hyperbolic photosynthesis–light response curve. Values of α and **P*** were computed from the curves of Stephens and Waggoner (1970). These values were expressed in PAR rather than total short-wave radiation and were corrected for the effectiveness of the leaf chamber incandescent lamps used by Stephens and Waggoner.

Dark respiration was added to the net photosynthesis data so that α and $\mathbf{P}^*$ describe curves for gross photosynthesis. $\mathbf{P}$ was chosen so as to force the hyperbolic curve to come to the experimentally determined maximum rates of photosynthesis at a PAR value of 560 W m^{-2} rather than at infinite PAR. If we set $\mathbf{P} = \frac{1}{2}\mathbf{P}^*$ in Eq. (7) and solve for α, then $\alpha = \mathbf{P}^*/\mathbf{I}_{\frac{1}{2}}$ where $\mathbf{I}_{\frac{1}{2}}$ is the PAR when $\mathbf{P} = \frac{1}{2}\mathbf{P}^*$. ($\alpha$ could also be determined from the initial slope of the leaf light response curve.) The leaf light-curve parameters of Goethalsia, Cecropia, and a lower canopy plant, Croton, are summarized in Table IV.

Table IV

Leaf metabolic parameters of three species used in simulating net photosynthesis of the Bosque de Florencia. Data derived from Stephens and Waggoner (1970).

Plant	Maximum photosynthesis (g m^{-2} h^{-1})	Initial slope (μg J^{-1})	Respiration rate (g m^{-2} h^{-1})	Reference temperature of respiration (°C)
Goethalsia	0.75	7.1	0.068	23
Cecropia	2.30	12.3	0.082	23
Croton	0.40	—	0.032	23

Figure 6 shows the light response curves of two of the three species in the Costa Rican forest. Simulations were run using both the Goethalsia curve and the Cecropia curve for the upper and middle stories of the forest. In both cases, the bottom four layers of the SPAM simulation were given maximum photosynthesis rates typical of the bottom-canopy vegetation (e.g., Croton). The simulations of the Goethalsia forest were intended to show what the actual forest was doing; whereas the simulations using the Cecropia data were intended to show photosynthesis rates in a pioneer forest of Cecropia. Lugo (1970) reported maximum photosynthesis rates for Puerto Rican Cecropia peltata seedlings that were about half of ours.

3. Diffuse Sky–Source Simulations

Runs were made assuming a ratio of diffuse (sky) to total radiation of 10%, 30%, and 60%. Observations at Ithaca, New York, U.S.A. by C. S. Yocum, L. H. Allen, and E. R. Lemon (unpublished, 1961) showed that sky radiation is richer in PAR than NIR under all sky/total ratios (see Vol. 1, p. 19). So the equivalent ratios of diffuse to total PAR were approximated by 11%,

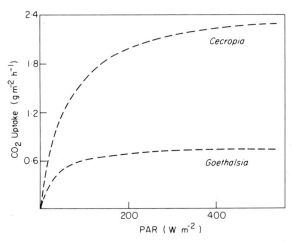

Fig. 6. Light response curves of *Goethalsia*, a successional dominant, and of *Cecropia*, a successional pioneer. Adapted from curves published by Stephens and Waggoner (1970). The ordinate
represents gross photosynthesis (net plus dark respiration) and the abscissa is photosynthetically
active radiation (PAR).

33%, and 66%. From total and diffuse (shade-band) Eppley pyranometer
records at Ithaca, New York, U.S.A., plots of percent diffuse radiation versus
total radiation were drawn at hourly intervals based on 15 days of records
from July 1 to August 21, 1970. These measurements were summarized in a
linear equation which described the diffuse/total ratio, x, in terms of the
total solar radiation, S_t, and the total solar radiation at 10% diffuse radiation,
S_t^*. Our linear approximation extrapolated to a fictitious 120% diffuse
radiation as $S_t \to 0$.

$$x = 1.2 + [(0.1-1.2)/S_t^*]S_t. \tag{8}$$

Rearranging, S_t becomes:

$$S_t = (1.2 - x)[S_t^*/(1.2-0.1)]. \tag{9}$$

Then the diffuse radiation, S_d, becomes:

$$S_d = xS_t. \tag{10}$$

The equations approximating the PAR portion of the solar spectrum are
(primes used on symbols):

$$x' = 1.3 + [(0.11-1.3)/S_t^{*\prime}], S_t' \tag{11}$$

$$S_t' = (1.3 - x')[S_t^{*\prime}/(1.3-0.11)], \tag{12}$$

$$S_d' = x'S_t'. \tag{13}$$

Our diffuse PAR extrapolated to a slightly higher S_d' of 130% as $S_t' \to 0$. The portion of total radiation in the PAR range was 48% (Moon, 1940), thus $S_t^{*'}$ was set at 48% of S_t^* in the simulations.

D. Results of Carbon Dioxide Exchange Simulations

1. *Distribution of Leaf Element Radiation Load*

Figure 7 shows the relative amount of leaf elements in each radiation load class at 1212 apparent solar time for three layers of the 15 simulated layers (layer 1, layer 4, and layer 9). Simulations were run for 10%, 30%, and 60%

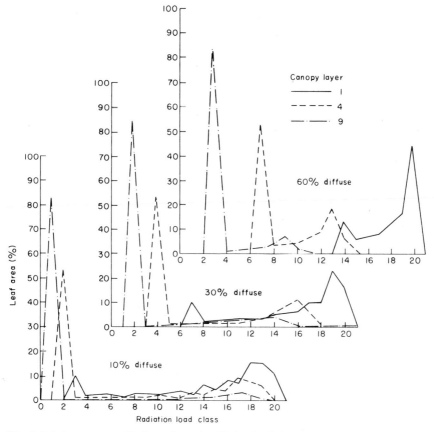

Fig. 7. Relative amount of leaf area in each radiation load class at 1212 apparent solar time, November 14, 1967, Turrialba, Costa Rica. The maximum PAR load is 540 W m^{-2} for 10% diffuse sky-source radiation, 430 W m^{-2} for 30% diffuse sky-source radiation, and 270 W m^{-2} for 60% diffuse sky-source radiation.

diffuse radiation to investigate the effects of sky sources. (In this section, as in the last section, 'diffuse' radiation implies light scattered from the sky.) The maximum possible PAR load was 540 W m^{-2} for the 10% case, 430 for the 30% case, and 270 for the 60% case. The maximum values included the sum of the diffuse load from above and reflected sources from below, as well as the direct-beam component.

The peaks at the left of each curve represent primarily the proportion of leaves that are shaded from direct solar rays. Within the top layer of the 15-layer simulation, only a small fraction of the leaf area is shaded, whereas over 80% of the leaf area elements is shaded in layer 9. In the 10% diffuse case, the direct-beam load covers radiation load classes 4 to 20. The various combination of 18 azimuth angle classes and 9 elevation angle classes give a distribution of leaf-to-sun angle classes from grazing angles to normal angles of incidence.

Sinclair (1971), Sinclair and Lemon (1973) and Allen and Lemon (1972) showed that a flat-plate traversing radiation sensor would yield two peaks in the radiation distributions, one peak for the shade, and one for sunflecks. When all azimuth and elevation angles are considered, the sunfleck peak becomes flattened and spread over a wider range of radiation loads.

In the 60% diffuse radiation case, the range of illumination loads in each layer is much smaller (Fig. 7). The leaf areas under the shade peaks are shifted upscale for two reasons; because a higher percentage of radiation is from the diffuse source (60%) and because the maximum radiation load is lower (270 compared with 540 W m^{-2}). This latter fact should be considered when comparing the leaf area distribution versus radiation load class in Fig. 7.

Figure 8 shows distributions of PAR versus radiation load class for the 10%, 30%, and 60% diffuse sources at 1212 apparent solar time. The information was obtained by multiplying the amount of leaf area in each of the 18 × 9 angle classes by the appropriate radiation load. Then these values were sorted into radiation load classes and summed for each layer. This figure is similar to Fig. 7 except that the peaks on the left side (shaded) are smaller and the peaks on the right side (sunfleck) are larger.

2. *Distribution of Photosynthesis in PAR Load Classes*

The radiation absorbed does not show the actual photosynthesis rates. Fig. 9 shows net CO_2 uptake distributions as a function of radiation load class. Since the curve relating photosynthesis to PAR for *Goethalsia* shown in Fig. 6 rises sharply in weak light and flattens in bright light, one would expect that leaves exposed only to diffuse sky radiation would exhibit higher photosynthesis rates than would be shown by the radiation load

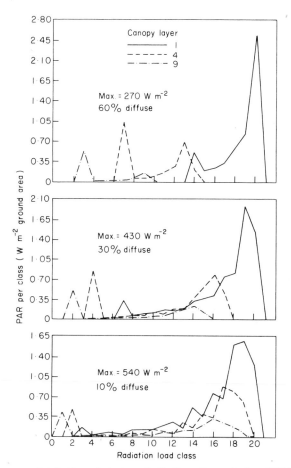

Fig. 8. Frequency distribution of photosynthetically active radiation (PAR) per unit ground area received by leaves in each radiation load class as a function of radiation load class. This figure represents the product of radiation load times relative leaf area of the radiation load classes. 1212 apparent solar time, November 14, 1967, Turrialba, Costa Rica.

curves of Fig. 8. This can be seen in Fig. 9. Figure 9 also shows that the net CO_2 uptake in weak light for 30% and 60% diffuse radiation is much greater than for the 10% diffuse case. There is graphical evidence that the canopy response to bright diffuse sky light is greater than for weak diffuse sky light. (See Table III for the energies.)

3. Daily Course of Photosynthesis

Figure 10 summarizes the daytime net CO_2 uptake for 10%, 30%, and 60% diffuse radiation. Information for the pioneer species, *Cecropia*, is also given.

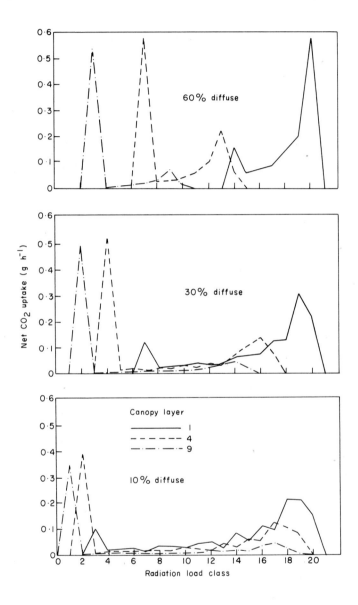

Fig. 9. Net photosynthesis per m^2 land area predicted by the SPAM model as a function of radiation load class for three canopy layers under three diffuse radiation source conditions. The maximum PAR load was 540 W m^{-2} for 10% diffuse, 430 W m^{-2} for 30% diffuse, and 270 W m^{-2} for 60% diffuse sky-source radiation. 1212 apparent solar time, November 14, 1967, Turrialba, Costa Rica.

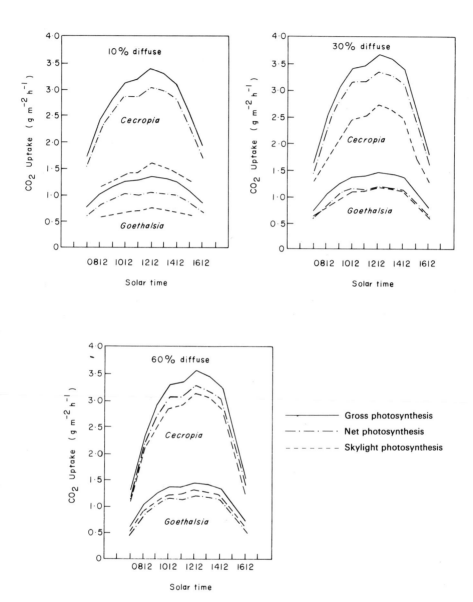

Fig. 10. Diurnal predictions of CO_2 uptake by *Goethalsia* and by *Cecropia* under 10 %, 30 %, and 60 % sky-source radiation conditions. $L = 3.3$. In addition to gross photosynthesis and net photosynthesis, the amount of photosynthesis that would be supported by skylight sources alone are shown. Simulation based on climatological data obtained at Turrialba, Costa Rica, November 14, 1967.

Gross photosynthesis, net photosynthesis, and skylight photosynthesis (photosynthesis which could have been supported by diffuse skylight sources *only*) are shown. Even in the 10 % diffuse case, about 55 % of the gross photosynthesis of *Goethalsia* could be supported by skylight only. Skylight would support about 45 % of the gross photosynthesis of the *Cecropia*.

The most significant fact which emerges from Fig. 10 is that the total photosynthesis rate of the whole canopy remains high even when radiation is 60 % diffuse despite the overall decrease in global radiation.

Figure 11 presents daily net photosynthesis at hourly intervals for each layer and the sum of the net photosynthesis over all 15 layers. The 60 % diffuse case is depicted. The lowest diurnal course represents the predicted

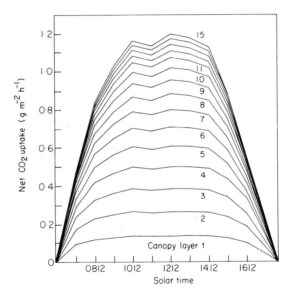

Fig. 11. Diurnal course of simulated net photosynthesis (60 % diffuse radiation) for each of the 15 layers of the simulated *Goethalsia* forest at Turrialva, Costa Rica, November 14, 1967. The series of curves, from bottom to top of the figure, represent the added contributions from layer 1 (at the top of the canopy) to layer 15 (at the bottom of the canopy). The area between the curves represents the simulated net photosynthesis of the respective layers.

photosynthesis from layer 1, the next highest curve for layer 1 plus 2, the third curve for layers 1 plus 2 plus 3, etc. The space between each curve represents the predicted photosynthesis for each of the layers. The spacings get thinner toward the top of the figure, as the net photosynthesis from the deeper canopy layers are added to the upper layer contributions. The four top increments are thinner still since these layers simulate photosynthesis of the shaded lower of the Costa Rican forest where maximum photosynthesis rates are least. The small dip in overall photosynthesis at 1112 occurred because real radiation at Turrialba on November 14, 1967, decreased during this hour due to temporary cloudiness.

E. Diurnal and Annual Carbon Dioxide Budgets

1. *Goethalsia, The Successional Dominant*

A diurnal CO_2 budget was prepared from the leaf photosynthesis and leaf respiration simulations (Table V). The SPAM simulations are a good guide for these components of the CO_2 exchange of the system, but give no information on twig, branch, trunk, and root respiration, and no information on decay of surface detritus or subsurface detritus. Müller and Nielsen (1965) reported the following annual production balance of an Ivory Coast forest based on dry matter and respiration measurements.

		tonnes ha^{-1} a^{-1}
Net biomass increment		9·0
(roots, trunks, branches, leaves)		
Respiration		39·1
roots	3·7	
branches, trunks	18·5	
leaves	16·9	
Biomass death and decay		4·4
roots	0·4	
branches, trunks	1·9	
leaves	2·1	
Gross production		52·5

The leaf area index of this forest was 3·16. The gross dry matter production would require an uptake of approximately 0·66 g CO_2 per m^2 leaf surface (one side) per hour for a period of 12 hours per day to support this gross production. On a unit land area basis the required CO_2 uptake rate would be about 2·1 g m^{-2} h^{-1}. Leaf respiration from Müller and Nielsen (1965) would be about 43% of the total living tissue respiration and about 39% of living tissue respiration plus biomass decay.

Kira et al. (1967), Kira and Shidei (1967), Kira (1969), and Kira and Ogawa (1971) reported the following dry matter budget for a Thailand evergreen seasonal forest near Khao Chong.

		tonnes $ha^{-1} a^{-1}$
Net biomass increase		5·3
(roots, trunks, branches, leaves)		
Respiration		94·7
roots	5·6	
branches, trunks	32·1	
leaves	57·0	
Biomass death and decay (litter)		23·2
roots	—	
branches, stems, etc.	11·4	
leaves	11·8	
Gross production		123·2

The analysis by Kira *et al.* (1967) showed that 60% of the respiration by living tissue of the Khao Chong forest was due to leaves. It also showed that leaf respiration was 48% of the respiration by living tissue plus detritus. Kira *et al.* pointed out that leaf respiration is approximately 50% of total respiration by living tissue in many tropical forests. We used this approximation in filling in the diurnal CO_2 budget of Table V. The night-time respiration of the leaves was estimated from the respiration rate computed at 0712 apparent solar time. This period had the lowest temperature of any of the simulation periods.

Respiration from other components of the forest, twigs, branches, trunks, and roots, were assumed equal to leaf respiration (Yoda, 1967).

Table V shows that the predicted net production was greatest for the 30% diffuse case, and was almost as large for 60% diffuse radiation. Photosynthetic efficiency based on PAR was computed at several stages in the diurnal CO_2 budget. During the 11-hour day the net photosynthetic efficiency was 0·96%, 1·28%, and 1·91% for 10%, 30%, and 60% diffuse radiation, respectively.

The average annual daily PAR (based on Turrialba climate station data) was 7·4 MJ m^{-2} day^{-1}, half way between 8·8 MJ m^{-2} day^{-1} for 30% diffuse radiation and 5·7 MJ m^{-2} day^{-1} for 60% diffuse. On the basis of these estimates and previous figures, we conclude that length of day may be more important for photosynthesis than irradiance in this type of forest. In effect, long day-lengths would simply extend the midday photosynthetic rates over a longer time span, since early morning and late afternoon irradiance near the equator would be very similar both for the longest and shortest days. The Smithsonian Meteorological Tables (List, 1958) show that day-length varies from 12 hr 43 min to 11 hr 32 min over the season at 10° N latitude. So the maximum difference is only 1 hr 11 min.

The net production listed in Table V, 3·6 g CO_2 per m^2 land per day, should be allocated to biomass increment, detritus, and grazing, but we have

Table V

Diurnal CO_2 budget, Bosque de Florencia, Turrialba, Costa Rica, November 14, 1967, based on SPAM simulations of *Goethalsia* photosynthesis. Photosynthesis and respiration in units of g CO_2 per m^2 land area per day

	Percent Diffuse Sky Source		
	10%	30%	60%
PAR ($MJ\ m^{-2}\ day^{-1}$)	10.7	8.8	5.8
Gross photosynthesis (11-h day)			
Upper canopy	10.5	10.9	10.7
Lower canopy	1.1	1.4	1.2
Total	11.6	12.3	11.9
Efficiency	1.22%	1.59%	2.33%
Net photosynthesis (11-h day)			
Upper canopy	8.3	8.9	8.8
Lower canopy	0.7	1.0	0.8
Total	9.1	9.9	9.7
Efficiency	0.96%	1.28%	1.91%
Leaf respiration			
Upper canopy (11-h day)	2.2	2.0	1.9
Lower canopy (11-h day)	0.3	0.3	0.3
Subtotal	2.5	2.3	2.2
Upper canopy (12-h night)	1.7	1.7	1.7
Lower canopy (13-h night)	0.3	0.3	0.3
Subtotal	2.0	2.0	2.0
Total	4.5	4.3	4.2
Net photosynthesis (24-h day)			
Upper canopy	6.7	7.2	7.2
Lower canopy	0.5	0.8	0.6
Total	7.2	8.0	7.8
Efficiency	0.75%	1.03%	1.52%
Respiration, trunk, stems, roots			
Day	2.5	2.3	2.1
Night	2.0	2.0	2.0
Total	4.5	4.3	4.1
Net production, biomass plus detritus	2.7	3.7	3.6
Efficiency	0.28%	0.47%	0.71%

no way to estimate this partition properly. The ratio of annual net biomass increment to detritus ranges from 2.0 for the Ivory Coast forest reported by Müller and Nielsen (1965) to 0.23 for the Khao Chong, Thailand forest reported by Kira *et al.* (1967). Lieth (1971) estimated herbivore consumption to be 8 to 12% of net production in rain-green forests and tropical rain forests. However, there were no large herbivores in this forest.

Table VI

Annual CO_2 and dry matter equivalent budget, Bosque de Florencia, Turrialba, Costa Rica, based on SPAM simulations of *Goethalsia* photosynthesis. Detritus estimates from Müller and Nielsen (1965). Units are tonnes ha^{-1} yr^{-1}. Average PAR = 7·4 MJ m^{-2} day^{-1} Dry matter = 0·61 × CO_2. Values for 60% diffuse radiation case from Table V were used in this summary

	CO_2	Dry matter	Efficiency (%)
Gross photosynthesis	45·8	27·9	1·93
Leaf respiration	16·3	9·9	
Net photosynthesis (24 h)	29·5	18·0	1·24
Respiration, other			
Day	9·0	5·5	
Night	7·2	4·4	
	16·2	9·9	
Net production (biomass and detritus)	13·3	8·1	0·56
Detritus (from Müller and Nielsen)	7·2	4·4	
Biomass increment	6·1	3·7	0·26
Weighted detritus	3·8	2·3	
Biomass increment	9·5	5·8	0·40

An annual CO_2 and dry matter budget was computed from the data of Table V. Values for 60% diffuse radiation were used in Table VI. Gross photosynthesis was adjusted by adding one hour of midday gross photosynthesis and subtracting 11 minutes of midday rates to adjust the November 14 rates to the maximum and minimum day-length rates, respectively. These were averaged and multiplied by 365 to give an annual value of 45·8 tonnes ha^{-1} of CO_2, or 27·9 t ha^{-1} of dry matter. Since much of the year is warmer than mid-November, we used the maximum computed leaf respiration rates of Table V (4·46 g m^{-2} h^{-1}) to yield annually 16·3 t ha^{-1} of CO_2 respired or 9·9 tonnes ha^{-1} of dry matter equivalent. Net photosynthesis defined as gross photosynthesis minus leaf respiration, was 29·5 t ha^{-1} of CO_2 and 18·0 t ha^{-1} of dry matter.

Respiration from other plant parts was estimated to be equivalent to the leaf respiration rate (Yoda, 1967), but was partitioned between day and night based on increased proportion of daylight and higher daytime temperatures.

The final summary yielded 8·1 t ha^{-1} yr^{-1} for biomass plus detritus increment. In terms of incident PAR, the efficiency of net production was 0·56%.

Biomass increment can be estimated from Müller and Nielsen (1965). If we assume a loss through detritus of 4·4 t dry matter ha^{-1} yr^{-1} the biomass gain by the Costa Rican forest would be $3.7 \, t \, ha^{-1} \, yr^{-1}$. On the other hand, if we weight the detritus by the gross production (or gross photosynthesis) of the Costa Rican forest relative to the Ivory Coast forest, the detritus becomes $2.3 \, t \, ha^{-1} \, yr^{-1}$ of dry matter, leaving a biomass increment of $5.8 \, t \, ha^{-1} \, yr^{-1}$.

The annual gross photosynthesis ($=$ gross production) computed for the Costa Rican forest, $27.9 \, t \, ha^{-1}$ of dry matter, was much lower than the $52.5 \, t \, ha^{-1}$ quoted by Müller and Nielsen (1965) for the Ivory Coast forest or the $123.2 \, t \, ha^{-1}$ given by Kira et al. (1967) for a Thailand tropical forest.

2. Cecropia, A Pioneer

Table VII presents a tentative diurnal CO_2 budget for the pioneer *Cecropia* based on SPAM predictions of photosynthesis for 10 one hour periods on November 14, 1967. The predicted net production is over four times that of *Goethalsia*. The same assumption that respiration of living tissue other than leaves would be about the same amount as leaf respiration was used. However, one would expect that more growth would occur with the larger photosynthesis rates of *Cecropia* and that more respiration due to growth would occur. McCree (1969) and Hesketh et al. (1971) found that respiration was more strongly related to photosynthesis than to biomass for annual crops with low biomass. Yoda (1967) pointed out that respiration rates were

Table VII

Cecropia diurnal CO_2 budget, based on climate data of November 14, 1967, taken at the Bosque de Florencia, Turrialba, Costa Rica. Photosynthesis and respiration units are $g \, CO_2$ per m^2 land area per day

	Percent diffuse sky source		
	10%	30%	60%
PAR ($MJ \, m^{-2} \, day^{-1}$)	10.7	8.8	5.8
Gross photosynthesis	27.5	29.2	27.5
Efficiency	2.89%	3.75%	5.38%
Net photosynthesis (10-h day)	24.8	26.6	25.1
Efficiency	2.60%	3.41%	4.91%
Leaf respiration (10-h day)	2.7	2.6	2.4
Leaf respiration (14-h night)	2.6	2.6	2.5
Net photosynthesis (24-h day)	22.2	24.0	22.6
Efficiency	2.33%	3.09%	4.43%
Respiration, trunks, stems, roots	5.4	5.2	4.9
Net production (biomass plus detritus)	16.8	18.8	17.7
Efficiency	1.76%	2.42%	3.47%

proportional to biomass in finer branches, and somewhat related to biomass for large branches and trunks. We did not have specific biomass information available on the Costa Rican forest, however.

The day-time efficiencies of net photosynthesis were 2·60%, 3·75%, and 5·38% for the 10%, 30%, and 60% diffuse light cases, respectively.

Table VIII

Cecropia annual CO_2 and dry matter budgets based on SPAM simulations. Detritus estimates from Müller and Nielsen (1965) and Kira *et al.* (1967). Average PAR = 7·4 MJ m^{-2} day^{-1} at Turrialba, Costa Rica. Estimates based on the 60% diffuse sky-source case of Table VII. Units are tonnes ha^{-1} yr^{-1}

	CO_2	Dry matter	Efficiency (%)
Gross photosynthesis (or gross production)	100·4	61·2	4·22
Leaf respiration (day)	10·0	6·1	
Net photosynthesis (day)	90·4	55·1	3·80
Leaf respiration (night)	9·5	5·8	
Net photosynthesis (24-h day)	80·9	49·3	3·40
Respiration (trunks, branches, roots)	19·6	11·9	
Net production (Biomass plus detritus)	61·3	37·4	2·58
Detritus (Müller and Nielsen)	7·2	4·4	
Biomass increment	54·1	33·0	2·27
Detritus (Kira *et al.*)	38·0	23·2	
Biomass increment	23·3	14·2	0·98

Table VIII is an annual CO_2 and dry matter budget for a hypothetical *Cecropia* forest. Predicted gross photosynthesis (or gross production) is 61·2 t dry matter ha^{-1} yr^{-1}, slightly more than that for the Ivory Coast forest, but only about one-half that of the Khao Chong, Thailand, forest. If the earlier assumption about living tissue respiration (excluding leaves) is correct, then this simulated forest would yield 37·4 t ha^{-1} of biomass plus detritus. In the table, two assumptions were made about detritus to obtain an estimate of annual biomass increment. First 4·4 t ha^{-1} were subtracted, based on evidence from Müller and Nielsen (1965), which left 33·0 t ha^{-1}. If this increment were correct, the forest would grow extremely fast. The second assumption was that 23·2 t ha^{-1} of detritus (based on Kira *et al.*, 1967) was subtracted from the net production. This would leave a biomass increment of 14·2 t ha^{-1}, still a substantial increment.

F. Photosynthesis of a Cambodian Forest

Hozumi et al. (1969b) predicted gross production (=gross photosynthesis) of an evergreen season tropical rain forest at Chékô, Cambodia based on the Monsi and Saeki (1953) light attenuation profile through the system. This forest structure was considered to consist of four layers, with the first layer ($L = 2.02$) composed of the tallest crowns, the second and third layers ($L = 3.78$) composed of mid-canopy species, and the fourth layer ($L = 1.57$) made up of the shade-tolerant ground layer smaller trees and herbaceous plants. Their photosynthesis summation formula predicted 52 g CO_2 m^{-2} day^{-1} gross production for the total canopy system ($L = 7.37$). This value is much larger than our prediction for a *Cecropia* forest (plus ground-level vegetation) with $L = 3.3$. Even for the top story, the Hozumi et al. (1969b) summation was 30 g m^{-2} day^{-1} for an L of only 2.02. The daily rate for the *Cecropia* simulations was about 29 g m^{-2} day^{-1}. The maximum rate of leaf photosynthesis for *Cecropia* was 2.3 g m^{-2} h^{-1}, whereas it was only 1.7 g m^{-2} h^{-1} for the top story trees (mainly *Dipterocarpus* cf. *dyeri* (chouteal) or *Anisoptera* cf. *glabra* (phdiek) of the Cambodian forest. Hozumi et al. (1969b) used a hyperbolic light response curve as we did in the *Cecropia* simulations. Their form of the hyperbolic equation (Monsi and Saeki, 1953) was:

$$\mathbf{P} = \frac{b\mathbf{S}}{1 + a\mathbf{S}}, \tag{14}$$

where b and a are constants.

The equation used by Saeki (1960) to give the average illumination on a leaf surface at a leaf area index of L below the top of the canopy was:

$$\mathbf{S} = \mathbf{S}(0)\mathcal{K} \exp(-\mathcal{K}L)/(1 - \tau), \tag{15}$$

where,

$\mathbf{S}(0)$ = short-wave radiation at top of canopy,

$\mathcal{K}$ = canopy light attenuation coefficient,

τ = leaf transmissivity.

Equation (15) was substituted into Eq. (14), and integrated with respect to L to yield gross photosynthesis.

$$\mathbf{P} = \frac{b}{\mathcal{K}a} \ln\left[\frac{(1 - \tau) + \mathcal{K}a\mathbf{S}(0)}{(1 - \tau) + \mathcal{K}a\mathbf{S}(0)\exp(-\mathcal{K}L)}\right]. \tag{16}$$

This equation basically predicts photosynthesis at the *average* radiation load that is predicted from the Monsi and Saeki (1953) exponential light-penetration equation. Photosynthesis light response curves, such as those in Fig. 6, are always nonlinear and are always convex upward (Saeki, 1960). Thus, it is apparent that photosynthesis computed from an average irradiance would be greater than that computed from a series of radiation load classes ranging from bright light to weak light. The SPAM model predictions should be more realistic since photosynthesis is predicted from a series of leaf exposures as was illustrated earlier in Figs 6, 7, and 8.

VI. SUMMARY AND CONCLUSIONS

The SPAM model depends primarily upon radiation interception and light response curves. These factors are one step closer to the photosynthetic act than bulk aerodynamic diffusion, and should be more reliable in this particular study. Waggoner (1969), Uchijima and Inoue (1970), and Sinclair et al. (1972) showed that bulk aerodynamic diffusion is generally not a limiting factor in CO_2 uptake.

The SPAM simulations gave distributions of leaf area, light interception, and photosynthesis as a function of radiation load. The model showed that stands with leaves which saturate in weak light are relatively insensitive to levels of illumination, especially since the diffuse (sky) component increases as the direct-beam radiation is scattered by haze or cloud. The simulations showed that skylight sources could support a large fraction of the total canopy photosynthesis, even on very clear days. The lower effectiveness of the blue end of the PAR spectrum (McCree, 1972) would decrease the apparent capability of the diffuse sky sources to promote photosynthesis, but it would still be remarkably effective.

SPAM simulations gave the diurnal course of photosynthesis for each of the 15 model layers of the system as well as the diurnal course for the whole system. For the *Goethalsia* forests, the efficiencies of PAR for gross photosynthesis was 1·22%, 1·59%, and 2·33% for the 10%, 30%, and 60% diffuse radiation cases, respectively. The efficiencies were 2·89%, 3·75%, and 5·38%, respectively, for the *Cecropia* simulations. Likewise, comparison of net production predictions of *Cecropia*, the fast growing pioneer, with *Goethalsia*, the persistent and eventual dominant, shows that *Cecropia* should outstrip *Goethalsia* in biomass accumulation. The gross daily photosynthesis of $29 \, \mathrm{g \, m^{-2}}$ ground area predicted by the SPAM model for *Cecropia* is more reasonable than the figure of $52 \, \mathrm{g \, m^{-2}}$ predicted by Hozumi et al. (1969b) for a Cambodian forest, especially since the maximum photosynthesis rates of leaves for the Cambodian forest were smaller $(1·7 \, \mathrm{g \, m^{-2} \, h^{-1}})$ than for *Cecropia* $(2·3 \, \mathrm{g \, m^{-2} \, h^{-1}})$. However, the Cambodian forest did have a leaf area index of 7·37 so perhaps it could support more photosynthesis than the Costa Rican *Cecropia* assuming $L = 3·3$. Kira *et al.* (1969) claim that clumped crowns and the low average leaf area density of the tall forests that they studied allowed more light to penetrate and hence larger leaf area indices to develop.

Another factor which has not been adequately handled by the SPAM modelling is the penumbra effect (Miller and Norman, 1971a; 1971b). Norman *et al.* (1971) compared the effects of computing photosynthesis at the bottom of a sumac canopy (*Rhus typhina* L.) and sunflower canopy (*Helianthus annuus* L.) under the following three assumptions:

(1) Photosynthesis based entirely on the average light transmission. (This predicted photosynthesis would be somewhat more than rates predicted by the method of Monsi and Saeki (1953) which are based on light absorbed per unit L.)

(2) Photosynthesis based on a diffuse background radiation plus geometric sunflecks. (This prediction is similar to the SPAM approach except that the whole range of radiation loads resulting from a distribution of leaf azimuth angles and leaf elevation angles was not taken into account.)

(3) Same as assumption (2) except that sunfleck penumbra was taken into account also.

For the sumac canopy, rate (1): rate (2) was 2·25 and rate (3): rate (2) was 1·52. The sunflower canopy with larger leaves showed less penumbral effect, the ratios being 2·01 and 1·12 respectively. Since Norman *et al.* (1971) did not take into account the complete range of leaf radiation loads based on an array of leaf angles, their computation under assumption (2) are lower than the SPAM computations would have been. Distributing direct-beam radiation over a range of leaf angles produces an effect on photosynthesis computations somewhat like spreading the direct-beam radiation over a wider range of leaf material due to the penumbra. Light is spread over the whole range of the photosynthetic response curve (Fig. 6). Adding penumbral effects to the SPAM model should increase the predicted photosynthesis somewhat, but probably not as much as the factors of 1·52 or 1·12 given above. However, with the large distances between canopy elements in a tall forest, penumbral effects should be greater than in a short crop. Also, the penumbral effects should be greatest under clear sky conditions, and least under heavy haze or cloud. With a uniform, umbrella-like upper canopy such as the Bosque de Florencia, the bottom of the forest may be bathed in pseudo 'enriched diffuse light' rather than well-defined gradations of sunfleck. With broken, irregular canopies as described by Kira *et al.* (1967) and Hozumi (1969a), the lower canopy sunflecks may appear as large, transient, more or less uniform zones.

Model predictions can set some reasonable limits for biomass accumulation as well as for other components of a forest system. However, both experimental and modelling efforts need to be maintained over long-term studies to make sure that our answers are reasonable and that our understanding of life processes in a forest system goes deeper than a tally sheet based on short-term observations.

REFERENCES

Allen, L. H., Jr. (1968). *J. appl. Meteorol.* **7**, 73–78.
Allen, L. H., Jr. (1973). 'Crop Micrometeorology: A. Wide-row light penetration; B. Carbon dioxide enrichment and diffusion'. Ph.D. Thesis, Cornell University, Ithaca, New York, U.S.A., 366 pp.
Allen, L. H., Jr. and Lemon, E. R. (1972). *Boundary-Layer Meteorol.* **3**, 246–254.
Allen, L. H., Jr., Jensen, S. E. and Lemon, E. R. (1971). *Science* **173**, 256–258.
Allen, L. H., Jr., Lemon, E. and Müller, L. (1972). *Ecology* **53**, 102–111.
Carnahan, B., Luther, H. A. and Wilkes, J. O. (1969). 'Applied Numerical Methods'. John Wiley and Sons, New York.
Duncan, W. G. (1971). *Crop Sci.* **11**, 482–485.
Duncan, W. G. and Barfield, B. J. (1970). *Trans. Am. Soc. Agr. Eng.* **13**, 246–248.
Duncan, W. G., Loomis, R. S., Williams, W. A. and Hanau, R. (1967). *Hilgardia* **38**, 181–205.
Gee, G. W. and Federer, C. A. (1972). *Water Resour. Res.* **8**, 1456–1460.
Hesketh, J. D., Baker, D. N. and Duncan, W. G. (1971). *Crop Sci.* **11**, 394–398.
Hozumi, K., Yoda, K., Kokawa, S. and Kira, T. (1969a). *Nature Life SE Asia* **6**, 1–56.
Hozumi, K., Yoda, K. and Kira, T. (1969b). *Nature Life SE Asia* **6**, 57–81.
Kira, T. (1969). *Malayan Forester* **32**, 375–384.
Kira, T. and Ogawa, H. (1971). *In* 'Productivity of Forest Ecosystems. Proc. Brussels Symp., 1969', pp. 310–321.
Kira, T. and Shidei, T. (1967). *Jap. J. Ecol.* **17**, 70–87.
Kira, T., Ogawa, H., Yoda, K. and Ogino, K. (1967). *Nature Life SE Asia* **5**, 149–174.
Kira, T., Shinozaki, K. and Hozumi, K. (1969). *Pl. Cell Physiol.* **10**, 129–142.
Kung, E. (1961). *In* 'Studies of the Three-dimensional Structure of the Planetary Boundary Layer' (H. H. Lettau, ed., Annu. Rept. to Meteorol. Dept., U.S. Army Electronic Proving Ground, Ft. Huachuca, Ariz.), pp. 27–35. Dept. of Meteorol., Univ. of Wisconsin, Madison.
Kung, E. C. and Lettau, H. H. (1961). *In* 'Studies of the Three-dimensional Structure of the Planetary Boundary Layer' (H. H. Lettau, ed., Annu. Rept. of Meteorol. Dept., U.S. Army Electronic Proving Ground, Ft. Huachuca, Ariz.) pp. 45–61. Dept. of Meteorol., Univ. of Wisconsin, Madison.
Kutzbach, J. E. (1961). *In* 'Studies of the Three-dimensional Structure of the Planetary Boundary Layer' (H. H. Lettau, ed., Annu. Rept, to Meteorol. Dept., U.S. Army Electronic Proving Ground, Ft. Huachuca, Ariz.), pp. 71–113. Dept. of Meteorol., Univ. of Wisconsin, Madison.
Lemon, E. R. (1967). *In* 'Harvesting the Sun: Photosynthesis in Plant Life' (A. San Pietro, F. A. Greer and T. J. Army, eds), pp. 263–290. Academic Press, New York.

Lemon, E., Allen, L. H. Jr. and Müller, L. (1970). *Bio Science* **20**, 1054–1059.
Lemon, E., Stewart, D. W. and Shawcroft, R. W. (1971). *Science* **174**, 371–378.
Lieth, H. (1971). *In* 'Perspectives on the Primary Productivity of the Earth', A.I.B.S. Symposium, Miami, Octover 24, 1971.
List, R. J. (1958). 'Smithsonian Meteorological Tables'. Smithsonian Institution, Washington.
Loomis, R. S. and Williams, W. A. (1969). *In* 'Physiological Aspects of Crop Yield' (J. D. Eastin, F. A. Haskins, C. Y. Sullivan, C. H. M. van Bavel, and R. C. Dinauer, eds), pp. 27–47. Amer. Soc. Agron. and Crop Sci. Soc. Amer., Madison.
Lugo, A. (1970). *In* 'A Tropical Rain Forest' (H. T. Odum, ed.) Chapters 1–7, pp. 81–102. U.S. Atomic Energy Commission Division of Technical Information, Oak Ridge, Tennessee.
McCree, K. J. (1969). *In* 'Prediction and Measurement of Photosynthetic Productivity, Proc. IBP/PP Tech. Meeting, Třeboň, 14–21 Sept. 1969' (I. Setlik, ed.) pp. 221–229. PUDOC, Wageningen.
McCree, K. J. (1972). *Agric. Met.* **9**, 191–216.
Miller, E. E. and Norman, J. M. (1971). *Agron. J.* **63**, 735–738.
Miller, E. E. and Norman, J. M. (1971b). *Agron. J.* **63**, 739–743.
Monsi, M. and Saeki, T. (1953). *Jap. J. Bot.* **14**, 22–52.
Monteith, J. L. (1965). *Ann. Bot. N.S.* **29**, 17–37.
Moon, P. (1940). *J. Franklin Inst.* **230**, 583–618.
Müller, D. and Nielsen, J. (1965). *Det Forstlige Forsgsvaesen Danmark.* **29**, 69–160.
Nichiporovich, A. A. (1961). *Soviet Pl. Physiol.* **8**, 428–435.
Nilson, T. (1971). *Agric. Met.* **8**, 25–38.
Norman, J. M., Miller, E. E. and Tanner, C. B. (1971). *Agron. J.* **63**, 743–748.
Owen, P. R. and Thompson, W. R. (1963). *J. Fluid Mech.* **15**, 321–334.
Robertson, G. W. and Russelo, D. A. (1968). 'Astrometeorological Estimator', *Ag. Met. Tech. Bul.* **14**, Plant Res. Inst., Canada Dept. of Agric. Ottawa, Ontario.
Saeki, T. (1960). *Bot. Mag. Tokyo* **73**, 55–63.
Shawcroft, R. W. (1970). 'Water Relations and Stomatal Response in a Corn Field', Ph.D. Thesis, Cornell Univ. Ithaca, New York, U.S.A. 127 pp.
Shawcroft, R. W. (1971). Tech. Rept. ECOM 2-68-I-7. *For* U.S. Army Electronic Command, Ft. Huachuca, Ariz., *by* Microclimate Investigations, U.S. Dept. of Agriculture, Ithaca, New York, 94 pp.
Shinn, J. H. (1971). 'Steady-state Two-dimensional Air Flow in Forests and the Disturbance of Surface Layer Flow by a Forest Wall', Ph.D. Thesis, Univ. of Wisconsin, Madison, Wisconsin, U.S.A.
Sinclair, T. R. (1971). 'An Evaluation of Leaf Angle Effect on Maize Photosynthesis and Productivity', Ph.D. Thesis, Cornell University, Ithaca, New York, U.S.A. 104 pp.
Sinclair, T. R., Knoerr, K. R. and Murphy, C. E. (1972). *Bull. Am. meteorol. Soc.* **53**, 1035.
Sinclair, T. R. and Lemon, E. R. (1973). *Solar Energy* **15**, 89–97.
Stearns, C. R. and Lettau, H. H. (1963). *In* 'Studies of the Effects of Variations in Boundary Conditions on the Atmospheric Boundary Layer' (H. H. Lettau, ed., Annu. Rept. to Meteorol. Dept., U.S. Army Electronics Research and Development Activity, Ft. Huachuca, Ariz.) pp. 115–138. Dept. of Meteorol., Univ. of Wisconsin, Madison.
Stephens, G. R. and Waggoner, P. E. (1970). *Bio Science* **20**, 1050–1053.

Stewart, D. W. (1970). 'A Simulation of Net photosynthesis of Field Corn', Ph.D. Thesis, Cornell Univ., Ithaca, New York, U.S.A. 132 pp.

Stewart, D. W. and Lemon, E. R. (1969). Tech. Rept. ECOM 2-68-I-6. *For* U.S. Army Electronics Command, Ft. Huachuca, Ariz. *by* Microclimate Investigations, U.S. Dept. of Agriculture, Ithaca, New York. 132 pp.

Swinbank, W. C. (1964). *Quart. J. R. met. Soc.* **90**, 119–135.

Uchijima, Z. and Inoue, K. (1970). *J. Agric. Met., Tokyo* **26**, 5–18.

Uchijima, Z. and Wright, J. L. (1964). *Bull. Nat. Inst. Agr. Sci., Japan* Ser. A., No. 11, 19–65.

Waggoner, P. E. (1969). *In* 'Physiological Aspects of Crop Yield' (J. D. Eastin, F. A. Haskins, C. Y. Sullivan, C. H. M. van Bavel, and R. C. Dinauer, eds) pp. 343–373. Amer. Soc. Agron. and Crop Sci. Soc. Amer., Madison.

Wit de, C. T. (1965). 'Photosynthesis of leaf canopies'. Versl. Landbouwk. Onderz. 663. Wageningen.

Woods, D. B. and Turner, N. C. (1971). *New Phytol.* **70**, 77–84.

Yoda, K. (1967). *Nature Life SE Asia* **5**, 83–148.

10. Citrus Orchards

J. D. KALMA and M. FUCHS

Division of Land Use Research, CSIRO, Canberra, Australia, and A.R.O.,
The Volcani Center, Bet Dagan, Israel

The objectives of this chapter are threefold. Firstly, a brief outline is given of some of the phenological and macro-climatic features of citrus. Secondly, recent studies on the micro-environment of citrus are reviewed, with particular emphasis on the transfer of radiation, momentum, latent and sensible heat. Finally, the practical importance of micro-climatology is discussed in sections on frost protection in citriculture and on means of controlling the solar radiation climate in citrus orchards.

I. PLANT CHARACTERISTICS AND MACRO-CLIMATIC FEATURES

Citrus is a typical mesophyte, growing in tropical and sub-tropical regions. The genus is characterized by glossy, evergreen leaves without any specialized

309

drought or heat-resisting qualities. Citrus root systems are shallow and spread moderately. Drought survival is in general limited (Hilgeman and Reuther, 1967). The various species of citrus all have their origin in tropical and sub-tropical South-east Asia and the Malay archipelago (Webber, 1943). Most present day commercial species are of sub-tropical origin, growing in climates ranging from humid to arid. The world's citrus belt lies between 40° N and 30° S and irrigation is common in many commercial production areas, concentrated at the belt's extreme latitudes. In contrast to the tropics where seasonal growth responds to rainfall (Cassin *et al.*, 1969), seasonal growth in the sub-tropics responds to temperature. In typical sub-tropical climates there is a cold-induced dormancy period of 2 to 4 months, which is usually followed by heavy flowering in early spring. The main crop ripens at one time, i.e. during autumn, winter or spring depending on variety and climate (Reuther and Rios-Castaña, 1969).

Mendel (1969) and Cassin *et al.* (1969) give the following screen temperature ranges for citrus growth : minimum, 12·5 to 13°C ; optimum, 23 to 24°C and maximum, 37 to 39°C. Accumulated temperatures (day degrees) above 12·5°C are regarded by Mendel (1969) and many others as the decisive factor in determining growth rate. Light seems to be of secondary importance under orchard conditions. Newman *et al.* (1967) however observed that for a wide climatic range of commercial citrus production areas in the U.S.A., net radiant heat load on the vegetative surface is a more important determinant for growth and yield than air temperatures. Cooper *et al.* (1963) reported on investigations into the main effects of climate on the phenology of commercial Valencia orange trees in four American States, finding air temperature, rainfall, and air humidity to be dominant environmental parameters.

To illustrate the wide range of climates in which citrus is grown, a climatic summary is presented in Table I for some of the world's centres of citriculture. Great differences exist in macro- and meso-climate between the various regions and as a consequence also in varieties, species and cultural practices.

Meso-climate affects citriculture in many ways. It affects the selection of orchard sites and the outlay and planting system which is adopted. It has also great bearing on the various cultural practices in the established orchard by which man attempts to modify the existing natural environment and to protect the crop from weather hazards. It affects the incidence of pests and diseases and the effectiveness of protective measures and control.

Recent micrometeorological studies in citrus are reviewed in the following section. Examples of the practical importance of such studies are given in the last two sections, by focusing on frost protection and on means of controlling the radiation climate.

Table I

Temperature, radiation, and rainfall for representative locations in some commercial citrus regions

Location		Latitude	Elevation m	Mean daily temperature (°C)		Mean daily solar radiation (MJ m^{-2})		Total rainfall (mm)	
				Jan	July	Jan	July	Nov–Apr	May–Oct
United States									
Brownsville	Tex.	25° 55′ N.	5	15·3	28·6	12·4	26·3	226	465
Tampa	Fla.	27° 57′	11	16·1	27·5	13·7	22·3	356	930
Jacksonville	Fla.	30° 20′	5	13·3	27·8	12·6	22·1	572	762
Yuma	Ariz.	32° 54′	43	12·5	33·1	12·1	27·5	48	38
Indio	Cal.	33° 44′	3	12·2	33·9	12·6	28·4	56	28
Riverside	Cal.	33° 57′	250	11·1	24·4	11·5	28·3	243	33
Fresno	Cal.	36° 46′	3	7·8	27·8	7·7	28·6	201	30
Japan									
Kagoshima		31° 34′ N.	5	6·6	26·8	9·9	21·4	737	1 600
Kumamoto		32° 49′	38	4·6	26·5	9·2	19·7	546	1 324
Osaka		34° 39′	7	4·5	26·6	9·4	16·4	456	903
Nagoya		35° 10′	51	2·9	25·7	9·9	17·7	492	1 054
Tokyo		35° 41′	4	3·7	25·1	10·4	18·7	519	1 044
Israel									
Beersheva		31° 14′ N.	270	11·6	26·2	13·7	30·8	191	9
Rehovot		31° 54′	50	13·0	25·8	12·6	29·7	512	18
Deganya		32° 43′	−200	13·6	29·7	13·1	30·8	369	15
Akko		32° 56′	10	13·6	25·7	12·6	28·8	548	29
Australia									
Richmond	N.S.W.	33° 36′ S.	22	23·1	10·3	23·4	9·5	447	266
Griffith	N.S.W.	34° 17′	128	24·2	8·9	29·3	10·1	172	214
Berri	S.A.	34° 17′	65	23·0	10·4	27·2	9·4	98	152
Swan Hill	Vic.	35° 22′	70	23·4	9·2	28·3	9·5	132	200

II. RECENT STUDIES OF THE MICRO-ENVIRONMENT OF CITRUS ORCHARDS

A. Radiative Transfer Processes

1. *Above-canopy Measurements*

The components of the radiation balance of the whole stand in a citrus plantation can be measured above the canopy of the trees, provided a representative area of the crop is sampled and provided also that radiative flux divergence between crop and radiation sensor can be disregarded.

There have been few reports of such studies. In many cases radiation studies form an integral part of evaporation studies (for example by Hashemi and Gerber, 1967) or studies on frost protection methods (e.g. Bartholic and Wiegand, 1969). In other studies, above-canopy measurements are used only as reference for within canopy measurements (e.g. Greene and Gerber, 1967).

Net radiation, R_n, is crop dependent and is rarely measured on a routine basis. Total short-wave radiation S_t however is frequently measured and can also be estimated fairly accurately. It is therefore understandable that many attempts have been made to arrive at functional, semi-empirical relationships between R_n and S_t. Stanhill *et al.* (1966) and Kalma and Stanhill (1969, 1972) reported on the diurnal and seasonal variation of the reflection coefficient, the heating coefficient and the net long-wave radiation at $S_t = 0$, by which R_n may be estimated from S_t (Monteith and Szeicz, 1961). On an annual basis ΣR_n for the citrus orchard at Rehovot, Israel was $0.56 \Sigma S_t$, and the relation for daily values (expressed in W m^{-2}) was

$$\Sigma R_n = 0.80 \Sigma S_t - 63.$$

2. *Within-canopy Measurements*

Studies of the radiation climate within a citrus canopy are hampered by horizontal heterogeneity. Monselise (1951) working with selenium barrier-layer photocells in grapefruit and orange plantations showed the difficulty of obtaining adequate numbers of reliable light measurements within the canopy of orchards. Since Monselise's work, experimenting with various sensors and integration techniques in orchards has continued till recently (see for example Maggs and Alexander, 1970) with little improvement in technology.

Green and Gerber (1967) measured net long-wave and net all-wave radiation at a number of heights in orange trees of different size, using Suomi-type net radiometers. Their measurements showed that light penetration increased with decreasing solar elevation and that 90% of all net radiation was absorbed in the top 30 to 90 cm of the canopy. Opposite results were obtained by Kalma and Stanhill (1969). Their measurements showed that absorption in the upper half of the canopy increased with decreasing solar elevation. Furthermore, R_n above the bare soil averaged 11% of the net radiation above the crop, i.e. $0.11\,R_n(0)$, varying from 0.16 at noon to 0.06 in the late afternoon. In a later paper (Kalma and Stanhill, 1972) these net radiation values were compared with values estimated from an exponential equation based on the one-dimensional model suggested by Cowan (1968a). Estimates and measurements were always within $\pm 0.1\,R_n(0)$.

From short-wave radiation measurements above and below the orange tree canopy, Kalma (1970) estimated that the mean transmission was 0.21, ranging from 0.36 in October to 0.13 in August. A similar value is likely for transmission of short-wave radiation reflected by the soil and, with mean soil and canopy reflection coefficients of 0.21 and 0.16 the mean short-wave absorption of the canopy was calculated to be 0.67 (see Fig. 1a). By contrast, a single layer mosaic of fresh orange leaves absorbed 49% of S_t and reflected 32% (see Fig. 1b). This absorptivity agrees with a value of 0.50 obtained

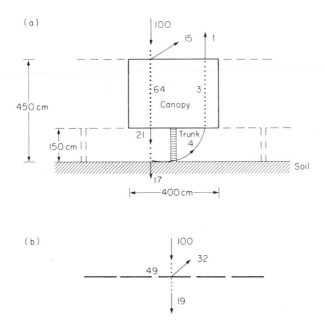

Fig. 1. The components of the short-wave radiation balance of the canopy as a whole (a) and of a single layer mosaic of fresh orange leaves (b) (after Kalma 1970).

for a layer of single leaves of 22 North American tree species (Birkebak and Birkebak, 1964). Thus, since absorption of the canopy as a whole is considerably greater than that of a single layer of leaves, the importance of radiation trapping in the canopy is clearly shown.

B. Momentum Transfer

Vertical fluxes of CO_2, heat, and water vapour above and within the canopy may be calculated from concentration profiles and diffusivities of the transported entity. The diffusivities of CO_2, heat and water vapour are frequently estimated by assuming them to be similar to the diffusivity or transfer coefficient for momentum.

Momentum transfer coefficients are mostly determined from measurements of wind speed at different heights above, within and below the canopy (see Vol. 1, Ch. 3). Spatially integrated measurements are rarely made. Usually small-scale irregularities are smoothed out, however, by computing average wind speed from total wind-run over intervals of say 5 or 10 minutes.

Brooks (1959) observed that, in stable conditions, the wind profile over a plantation of young trees was best described by a power law, and under conditions of neutral stability by the familiar logarithmic expression. Adopting a logarithmic profile he reported that for trees about 3·2 m high, values of the two steady-state wind profile parameters z_0 and d, were 31 and 256 cm respectively. Kalma and Stanhill (1972) give values for an orange plantation with almost complete cover and an average tree height of 4·35 m. They report mean values for 72 profiles of 40 and 286 cm for z_0 and d. The temperature gradient for 58 profiles was less than $0·15°C\,m^{-1}$. No relation was observed between z_0 or d and measured temperature gradients, nor was there any effect of wind direction. On some occasions a direct relationship between wind speed and z_0 was observed and an inverse one with d. For an orange plantation with 75% cover and an average tree height of 4 m, Bartholic and Smithey (personal communication, 1968) found $z_0 = 40$ cm and $d = 250$ cm.

Empirical relationships between z_0 and d and crop height such as those provided by Tanner and Pelton (1960), Kung and Lettau (1961), and Szeicz *et al.* (1969) for z_0 and by Cowan (1968a) and Stanhill (1969) for d, are adequate for citrus plantations which cover more than 75% of the underlying ground surface.

Although reported measurements of wind flow within orchard canopies are scarce, most results show that especially in younger, more open orchards, spatial differences in horizontal wind speed may be considerable (e.g. Brooks, 1959). It is because of this lack of horizontal homogeneity that many one-dimensional approaches to transfer of water vapour, heat, or CO_2 fail. Kalma (1970) presented data indicating that wind speeds below an orchard canopy of 90% ground cover, in vertical projection, were actually faster than wind speeds in the canopy above. This result is similar to Allen's observations (1968) in a larch plantation. Cowan (1968a) assumed in his model that eddy diffusivity varied linearly with wind speed and that leaf area per unit volume of space was constant with height. Agreement between this theory and observations in the canopy is reasonably satisfactory but wind speeds in the trunk space were twice the computed values.

C. Transfer of Latent and Sensible Heat

Irrigation plays an important role in many commercial citrus areas and much attention has been given to determining timing and design criteria and irrigation (Till, 1969). Not surprisingly, a great variety of approaches have been used to determine water loss from orchards. Tanner (1968) discussed many such methods in great detail.

A unique long-term experiment on the water relations of citrus is in progress in a 390 ha watershed in California. Reports by Davis and Grass (1966), Davis et al. (1969) and Bingham et al. (1971) contain results obtained thus far. With 90% of the area planted with citrus, the watershed may be viewed as a macrolysimeter, because all sources and losses of water (other than evapotranspiration) are measured. The mean annual evaporation from vegetation and soil over 9 years was 720 mm, varying from 22 in January to 102 mm in August. For the 1967 to 1971 period, evapotranspiration per year was 87% of that computed with the Blaney and Criddle method and 38% of the total evaporation from a screened Class A pan.

Kalma (1970) adopted a water balance technique in an orange plantation in Israel. At two-weekly intervals over two years he measured soil-water content in a 4 × 4 m plot between four trees of similar shape, size and age, down to 2·8 m using a dense grid of neutron scattering access tubes. Irrigation and rainfall reaching the ground were also intensively measured. Deep drainage was estimated from successive moisture profiles. Mean annual evaporation over the two year period was 840 mm or about 54% of the total evaporation from a screened Class A pan. Daily rates varied from 1·0 mm in midwinter to 4·4 mm in midsummer.

Van Bavel et al. (1967), using a similar water balance technique for two years in a mature orange plantation in Arizona, found an average annual evaporation of 1080 mm, with daily rates ranging from 0·9 mm in midwinter to 5·3 mm in midsummer. Water loss from crop and soil was 66% of the annual evaporation from the screened Class A pan.

Recent examples of work in citrus with the combination method are studies by Van Bavel *et al.* (1967), Hashemi and Gerber (1967) and Kalma (1970). This approach requires information and assumptions about the state of the surface and introduces the concept of potential evaporation. Van Bavel (1966) defined potential evaporation as the rate of vaporization from a wet surface which imposes no restrictions on the diffusion of water vapour. This definition is based on the concept of surface resistances introduced by Penman and Schofield (1951). It indicates that under non-potential conditions there is an additional resistance to water vapour (see for example Slatyer and McIlroy, 1961; Monteith, 1963, 1965).

Internal resistances to water loss have been computed by various authors from potential evaporation estimated by the combination method and determinations of actual water use (Van Bavel *et al.*, 1967; Hilgeman and Reuther, 1967; Stanhill, 1972). Resistances were generally related to the soil-water status. Tanner (1968), however, states that any agreement between such estimates of internal resistance and measurements of leaf diffusion resistance is fortuitous. Tanner submits that canopy models are needed which consider the spatial distribution of transfer coefficients and sources and sinks of latent and sensible heat, in contrast to the plane models proposed by Monteith (1965) and others, which have been frequently used in recent years.

Measurements of leaf diffusion resistance have been reported by Van Bavel *et al.* (1967) and Hilgeman *et al.* (1969). Mean midday values observed for the lower epidermis ranged from 5 to 18 s cm^{-1}, which is higher than for other field crops. Citrus leaves have no or very few stomata in the upper epidermis, the resistance of which is typically about 50 s cm^{-1}.

The atmospheric resistance between crop surface and the free air is assumed to be the same for latent and sensible heat. Artificial leaves were used by Kalma (1970) to measure atmospheric resistance at various heights in the orange tree canopy. These measurements were compared with values computed from aerodynamic characteristics and wind speed measurements following Monteith (1965). Agreement was best when the need for atmospheric stability correction of the aerodynamically computed resistances was least. The same study also contains data on the diurnal variation in total diffusive plant resistances calculated from Bowen ratio measurements of actual water loss and calculations of potential evaporation by the combination method made over half-hourly periods. Total diffusive resistances varied between 0·5 and 4 s cm^{-1} for most of the daylight hours. Atmospheric resistance was about one-tenth of the internal plant resistances.

Hilgeman *et al.* (1969) also reported measurements of apparent trans-piration by the cut-leaf method and determinations of the internal water deficit of leaves. On summer days in Arizona, apparent transpiration increases rapidly till 1000 h, remains high to 1600, and then decreases rapidly. Similar trends were observed in the internal leaf-water deficit and opposite changes in the leaf-diffusion resistance.

As pointed out before, one-dimensional models, such as that described by Brown and Covey (1966), have been used to simulate the vertical transfer of latent and sensible heat, momentum or CO_2. These models assume horizontal homogeneity. The difficulty of obtaining spatially differentiated i.e. three-dimensional measurements is well known. The vertical transfer of latent and sensible heat has been determined in various crops using an energy budget technique based on energy and mass conservation. From profiles of net radiation, air temperatures, air vapour pressures and measurement of soil heat flux, exchange coefficients were determined by Kalma (1970) for an orange plantation, with an average tree height of 4·25 m. The discordance of the results was too large to be accounted for by the relative errors in measurement of dry- and wet-bulb temperatures, net radiation and soil heat flux. It was assumed that the lack of horizontal homogeneity was by far the most important factor in the failure of the energy budget technique in the orchard canopy.

III. FROST PROTECTION IN CITRICULTURE

A. Frost Types

Protection from advective frosts, which are accompanied by winds below 0°C, has been largely ineffective. The present discussion will therefore be confined to radiation frosts, which occur under clear, calm, and dry atmospheric conditions, with the temperature of the invading air masses generally above freezing point. These frosts occur especially at night, when the loss of radiant heat from the surface is greater than the radiative flux towards the vegetative surface. This net radiant loss causes a temperature inversion near the surface. The resulting stable stratification of the air lessens the convective heat exchange between air and leaves and sustains the critically low leaf temperatures.

B. Protection Methods

Blanc (1963) distinguished between passive and active frost protection methods. Passive methods in citriculture are generally very important. They are concerned with ways of totally avoiding frost damage, for example by area and site selection, by choice of planting system, varieties and root stocks and by breeding for cold hardiness. Active methods comprise artificial measures applied immediately before and during the period of actual frost, such as overhead sprinkling, orchard heating and wind machines. (In California alone several millions of dollars are spent annually on active methods with a total investment in equipment of 15 to 20 million dollars). The effectiveness of active methods is difficult to assess because these methods modify a dynamic micro-environment. It is apparent from the literature that more information is needed in many citrus areas on the local suitability of these methods. Extrapolation in space and in time of the results of a few specific experiments has been virtually impossible. Much more environmental research is needed.

Orchard heaters arrest the cooling of the tree canopy, both by increasing the air temperature and hence the flux of sensible heat towards the leaves and by adding radiant heat from the heater to the energy balance of the leaves. When air temperature increases with height the turbulence generated by air jets from wind machines (or helicopters) causes the warmer air aloft to mix with the colder air in the orchard. It enhances the flux of sensible heat towards the leaves by increasing the turbulent transfer coefficient for sensible heat. In overhead sprinkling during frost, the release of latent heat in the freezing of water is utilized to offset the heat loss to the environment.

It is difficult to assess the economic value of frost protection, because the cost of protection goes by the hectare, but the net return comes only from the fruit saved. One must therefore consider (1) the quality and quantity of the product; (2) the frequency and severity of damaging frosts; and (3) the cost of protection per degree temperature rise.

C. Environmental Studies on Frost Protection for Citrus

Bartholic and Wiegand (1969) state that environmental studies on frost protection by orchard heating should have three objectives. First, existing micrometeorological relationships should be further developed and evaluated. Second, models should be developed to predict expected minimum temperatures. Third, a methodology should be provided for calculating the amount of heat required to keep citrus at a safe temperature, in a given microenvironment. This applies also to protection by overhead sprinkling and generating turbulence with wind machines.

Brooks et al. (1952) assumed that latent heat exchange could be neglected and that 75% of the radiant heat loss in an 'infinitely' large orchard comes from the soil, 19% from the overhead air and 8% from the orchard air and trees.

Crawford (1964) treated the orchard as enclosed in a box, for which the heat exchange across the sides has to be calculated. His simplified solution of the energy balance of the heated orchard assumes steady state conditions and is independent of the thermal properties and initial temperatures of air, trees and soil.

In contrast to Brooks *et al.* (1952), he neglected convection from the air, but included advection of upwind cold air and omni-directional flow induced by the buoyancy differences between the heated and unheated orchards. Nomograms permit rapid computation of heating requirements, which came to within 5% of measurements in a limited number of field studies.

Crawford's techniques were re-examined by Gerber (1969), who developed a method of estimating heating requirements, using heat transfer concepts as earlier suggested by Turrell and Austin (1965) and Businger (1965). Gerber adopted Crawford's approach to the advection problem. He showed how the heat required for maintaining individual objects at the lowest tolerable temperature may be estimated and he presented analytical expressions for use with the orchard layer as a whole. Gerber's method showed that the transition from individual leaves to the orchard layer as a whole with its specific crop architecture creates great difficulties which can be solved only by the introduction of empirical crop constants. The solution for an incomplete crop cover clearly illustrates this point.

Similarly, for overhead sprinkling as the frost protection method, Businger (1965) and Harrison *et al.* (1971) discussed the energy requirements of individual leaves, but found extension to the whole crop difficult. Run-off, uneven distribution and wind drift are uncertain quantities. Harrison *et al.* (1971) provided tables and figures for determining sprinkling rates from wind speed, minimum temperatures and ambient vapour pressure. The relationship of these rates and the energy balance of individual leaves and the whole crop is unfortunately not fully explained.

Crawford (1965) stated that for the planning of frost protection by wind machines, various generalizations are needed. Schultz (1962, 1969) clearly showed how the effectiveness of wind machines increases with the strength of the inversion, while a close connection is observed between free air subsidence, with adiabatic warming, and the inversion strength near the ground. Schultz (1969) also revealed that nightly air-drainage as a local parameter, may be of great importance for the formation of strong inversions. Empirical relationships between total area affected by the wind machine and its thrust (which is directly related to the jet produced) were presented by Crawford (1964) and Bates (1971/1972). Protection with wind machines obviously requires a permanent installation which is designed on the basis of local experience. Unlike orchard heating and, to a lesser extent, overhead sprinkling, this protection method does not need short-term solutions for operating. An increased interest has recently been shown in combinations of heaters and wind machines. Crawford (1965) stated that such combinations provide better protection than one method alone.

It is apparent from this section on frost protection in citrus that since the study of Brooks *et al.* (1952) much effort has been devoted to the various methods, but that few real advances will occur until an integrated approach to the problem is adopted.

IV. MEANS OF CONTROLLING THE SOLAR RADIATION CLIMATE IN CITRUS ORCHARDS

Planting distances in most citrus districts vary widely. They depend on variety, rootstock species, soil fertility, length of growing season and several management considerations. Planting distance strongly affects the radiation climate of the orchard and therefore its productivity.

Yield and fruit quality of deciduous and evergreen fruit trees are related to incoming global radiation (Kvåle, 1969; Newman *et al.* 1967). The relation between light interception and yield (and quality) has rarely been investigated however, in sharp contrast to the attention the manipulation of tree canopy and tree density has generally received in forestry.

The practices of pruning citrus trees, and of cultivating and mulching the soil beneath them modify the absorption of solar radiation by the canopy and the soil. As pointed out earlier, present experimental studies of the distribution of solar radiation within citrus orchards cannot provide guidelines for these practices because of the large spatial variability of the canopy. Theoretical models of the scattering of solar radiation based on the optical and geometrical properties of the leaves (de Wit, 1965; Cowan, 1968b, Isobe, 1969) have rarely been applied in citrus orchards mainly because of extensive sampling problems (Monselise, 1951). However, technological advances have made canopy measurements by hemispherical photographs much more convenient (Anderson, 1971; Bonhomme and Chartier, 1972). The gap frequency distribution as a function of solar elevation and azimuth, easily obtained from these photographs, can be used to assess the absorption and scattering of solar radiation by the foliage and by the soil underneath the vegetation (Fuchs, 1972).

The division of the complex orchard canopy into a uniformly scattering leaf zone and an optically passive trunk zone implied by the analysis does not define the profile of solar radiation absorption within the canopy. However, this limitation is serious only in theoretical studies of canopy photosynthesis. In practice, the simplified analysis permits the gross effects of changes in canopy density and soil reflectivity on the solar radiation balance to be predicted.

The azimuthal and altitudinal distribution of the canopy elements obtained by Kalma (1970) from a series of 12 hemispherical photographs of the sky from the ground through the citrus orchard canopy illustrates such an approach.

A. Interception of Solar Radiation

The interception of direct solar radiation by the citrus trees does not vary much with solar elevation. The measurements shown in Fig. 2 indicate that

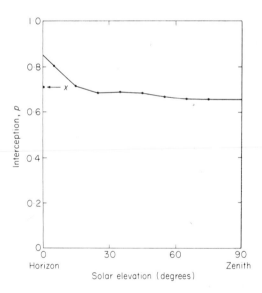

Fig. 2. Interception, p, of direct solar radiation by a citrus orchard canopy as a function of solar elevation. The hemispherical interception for diffuse radiation, x, equals $\int_{h=0}^{\pi/2} p \sin h \, dh$ when h is the angle of elevation.

the canopy is open. The relatively small values of the interception at low solar elevation are the result of the strict tree alignment in a 4×4 m planting pattern which leaves wide regular pathways where solar radiation can penetrate even at small angle of incidence (Fig. 3).

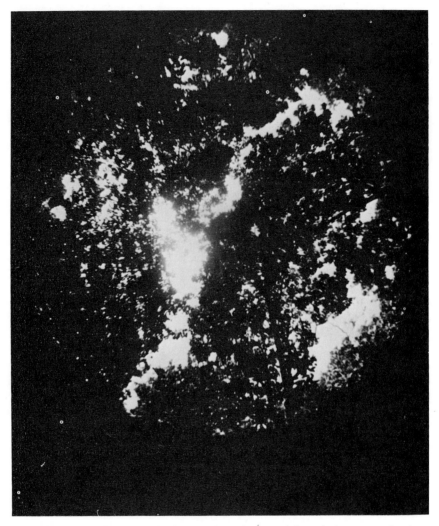

Fig. 3. Hemispherical photograph of the sky from the ground through the canopy of a citrus orchard, illustrating the continuity of gaps due to tree alignment.

The average daily fraction of direct solar radiation intercepted by the trees does not change significantly during the year and is practically equal to the hemispherical interception of isotropic diffuse radiation. This similarity allows a single analysis of the time-averaged direct and diffuse solar radiation. Similarly, the hemispherical interception for diffuse radiation can be used as a measure of the cover density.

Kalma (1970) indicated that the average leaf area index for his citrus orchard was 6·5, so the amount of incoming radiation scattered downward by the canopy must be small. The solar radiation reaching the soil is essentially transmitted through gaps. Accordingly, the relative transmission of diffuse solar radiation was estimated to be 0·29 which is somewhat larger than the average value of 0·21 measured by Kalma.

Following Fuchs (1972), it was found that the solar reflectivity of the foliage was 0·22, using Kalma's average value of 0·16 for the reflectivity of the orchard canopy and a reflectivity of 0·12 for the bare soil.

B. Changing the Reflectivity of the Soil and the Cover Density

Artificial mulches can be used to change the solar reflectivity of the soil over a wide range and values below 0·05 can be obtained with various black plastic coatings. Stanhill (1965) reported that a magnesium carbonate dressing increased the soil reflectivity to 0·65. Figure 3 shows how the change of soil reflectivity ρ_s modifies the absorption of solar radiation by the soil underneath a citrus orchard canopy. The curve with a hemispherical interception of 0·7 corresponds closely to the condition of the orchard studied by Kalma. The inverse relationship between soil reflectivity and absorption is nearly linear and the slope of the lines depends on the cover density. The comparison between the curves, corresponding to the different values of hemispherical interception, indicates that an increase of soil reflectivity may offset the effect of decreasing the canopy cover density. For example, the effect of changing the hemispherical interception from 0·7 to 0·5 may be compensated by an increase of soil reflectivity from 0·12 to 0·48.

On the other hand, increasing the soil reflectivity increases the absorption of solar radiation by the canopy (Fig. 5) as a result of the increase of upward solar radiation from the soil surface implied in Fig. 4. Comparison of the slopes of corresponding lines in Figs 4 and 5 indicates that when ρ_s increases, only a fraction of the additional reflection from the soil is absorbed by the canopy: the rest is retransmitted through the gaps in the canopy. Consequently, the solar radiation absorption by the vegetation system, i.e. soil and canopy combined, decreases when the soil reflectivity is increased.

Modification of the cover density in citrus groves is restricted by considerations other than the solar radiation balance. Travel space for agricultural implements limits the upper density of tree population and economic factors set a lower limit. Within these limits there is scope for variation of cover density as occurs of course naturally with the growth of the trees.

Figure 5 also shows that the effect of increasing the cover density on the absorption of solar radiation by the canopy can also be obtained by increasing

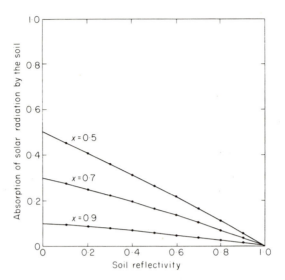

Fig. 4. Effect of the soil reflectivity on the mean absorption of solar radiation by the bare soil beneath a citrus orchard at various cover densities. The cover density is measured by x, the hemispherical interception.

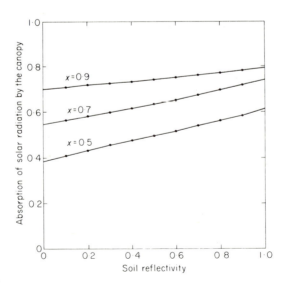

Fig. 5. Effect of the soil reflectivity on the mean absorption of solar radiation by the canopy of a citrus orchard at various cover densities, measured by x, the hemispherical interception.

the reflectivity of the soil. For example, the absorption of solar radiation by the foliage of a citrus orchard intercepting 50% of the incoming solar radiation on a soil with $\rho_s = 0.8$, will be the same as for a canopy with 70% interception but with $\rho_s = 0.1$. However, the absorbed solar radiation will affect the orchard differently as in the first situation the absorption is by a smaller number of leaves and through a larger participation of the lower leaves.

Increasing the interception of solar radiation by the foliage lowers the flux reaching the soil surface, and consequently the absorption by the soil. For small values of the soil reflectivity (Fig. 6) the relationship between

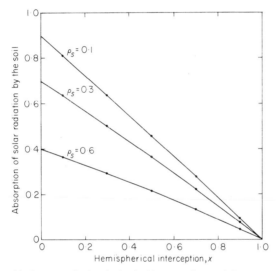

Fig. 6. Relationship between the hemispherical interception and the mean absorption of solar radiation by the bare soil beneath the canopy of a citrus orchard. The parameter ρ_s is the reflectivity of the soil.

interception and absorption is nearly linear. At large values of ρ_s however, the curvature of the lines increases, indicating that multiple reflections between foliage and soil significantly increase the absorption of solar radiation by the soil.

The dependence between absorption of solar radiation by the canopy and hemispherical interception by the canopy is also modified by the soil reflectivity (Fig. 7), but less markedly.

The maximum absorption by the canopy is reached at full cover density regardless of the soil reflectivity as a consequence of the very small downward scattering of radiation by the citrus foliage. The increased absorption

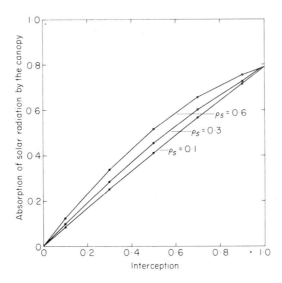

Fig. 7. Relationship between the hemispherical interception and the mean absorption of solar radiation by the canopy of a citrus orchard. The parameter ρ_s is the reflectivity of the soil.

of solar radiation by the canopy noted in Fig. 5 will occur for interception values in the open interval between 0 and 1. However, the maximum absolute increase of absorption is at mid-range values of the interception. It can be shown from the equation of absorption by the canopy presented by Fuchs (1972) that the maximum increase of absorption by citrus orchard foliage resulting from an increase of soil reflectivity (ρ_s) occurs when the hemispherical interception equals $(2 \text{ to } 0 \cdot 22\,\rho_s)^{-1}$.

In conclusion it appears that the radiation climate of the citrus orchards can be controlled by manipulation of the cover density and the soil reflectivity. Both operations interact and the optimum effectivities of any one of them depends upon the other.

REFERENCES

Allen, L. H. (1968). *J. appl. Met.* **7**, 73–78.
Anderson, M. C. (1971). *In* 'Plant Photosynthesis Production. Manual of Methods.' (Z. Sesták, J. Catský and P. G. Jarvis, eds), pp. 412–466. Junk, The Hague.
Bartholic, J. F. and Wiegand, C. L. (1969). *Proc. 1st int. Citrus Symp.* **2**, 583–592.
Bates, E. M. (1971/1972). *Agric. Met.* **9**, 335–346.
Bingham, F. T., Davis, S. and Shade, E. (1971). *Soil Sci.* **112**, 410–418.
Birkebak, R. and Birkebak, R. (1964). *Ecology* **45**, 646–649.

Blanc, M. L. (1963). 'Protection against Frost Damage'. W.M.O. Technical Note No. 51, Geneva, Switzerland.

Bonhomme, R. and Chartier, P. (1972). *Israel J. agric. Res.* **22**, 53–61.

Brooks, F. A. (1959). 'Physical Microclimatology'. University of California, Davis, California.

Brooks, F. A., Kelly, C. F., Rhoades, D. G. and Schultz, H. B. (1952). *Agric. Engn.* **33**, 74–78, 143–147, 154.

Brown, K. W. and Covey, W. (1966). *Agric. Met.* **3**, 73–96.

Businger, J. (1965). *Met. Monogr.* **6**, 74–80.

Cassin, J., Bourdeaut, J., Fougue, A., Furon, V., Gaillard, J. P., LeBourdelles, J., Montagut, G. and Moreuil, C. (1969). *Proc. 1st int. Citrus Symp.* **1**, 315–323.

Cooper, W. C., Peynado, A., Furr, J. R., Hilgeman, R. H., Cahoon, G. A. and Boswell, S. B. (1963). *Proc. Am. Soc. hort. Sci.* **82**, 180–192.

Cowan, I. R. (1968a). *Quart. J. R. met. Soc.* **94**, 523–544.

Cowan, I. R. (1968b). *J. appl. Ecol.* **5**, 367–379.

Crawford, T. V. (1964). *J. appl. Met.* **3**, 750–760.

Crawford, T. V. (1965). *Met. Monogr.* **6**, 81–87.

Davis, S. and Grass, L. B. (1966). *Trans. Am. Soc. agric. Engrs.* **9**, 108–109.

Davis, S., Bingham, F. T., Shade, E. R. and Grass, L. B. (1969). *Proc. 1st int. Citrus Symp.* **3**, 1771–1777.

Fuchs, M. (1972). *In* 'Optimizing the Soil Physical Environment toward Greater Crop Yields' (D. Hillel, ed.) pp. 173–191. Academic Press, New York and London.

Gerber, J. F. (1969). *Proc. 1st int. Citrus Symp.*, **2**, 545–550.

Greene, B. A. and Gerber, J. F. (1967). *Proc. Am. Soc. hort. Sci.* **90**, 77–85.

Harrison, D. S., Gerber, J. F. and Choate, R. E. (1971). 'Sprinkler Irrigation for Cold Protection'. Circular 348, Florida Coop. Ext. Service, Univ. of Florida, Gainesville, Florida.

Hashemi, F. and Gerber, J. F. (1967). *Proc. Am. Soc. hort. Sci.* **91**, 173–179.

Hilgeman, R. H. and Reuther, W. (1967). *In* 'Irrigation of Agricultural Lands' (R. M. Hagan, H. R. Haise and T. W. Edminster, eds), pp. 704–718. American Soc. Agronomy, Madison, Wisc.

Hilgeman, R. H., Ehrler, W. L., Everling, C. E. and Sharp, F. O. (1969). *Proc. 1st int. Citrus Symp.* **3**, 1713–1723.

Isobe, S. (1969). *Bull. Nat. Inst. Agric. Sci. (Japan)* A**16**, 1–25.

Kalma, J. D. (1970). 'Some Aspects of the Water Balance of an Irrigated Orange Plantation'. Ph.D. Thesis, published by the Volcani Institute of Agricultural Research, Bet Dagan, Israel.

Kalma, J. D. and Stanhill, G. (1969). *Sol. Energy* **12**, 491–508.

Kalma, J. D. and Stanhill, G. (1972). *Agric. Met.* **10**, 185–201.

Kung, E. C. and Lettau, H. H. (1961). 'Regional and Meridional Distributions of Continental Vegetation Cover and Aerodynamic Roughness Parameters'. Ann. Rep. Ft. Hauchuca, Contract DA-36-039-SC-80282.

Kvåle, A. (1969). *Acta Agric. Scand.* **19**, 229–239.

Landsberg, J. J. and James, G. B. (1971). *J. appl. Ecol.* **8**, 729–741.

Maggs, D. H. and Alexander, D. McE. (1970). *J. appl. Ecol.* **7**, 639–646.

Mendel, K. (1969). *Proc. 1st int. Citrus Symp.* **1**, 259–265.

Monselise, S. P. (1951). *Bull. Res. Coun. Israel.* **1**, 36–53.

Monteith, J. L. (1963). *In* 'Environmental Control of Plant Growth' (L. T. Evans, ed.), pp. 95–112. Academic Press, New York.

Monteith, J. L. (1965). *Symp. Soc. exp. Biol.* **19**, 205–234.

Monteith, J. L. and Szeicz, G. (1961). *Quart. J. R. met. Soc.* **87**, 159–170.

Newman, J. E., Cooper, W. C., Reuther, W., Cahoon, G. A. and Peynado, A. (1967). *In* 'Ground Level Climatology' (R. H. Shaw, ed.), pp. 127–147. A.A.A.S., Washington, D.C.

Penman, H. L. and Schofield, R. K. (1951). *Symp. Soc. exp. Biol.* **5**, 115–129.

Reuther, W. and Rios-Castaña, D. (1969). *Proc. 1st int. Citrus Symp.* **1**, 277–300.

Schultz, H. B. (1962). *Proc. 2nd int. Congr. Biomet.* 614–629.

Schultz, H. B. (1969). *Proc. 1st int. Citrus Symp.* **2**, 571–573.

Slatyer, R. O. and McIlroy, I. V. (1961). 'Practical Microclimatology' CSIRO, Melbourne, Australia.

Stanhill, G. (1965). *Agric. Met.* **2**, 197–203.

Stanhill, G. (1969). *J. appl. Met.* **8**, 509–513.

Stanhill, G. (1972). *Proc. 18th int. hort. Congr.* Tel Aviv.

Stanhill, G., Hofstede, G. J. and Kalma, J. D. (1966). *Quart. J. R. met. Soc.* **92**, 128–140.

Szeicz, G., Endrödi, G. and Tajchman, S. (1969). *Wat. Resour. Res.* **5**, 380–394.

Tanner, C. B. (1968). *In* 'Water Deficits and Plant Growth', (T. T. Kozlowski, ed.), Vol. 1, pp. 73–106. Academic Press, New York.

Tanner, C. B. and Pelton, W. L. (1960). *J. geophys. Res.* **65**, 3391–3413.

Till, M. R. (1969). *Proc. 1st int. Citrus Symp.* **3**, 1731–1747.

Turrell, F. M. and Austin, S. W. (1965). *Ecology* **46**, 25–34.

Van Bavel, C. H. M. (1966). *Wat. Resour. Res.* **2**, 455–467.

Van Bavel, C. H. M., Newman, J. E. and Hilgeman, R. N. (1967). *Agric. Met.* **4**, 27–37.

Webber, H. J. (1943). *In* 'The Citrus Industry', (H. J. Webber and L. D. Batchelor, eds), Vol. 1, pp. 1–40. Univ. Calif. Press, Berkeley, and Los Angeles.

Wit de, C. T. (1965). 'Photosynthesis of Leaf Canopies'. *Agric. Res.* Rept. 663. Centre for Agricultural Publications and Documentation, Wageningen.

11. Swamps

E. LINACRE

School of Earth Sciences, Macquarie University, Sydney, Australia

I. INTRODUCTION

A. Terminology

The word 'swamp' has been applied to the forested Everglades area of Florida, to marshes, bogs, reed fields, and also to Canadian muskeg. These areas differ greatly from each other in the extent to which the water surface is exposed, in their aerodynamic roughness, in the absorption of radiation, and in the rate of carbon dioxide intake. Considerable literature on 'swamps' comes from Russia, where unfortunately the same word (*boloto*) refers to mire, fen, and bog (Dooge, 1972). So the terminology used in the literature is confusing. Here, a 'mire' is any vegetated area with a wet organic substrate, one kind being a 'swamp', where there is standing surface water. The particular form of the swamp depends on the extent to which the water is stagnant, permanent, saline, acidic, or deep, and the underlying soil inorganic or peat (Godwin *et al.*, 1941; Anon, 1962; Bulavko, 1971; Krolikowska, personal communication 1972). In the present survey, attention is focussed on reed fields in particular, because most of the available literature refers to this kind of swamp. However, many of the generalizations apply also to other kinds.

B. Vegetation

Reed fields may contain species of the following genera: *Scirpus*, *Typha* *Phragmites*, *Cyperus* and/or *Juncus*. Illustrations of many of the various canopies have been published by Stewart and Kantrud (1972). Species that grow in temperate climates are characterized by seasonal growth: new foliage grows in spring and dies in autumn (Young and Blaney, 1942) so that a mature stand will usually contain dead foliage, remaining from previous years.

Scirpus and *Typha* species are known as common bulrushes. They abound along water courses and on the edges of ponds in America (e.g. Christiansen and Low, 1970), where these species are called 'tules'. The stomata of *Scirpus validus*, for instance, are usually wide open (Eisenlohr, 1966a), allowing free transpiration.

Typha species are also commonly referred to as 'cattails' or hardstem bulrush (Yeo, 1964). They occupied 70% of the Australian swamp studied by Linacre *et al.* (1970). In Arizona the leaves of these species grow to 2 metres high between March and July, and then begin to die in October (McDonald and Hughes 1968). *Typha* grows best in only a few centimetres of water, depending on flow rate and aeration.

Phragmites communis is the common reed (Garter *et al.*, 1962), which may grow up to 6 m high (Gel'bukh, 1964) above water as much as 2 m deep (Young and Blaney, 1942). In Uganda, *Phragmites mauritanius* grows to 8 m high. Haslam (private communication, 1972) reported the height of foliage in English *Phragmites* stands as between 65 and 250 cm, with, say, 300 shoots in each square metre of the water surface for the lower growth, and about 100 per m^2 for the taller. A 1 m^2 area can contain 1·7 to 3·4 m^2 of leaf (Eisenlohr, 1966a), i.e. the leaf area index (L) is 1·7 to 3·4. Gel'bukh (1964) reported values of leaf area index for such reeds in Russia of 0·35 in May, 6·6 in August, and 1·8 in October. Thompson (private communication, 1973) referred to values of L over 10 in Africa.

Phragmites is discouraged by increased salinity of the water and substrate (Bird, 1961), and *Typha* and *Scirpus* require only slight brackishness. The salinity is measured in terms of the electrical conductivity of the water (Stewart and Kantrud, 1972: D19), and is affected indirectly by atmospheric conditions. Rain decreases the concentration, whilst evaporation and stagnation increase it.

Cyperus Papyrus, commonly called papyrus, thrives in Uganda, for example (Lind, 1956).

Juncus species are rushes (Young and Blaney, 1942).

C. Extent

Swamps may cover 2·3 % of the globe's land area—a figure given by Bulavko (1971) for 'marshes'. However 'marshes' include swamps and also areas with waterlogged soils of low organic content and, say, a cover of short grasses. About 9 % of the U.S.S.R. is marshland, notably in Byelorussia. Bavina (1971) mentioned 70 % of the world's swamps as being within U.S.S.R. but this figure does not refer to reed fields, as it is bogs that occupy tens of millions of hectares in Russia (Romanov, 1956).

Luther and Rzoska (1971) described over 30 inland wetlands. There are many thousands of acres of reed swamps in each of Rumania, Bulgaria, Sudan, and Algeria, as well as in Holland (Zonderwijk, 1962) and Uganda (Lind, 1956). Papyrus swamps cover 6·5 % of the total area of Uganda, including 60 km^2 associated with Lake Edward (K. Thompson, personal communication, 1973). Even in England there are some 5000 hectares in the Fens and 5000 in Kent. Robinson (1949) mentioned many thousand hectares of tules and phreatophytes (the latter depend on subterranean water) in Arizona and Missouri. In North Dakota, there are several hundred thousand natural ponds, most of which are at least partially covered by reeds (Shjeflo, 1968). Also many hundred hectares of swampland in New Jersey influence the pattern of flow of the Passaic River (Vecchioli *et al.*, 1962).

D. Significance

Some half million people inhabit the Sudanese swamps, and elsewhere such areas provide reeds for thatching, shelter for wild fowl, and peat. In the Danube delta, *Phragmites* is used on a large scale for paper production.

But swamps in general have been regarded as undesirable, requiring reclamation. They hinder communication and river flow, and tend to be
unproductive in an economic sense; they are associated with mosquito-
borne diseases. It has been thought that the foliage of reed fields acts like
wicks, increasing the loss of water by evaporation (e.g. Stearns and Bryan,
1925), thus depleting the amount available for irrigation farming, say
(Linacre et al., 1970). The last point is examined later in this chapter.

The relative inhospitality of swamps has deterred scientific examination
of their characteristics, but their extent and their influence downstream, and
the potential fertility of drained-swamp soils, are now attracting more
attention.

E. Previous Studies

The interaction of swamp and the atmosphere has received scant consideration, partly because of the difficulty of access. Another reason is the horizontal
structure, which is neither uniformly regular nor random, but clumped
(Ondok, 1970), and therefore not easy to represent. So the absorption of
incoming radiation, its attenuation within stands, and the optical characteristics of swamp foliage do not appear to have been elucidated yet. New
information will be available from Thompson and Gaudet (in press) and
from S. F. Phillips, who is studying the micrometeorology of a stand of
Phragmites communis in Manitoba. There is no published information on
the aerodynamic features of swamps or their foliage.

The main concern related to meteorology has been to determine the
evaporation rates of swamps, in particular for comparison with the rate of
evaporation from a lake in a similar environment. This is an important
matter to those considering the draining or clearing of swamps. If a swamp
loses appreciable water by evaporation, it represents a waste in an arid area,
whereas if it *reduces* the evaporation from a lake that would otherwise
exist, then it conserves water, and would have real value in much of Africa,
for example.

II. METHODS OF MEASURING EVAPORATION

A. Water-balance Method

The balance equation relating water outflow to inflow and change of storage
allows determination of the evaporation rate (E), as follows

$$E = P + I + \Delta H - O - S \tag{1}$$

where P is the precipitation, I the inflow, ΔH the fall in water level, O the outflow, S the seepage into the soil beneath. One complication is the change of water-surface area with change of water level (Christiansen and Low, 1970), and another is the change of water content of the soil which is exposed as the level falls. The seepage S is not easy to determine, and the accuracy of measuring precipitation is less than is commonly assumed, on account of spatial variation and the unrepresentativeness of gauge collection (Hutchinson, 1971). Even the change of water level (ΔH) is not simple to measure, because the hydrodynamic impedances of a large swamp, and the effect of wind may prevent uniformity of level. So ΔH is measured at, say, four places to find an average change of level (Christiansen and Low, 1970). The summation of all the errors in measuring $P, I, \Delta H, O$, and S leads to an appreciable error in the implied evaporation rate E. For example, an attempt to estimate the evaporation from a large Australian swamp by a water balance gave nonsensical results, largely because it was not possible to measure accurately the inflow and outflow (Linacre *et al.*, 1970).

The water-balance procedure has been refined to allow determination of seepage S, and to separate the evaporation from the water surface in a stagnant swamp from the transpiration from the foliage. Eisenlohr (1966a, 1966b) assumed that a form of the Dalton evaporation formula is valid for a swamp, namely,

$$E = Nu(e_w - e)$$

where N is a constant to be determined, u is the wind speed at some arbitrary height, e_w is the saturated water-vapour pressure at the temperature of the evaporating surface, and e is the vapour pressure at an arbitrary height. Meteorological measurements are taken in dry periods when precipitation is zero. Then, for stagnant water (for which $I = O =$ zero), the change of level ΔH equals the sum $(E + S)$ i.e. evaporation plus seepage. It is assumed that the seepage rate is constant, and mean values of $u(e_w - e)$ are graphed against measurements of ΔH over a few days. The slope of the graphed line is N, and the axis-intercept is S. This procedure was used by Shjeflo (1968), who calculated evaporation rates from potholes in North Dakota of about 2 mm/day.

The value of N fluctuates during the growing season of the vegetation, being 50% higher in summer than in spring, for instance. This increase has been attributed to the growth of vegetation, and the lowest value (N_w) has been ascribed to evaporation from the water surface alone (Eisenlohr, 1966b). The difference ($N - N_w$) may be attributed to evaporation from the foliage, and proves to be proportional to vegetation height (Eisenlohr, 1966a).

This ingenious procedure merits further examination, though it has weaknesses. The seepage rate may not be constant. It may not be appropriate to take only water-surface temperature to determine e_w, if leaves at different temperatures are also evaporating. Also, it may not be proper to assume a constant value for N_w, since canopy development will presumably reduce the evaporation from the water. Finally, the conditions in which ΔH is measured (i.e. with no rain) are those likely to promote the supply of energy by advection, so the values of N relate only to this special circumstance.

B. Swamp Vegetation in Tanks

An alternative to taking the water balance of an entire swamp is to study a representative sample, enclosed in a tank. The tank might be an oil-drum, with its top removed, set flush in the flooded ground, but with the rim above water level. The amount of water required to maintain its level within the tank at a fixed height is a measure of the evaporation from the swamp plants growing in the soil in the base of the tank. Duplicate tanks provide a check on leaks, which would alter the apparent evaporation rate (Dooge, 1972).

Gibbs and Partners (1956) inferred values of swamp evaporation rate from the arbitrary formula of Olivier and the well-known Thornthwaite formula. Monthly values differed by up to 40 % from the evaporation rates obtained from an experimental tank near Shambe in the Sudd region of the Sudan in 1947.

C. Bowen Ratio Method

A superior method of determining the evaporation rate from any uniform terrain involves the 'Bowen ratio' equation, and was used by Bavina (1967) and Rijks (1969) in the case of a swamp. The 'Bowen ratio' (β) is the ratio $C/\lambda E$, where C and λE are energy fluxes of sensible and latent heat, respectively, λ is the latent heat of evaporation of unit mass of water (about 2450 J/g), and E is the evaporation rate. The sum ($C + \lambda E$) is the outgoing energy from the swamp to the atmosphere, and roughly equals the incoming net radiation flux $\mathbf{R}_n$, which can be measured directly, or estimated (Linacre 1969b).

$$\mathbf{R}_n = \mathbf{C} + \lambda \mathbf{E} = \lambda \mathbf{E}(1 + \beta) \qquad (2)$$

Hence

$$\mathbf{E} = \mathbf{R}_n/\lambda(1 + \beta). \qquad (3)$$

The Bowen ratio (β) represents the partition of the available energy, and can be inferred from measurements of dry-bulb and wet-bulb temperatures at two heights. Values of evaporation rate from Estonian bogs obtained in this way by Bavina (1967) were as much as 30 % less than tank measurements. Bowen ratio estimates of evaporation are more likely to be accurate than are tank measurements, because they involve undisturbed natural vegetation, and average the evaporation from the substantial area which affects the air temperatures measured at, say, 1 metre and 3 metres above the foliage.

D. Other Methods

Similar advantages also apply to the eddy-correlation method used by Linacre et al. (1970). Electric currents, generated by a light propeller driven by vertical air motions, are linked electronically to a temperature sensor, so that the heat flux C is measured, and λE is then obtained by subtraction from the measured net radiation R_n. The instrument is free of the errors of measuring small temperature differences, associated with the Bowen ratio method.

A different method of measuring the evaporation from individual reed leaves has been used by Krolikowska (1971). This is the botanists' well-known method of repeatedly weighing an excised leaf or shoot about a metre in length. However, such a method gives figures impossible to relate to the evaporation of a canopy, and even an intact leaf will transpire differently because of a different internal water pressure and hence stomatal condition, different attitude with respect to sun and wind, and so on. Gel'bukh (1964) obtained larger weight losses from excised reeds in 1954 than 1955 despite hotter and drier conditions in the later year. He also measured greater evaporation than from reeds growing in tanks.

This brief survey indicates that the best methods of measuring the evaporation from extensive swamps are the Bowen ratio and the eddy-correlation methods.

III. EVAPORATION RATES

A. Effect of Advection

Swamps may be either extensive areas, or else narrow strips lining a pond or river, for example. The evaporation rates of the two kinds differ greatly during dry periods, when the environs are arid. Evaporation from plants at the edge of an extensive area, or along a narrow strip of swampland, is enhanced by the dryness and warmth of the oncoming wind, i.e. by the advection of energy. The advected heat supplements the incoming net radiation, providing more energy for evaporation. In fact, advection affects all swamp evaporation except in the middle of a large swamp, where the air is remote from the influence of the surroundings. Advection specially affects the evaporation of an oasis, such as a tank of swamp vegetation set in a bare field (Young and Blaney, 1942; National Resources Planning Board, 1942). An exposed tank of triangular-stemmed bulrushes at Victorville in California evaporated water three times faster than a tank set within a swamp (Turner, personal communication). A tank containing *Scirpus*, exposed to desert conditions 6 metres outside a swamp in California, evaporated 159 cm whilst two tanks with similar vegetation located 8 m *inside* the swamp each lost only 35 cm (Young and Blaney, 1942). These figures demonstrate the large effect of advection on evaporation. Lang *et al.* (1972) found that over half the evaporation energy came from advection, in the case of an Australian rice crop extending several hundred metres.

Unfortunately most of the measurements on evaporation from swamp vegetation have been made with tanks of plants set conveniently only a few metres within the edge of a swamp, or within the fringe of swamp around a lake (e.g. National Resources Committee, 1938; McDonald and Hughes, 1968), so values obtained there, whilst relevant to similar strips, are greater than those from an extensive swamp. The Russian authorities recognize the importance of the boundary effects, in recommending the placing of stations for hydrological measurements at the centre of the swamp (Anon, 1961).

B. The Clothes-line Effect

Advection comprises the 'oasis effect' due to horizontal differences of wetness in a region, and the 'clothes-line effect' due to differences of geometry. The latter creates increased evaporation resulting from ventilation *through* the canopy. It occurs at the edge of a swamp, or in an isolated stand of the vegetation, or through the protuberant part of especially vigorous growth. Clothes-line evaporation through an isolated tank of cattails at Carlsbad in New Mexico may have partly accounted for water loss greater than from the exposed water surface within a U.S. Class A pan evaporimeter (National Resources Planning Board, 1942). Also, the taller foliage resulting from more vigorous growth of tules in one tank in a small swamp at Mesilla Dam in New Mexico (National Resources Committee, 1938) evaporated 136 cm during July—December, compared with 85 cm from a tank of smaller plants. Similarly, McDonald and Hughes (1968) showed that the water loss from one tank of cattails in an Arizona swamp was up to 50 % greater than from two adjacent tanks of cattails, 69 to 90 cm less tall. These results show the importance of measurements on typical plants, whereas tules grown in tanks tend to differ in growth from the undisturbed, freely ramifying plants of the surrounding swamp (Young and Blaney, 1942).

C. Effect of Climate

Figure 1 shows the expected maximum evaporation rates in high summer, corroborated by Blaney and Hanson (1965). The summer maximum is due partly to the relative lushness of the vegetation between July and September (Eisenlohr, 1966b).

The importance of advection is illustrated by Fig. 1, where the order of evaporation rates is the order of the aridity of the regions. The evaporation

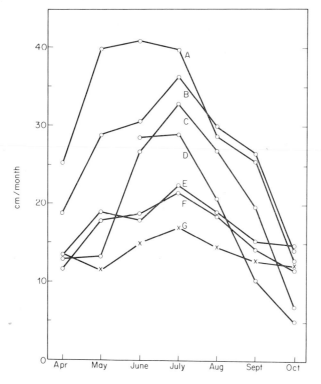

Fig. 1. Tules evaporation. (A) Mitty Lake, Ariz. (McDonald and Hughes, 1968). (B) Victorville, Calif. (Blaney, 1959). (C) Albuquerque, N. Mex. (Blaney, 1959). (D) Alamoso, Col. (Blaney, 1959). (E) San Luis Rey, Calif. (Blaney, 1959). (F) Bonsall, Calif. (Blaney, 1959). (G) San Bernardino, Calif. (Blaney, 1959).

from tules in the Californian desert is typically 196 cm/year, but nearer the coast only 147 cm/year (Blaney, 1961).

D. Effect of Kind of Vegetation

Experiments with various species in different tanks at Santa Ana gave monthly-mean evaporation rates as follows (Young and Blaney, 1942):

	July	January
roundstem	75	6 cm/month
triangular tules	69	7
cattails	47	6

Likewise at King Island in the Delta, California, Young and Blaney (1942) found:

	August	January
tules	39	5 cm/month
cattails	37	3

Bearing in mind the impreciseness of this kind of measurement, it appears that differences between the plant types are relatively unimportant, as also observed by Parshall in 1937 (Young and Blaney, 1942). Advection was presumably responsible for the very large values and the differences between evaporation rates for tules and cattails reported by Robinson (1952) for different sites in Idaho namely, 107 cm and 154 cm, respectively, for 22 June to 30 September.

E. Effect of Canopy Structure

It was assumed in a Ugandan report (Gibb and Partners, 1956) that shade and mulching provided by papyrus would *reduce* the evaporation from the wet soil beneath. This assumption was based, they said, on Migahid's observations that 'water lost by evaporation from a free water surface inside a swamp was about 20% of the loss from a free water surface in the open'. The logic of this is not clear.

Canopy density affects the wind flow and light penetration through the vegetation, and hence influences evaporation from the water beneath (Gel'bukh 1964). A denser canopy may be expected to increase transpiration, at least in the range of small values of the leaf area index, but will reduce the water-surface evaporation (Shjeflo, 1968; Eisenlohr, 1966a). In this connexion, the results of Rijks (1969) show that swamp evaporation increases with wind, though the relationship between the two may be complicated by an effect of wind on canopy roughness, as in the case of rice (Evans, 1971).

The contribution to the total evaporation rate made by water beneath a swamp canopy is indicated by measurements on cattails and tules in tanks in a swamp in California (Young and Blaney, 1942). Evaporation in August, was 37 cm when the water table was just at ground level, but 45 cm when the plants stood 25 cm in water.

F. Relationship to Incoming Radiation

A water-budgetting procedure was used by Eisenlohr (1966b) on a typical North Dakota pothole, i.e. a pond of several hectares, fringed or (partly) covered by bulrush or cattails (Shjeflo, 1968). In the summer season, May to September, when most growth and evaporation occur, Eisenlohr found a precipitation inflow of 39·9 cm, a decrease of level of 14·9 cm (with consequent reduction of evaporating area), run-on from land around of 6·9 cm, seepage outflow of 11·7 cm, and evaporation of 50·3 cm. Dominance of the evaporation was also reported by Shjeflo (1968) in similar circumstances for an annual precipitation of 36 cm, spring snowmelt 24 cm, run-on of 16 cm, seepage 18 cm, evaporation 63 cm. In other words, at the height of summer the evaporation rate was around 13 cm/month.

At a latitude of 50° N in June, the extra-terrestrial radiation intensity is 475 W m^{-2} and the day-length (N) is 16·3 hours (Linacre, 1969b). Hence the ground-level global radiation (S_t) is 165 W m^{-2}, assuming Angstrom's modified formula (Linacre, 1969a), and a bright-sunshine duration (n) of 10 hours daily. Then the net radiation intensity (R_n) is given as follows (Linacre, 1969b):

$$R_n = (1 - \rho)S_t - 1·1\left(0·2 + 0·8\frac{n}{N}\right)(100 - T) \text{ W m}^{-2} \tag{4}$$

where ρ is the surface albedo, and T is the mean temperature (°C). If T is 18°C, and ρ is 22% (Bray, 1962), R_n is 145 W m^{-2}.

Evaporation at 13 cm/month, say, absorbs heat at the rate of 120 W m^{-2} which corresponds to almost all of the incoming net radiation energy (R_n). Thus much faster rates (e.g. Fig. 1) would necessarily imply considerable advection.

A relationship between the evaporation (E) from *bogs* and incoming net radiation was derived empirically by Romanov (1962):

$$\lambda E = aR_n + b \tag{5}$$

where λ is the latent heat of evaporation of water and a, b are nominally constants, though b is said to represent advection, normally small enough to be disregarded (Bavina, 1971). The factor 'a' depends on the depth of the water table in the bog, with values up to 0·82 when λE and $\mathbf{R}_n$ are in the same units (Romanov, 1956), or 0·7 if corrected values of $\mathbf{R}_n$ are used (Bavina, 1967). Such a value from Russia is in good accord with the deductions of Priestley and Taylor (1972) that the ratio $\lambda E/\mathbf{R}_n$ for any extensive well-watered region should be 0·71 at 10°C, 0·86 at 20°C, etc. In the relatively warm climate of Uganda, Rijks (1969) obtained 0·7 for the ratio of $\lambda E/\mathbf{R}_n$, whilst for a *Typha* marsh in Canada, Bray (1962) deduced values which imply 0·49.

For *rice*, the ratio $\lambda E/\mathbf{R}_n$ is about 1·7 (Evans, 1971), implying substantial advection of energy in the air flow, to supplement $\mathbf{R}_n$ and thus increase the latent-heat flux λE (see also p. 51). Such advection is plausible with measurements only 60 metres from the edge of the crop. Measurements on small plots can lead to large errors when extrapolated to extensively irrigated areas (Lourence *et al.*, 1970).

G. Comparison of Tank and Pan Evaporation Rates

Measurements with a tank of swamp plants 3 m within a swamp, and a pan of water 5 m outside gave mean values at Isleta, New Mexico, of 170 cm and 111 cm respectively for June to November (National Resources Committee, 1938). Likewise at Parma, Colorado, a tank of tules evaporated water 1·26 times faster than a pan of water (Young and Blaney, 1942). Christiansen and Low (1970) also reported that cattails evaporated more than a pan evaporimeter, especially in summer. Measurements on tules at San Luis Rey and at Victorville showed *annual* evaporation rates roughly equal to pan evaporation (Phreatophyte Subcommittee of Pacific Southwest Inter-agency Committee, 1957, based on work by Blaney, Taylor and Young, 1933; Muckel and Blaney, 1945) but evaporation from tules was faster in *summer*. In Russia too, a tank of reeds, area 0·3 m^2, evaporated more than an adjacent pan of water (Gel'bukh 1964). The excess evaporation from reeds is presumably due to a combination of edge and clothes-line effects influencing the vegetation, and the sheltering of the pan evaporimeter adjacent to it.

H. Comparison of Swamp and Lake Evaporation Rates

Pan evaporimeters are samples of lakes, known to evaporate water faster, on account of advection, exposure of the sides to sunshine, and so on. For a U.S. Class A pan, the ratio of pan to lake evaporation is commonly taken to be about 1·4, so rates of swamp evaporation faster than pan evaporation must exceed lake evaporation rates even more. Thus, early measurements on swamp and pan evaporimeters led to the conclusion that the vegetation substantially augmented the water loss (Gel'bukh, 1963).

This conclusion might be expected also from considerations of surface roughness (Linacre *et al.*, 1970) and thermal response. The small weight of a canopy, compared with an equal volume of water, makes it heat more rapidly in the day, which presumably promotes evaporation. Indeed, summertime measurements by Young and Blaney (1942), at Prado in California, showed an earlier start of evaporation from tules each day, but a more rapid decline in the early afternoon presumably due to the relatively small thermal capacity of the canopy. In addition, the evaporation from the water surface within a swamp is expected to be faster because shallow water allows more rapid heating. Shjeflo (1968) observed an increased evaporation rate from shallow potholes in hot days. In an Australian swamp, the water surface during day-time was about 1°C *hotter* than the air above it in some summertime experiments, whereas a nearby lake surface was 3 to 5°C *cooler* (Linacre *et al.*, 1970). The swamp water would probably have been even hotter in the absence of the reeds, because of its shallowness and turbidity (Rose and Chapman, 1968).

Gel'bukh (1964) used an empirical formula for lake evaporation and ignored the net-radiation flux. Also reed-field evaporation was calculated from an expression depending solely on the air dryness, irrespective of wind speed or net radiation. The ratio of the two expressions of unexplained origin, each containing empirical numerical constants, implies a swamp evaporation greatly in excess of lake evaporation. This deduction by Gel'bukh was justified by the notion that vegetation absorbs more solar energy than water, whereas in fact it will retain *less*, having a greater reflectivity.

More recently, Rijks (1969), working at the middle of a Ugandan swamp of papyrus (*Cyperus papyrus*) 500 m wide and 3·8 m high, obtained daily values of evaporation in the range 0·38 to 0·81 times the calculated lake evaporation rate. The ratio was about 0·48 when lake evaporation was between 2 and 4 mm/day, 0·61 for 4 to 6, and 0·66 for 6 to 7 mm/day, reflecting the dependence of $\lambda E/R_n$ on temperature mentioned earlier. Rijks concluded that the dead foliage in an old stand of papyrus acts as a major impedance to the evaporation process. He suggested that Penman's analysis of Migahid's observations (Penman, 1963), pointing to equality of water evaporation and swamp vegetation evaporation, was due to the plants in that experiment being luxuriant and freshly transplanted.

The evaporation of bogs is less than that of open water but approaches equality during wet summer months in Germany, according to Eggelsmann

(1963). In contrast, Bulavko (1971) reported that evaporation from a water-logged marsh is often 20 to 40% more than from a lake. In neither case was the evidence given. Bay (1967) measured the monthly evaporation from two Minnesota bogs of about 9 hectares and found that the rates during May to October varied between 88 and 121% of lake evaporation, estimated by an approximate formula. In short, rough equality with lakes is probably the most reasonable inference for bog evaporation.

A recent paper by Thompson and Boyce (1972) on the evaporation from South African sugar cane is relevant to the case of reed swamps, as the geometry is similar. With well-irrigated crops, the measured evaporation rate was about the same as that from a Class A pan evaporimeter, i.e. more than from a lake.

Hall et al. (1972) used the water-balance equation (1) to calculate the evaporation from 3 ha of mixed vegetation (not reeds) within a 4-ha swamp in New Hampshire. During a June—August period, precipitation was 20 cm, inflow 11 cm, and the fall in water level 35 cm. The submerged peat associated with the swamp was reckoned to contain 30% by volume of water, and the evaporation rate from the 1 ha of clear water was calculated from the well known Penman equation as 20 cm. These figures imply an evaporation rate from the vegetation of 36 cm, i.e. appreciably more than from the open water.

The only comparisons of the *measured* evaporation of a swamp and the *measured* evaporation of open water in the same climate were by Eisenlohr (1966a) and Linacre et al. (1970). The former found the average evaporation loss for May–October was 60 cm for vegetated patches and 67 cm for pot-holes without vegetation. In other words, the implication is that swamp vegetation *reduced* evaporation. The measurements of Linacre et al. (1970) were in the middle of an Australian swamp of 3000 ha and, simultaneously, a 240 ha lake 16 km away. The ratio of evaporation rates in summer depended on whether the surrounding country was wet or dry. When the country was dry, the swamp evaporation over four day-time hours was only about a third that of the lake, whereas, after heavy rain had wetted the entire region, lake evaporation was reduced to about that of the swamp. This change emphasizes the importance of the oasis effect on the relatively small lake, and differences between lake and swamp vegetation are trivial in comparison.

In other words, it might be concluded that swamp vegetation *enhances* evaporation on the basis of Gelbukh's measurements, considerations of surface roughness and thermal response, analogous studies on sugar cane, and the measurements by Hall et al. But the contrary and more sound conclusion was reached in the later, more sophisticated field measurements by Shjeflo et al. In any case, the effects of advection can be overwhelming.

When advection is negligible, it is possible to use a simple formula to estimate swamp evaporation, derived from that of Penman. A formula for the evaporation rate for a lake given by Linacre (1969a) reduces to the following:

$$\lambda E_0 = \frac{10 \mathbf{R}_n + 100(T_{max} - T_{min})}{22 - 0.15(T_{max} + T_{min})} \tag{6}$$

assuming a wind speed of 3 m s^{-1}, where λ is the latent heat of evaporation, $\mathbf{R}_n$ is given by Eq. (4) and T_{max} and T_{min} are respectively the daily maximum and minimum temperatures at 2 m, say. If $\mathbf{R}_n$ is 145 W m^{-2}, T_{max} is 24°C and T_{min} is 12°C, λE_0 is 100 W m^{-2}, i.e. E_0 is 11 cm/month. This resembles the 13 cm/month mentioned in Section III.F.

IV. THE RATE OF GROWTH OF SWAMP PLANTS

Thompson (1973, personal communication) calculated large apparent efficiencies of over 5 % for the conversion of incident radiation in the photo-synthesis of tropical papyrus. However, the fast growth rate is not necessarily a result of rapid photosynthesis, but may simply represent a remobilization of reserve material stored within the submerged parts. This is the case for reeds growing in Czechoslovakia (Dykyjova et al., 1970).

Considerable plant-physiological information is available about the photosynthesis of individual swamp plants (e.g. McNaughton and Fullem, 1969), as distinct from natural stands. The intake of carbon dioxide in photosynthesis by *Phragmites communis* in Austria has been examined by Burian (1971). The growth of swamp vegetation is favoured by abundant water and by vertical leaf orientation, which reduces intense heating by the sun and mutual shading of the leaves. Also swamps often occur in regions of much sunshine and warmth. As a consequence, the rate of net dry-weight production is as high as 2.7 to 3.3 kg m^{-2} per year for *Spartina alterniflora* in Georgia or *Phragmites communis* in Long Island, and 6.4 kg m^{-2} per year for cultivated *Scirpus americana* in Germany (Keefe, 1972). In energy terms, these rates may represent up to 1.5 % of the incident radiation. The figure for *gross* photosynthesis, which makes no allowance for respiration, may be up to 6 % in the case of *Spartina*. For *Typha* marsh, Bray (1962) calculated that the overall dry-matter production accounted for 1.1 % of the incoming radiation.

V. CONCLUSIONS

A thorough search of the (English) literature has demonstrated the paucity of knowledge on the physics of swamps. Most work has been on evaporation rates, but the *process* of evaporation from a reed field, say, remains a matter of ignorance. It is clear that advection of energy from outside a small swamp greatly increases evaporation rates, as in the case of a small body of water. This complicates the comparison of swamp and lake evaporation. Compared with regional climate and local advection, the presence of swamp vegetation and its type are relatively minor influences on evaporation rates, at least whilst the vegetation is actively growing.

ACKNOWLEDGMENTS

Several dozen experts from many countries kindly supplied information for this survey. In particular, I thank Dr Julian Rzoska (International Biological Program, Central Office, London) for advice, and also Dr David Mitchell (University of Rhodesia) and Professor Samuel Snedaker (Center for Aquatic Sciences, Gainesville, Florida); Mr Harry Blaney for helpful discussions and guidance through the early literature held at the Water Resources Center at the University of California, Los Angeles; Commonwealth Bureau of Soils (Commonwealth Agricultural Bureaux, Harpenden, England) for compiling an annotated bibliography (number 1546, June 1972) on Soils, Vegetation and Reclamation of Swampy Areas (1967 to 1971): The Director of the Water Resources Scientific Information Service (Office of Water Resources Research, U.S. Dept. of the Interior) for a computer scan and printout of abstracts on all holdings concerning swamps; Dr Don Boelter (US Dept. of Agriculture, Grand Rapids, Minnesota), Dr Joseph Ondok (Czechoslovak Academy of Sciences), Dr S. S. Khodkin (International division, Hydrometeorological Service of USSR) for extensive source material; and Dr Keith Thompson of Makere University (Uganda) for expert comments on a draft of this paper.

REFERENCES

Anon. (1961). Instructions for Hydrometeorological Stations and Posts, No. 8: Manual of Hydrological Observations in Swamplands. Meteorological Translation No. 6, of Department of Transport, Canada, p. 25.
Anon. (1962). Project Mar: Conference on the Conservation and Management of Temperate Marshes, Bogs and other Wetlands, Les Saintes Maries de la Mer, 12–16 Nov. 1962, **2**, 10.
Bavina, L. G. (1967). *Soviet Hydrology* **4**, 348–370. (Am. Geophys. Union, 1968, translated from *Trudy GGI*: 69–96, 1967).
Bavina, L. G. (1971). Water Balance of Swamps and its Computation. Symp. on World Water Balance, Internat. Assoc. Sci. Hydrol. Unesco, Vol. 2, pp. 461–466.

Bay, R. R. (1967). Evapotranspiration from two peatland watersheds. Geochemistry, Precipitation, Evaporation, Soil-moisture, Hydrometry, Gen. Assembly, Bern, Sept.–Oct., pp. 300–307.

Bird, E. (1961). *Vict. Nat.* **77**, 261–268.

Blaney, H. F. (1959). Monthly coefficients for consumptive use of water formula ($u = kf$) and use of water by phreatophytes. Meteorological Meeting, Tel Aviv, May 6th' 9, 12.

Blaney, H. F. (1961). *J. Irrig. Drain. Div. Proc. Am. Soc. Civil Engrs*, Sept., 37–46.

Blaney, H. F. and Hanson, E. G. (1965). Consumptive use and water requirements in New Mexico, Tech. Rept. 32, N. Mex. State Engineer, Santa Fe, p. 43.

Blaney, H. F., Taylor, C. A. and Young, A. A. (1933). Water losses under natural conditions in wet areas in Southern Calif., State Dept. of Public Works, Div. of Water Resources, Bull. 44.

Bray, J. R. (1962). *Science*, **136**, 1119–1120.

Bulavko, A. G. (1971), *Nature Res.* **7**, 12–14.

Burian, K. (1971). *Hidrobiologia* **12**, 203–218.

Christiansen, J. E. and Low, J. B. (1970). 'Water Requirements of Waterfowl Marshlands in Northern Utah,' 19 Utah Division Fish and Game, Wildlife Management Inst., Bureau of Sport Fisheries and Wildlife.

Dooge, J. (1972). 'The water balance of bogs and fens'. Paper to Internat. Symp. Marshridden Areas, Minsk, July.

Dykyjova, D., Ondok, J. P. and Priban, K. B. (1970). *Photosynthetica*, **4**, 280–287.

Eggelsmann, R. (1963). Die potentielle und aktuelle evaporation eines Seeklimahochmoores. Internat. Assoc. Sci. Hydrol. Publication No. 62, pp. 88–97.

Eisenlohr, W. S. (1966a). *Water Resour. Res.* **2**, 443–453.

Eisenlohr, W. S. (1966b). Determining the water balance of a lake containing vegetation. Internat. Assoc. Sci. Hydrol. Publication No. 70, Hydrology of Lakes and Reservoirs, **1**, 91–99.

Evans, G. N. (1971). *Agric. Met.* **8**, 117–127.

Garter, H. P., van Dijk, J., Docter, D. A. and Westhoff, V. (1962). Conservation and management of the Netherlands lowland marshes. Project Mar: Conference on the Conservation and Management of Temperate Marshes, Bogs and other Wetlands, Les Saintes Maries de La Mer, 12–16 Nov. 1962. **1**, 248–259.

Gel'bukh, T. M. (1963). Evaporation from overgrowing reservoirs. Internat. Assoc. Sci. Hydrol., Gen. Assembly, Berkely, 19–31 Aug. 1963.

Gel'bukh, T. M. (1964). *Trudy GGI*, **92**, 152–174, (translated by D. B. Kringold in *Soviet Hydrol.* **4**, 363–382).

Gibb, A. and Partners (Africa). (1956). Water Resources Survey of Uganda 1954–5, Applied to Irrigation and Swamp Reclamation: Supplementary Report on Swamp Hydrology, pp. 9–12.

Godwin, H. *et al.* (1941). *Chron. Bot.* **6**, 260–61.

Hall, F. R., Rutherford, R. J. and Byers, G. L. (1972). The influence of a New England wetland on water quantity and quality. Univ. of New Hampshire, Durham, N.H., Water Resource Center, Research Department Rept. 4, 51 pp.

Hutchinson, P. (1971). *Weather* **26**, 366–71.

Keefe, C. W. (1972). *Contrib. Mar. Sci.* **16**, 163–181.

Krolikowska, J. (1971). *Pol. Arch. Hydrobiol.* **18**, 347–358.

Lang, A. R. G., Evans, G. N. and Ho, P. (1972). C.S.I.R.O. Irrigation Research Division Annual Report 1971–72, pp. 8–9.

Linacre, E. T. (1969a). *J. Irrig. Drain. Div. Proc. Am. Soc. Civil Engrs* **95**, 348–352.

Linacre, E. T. (1969b). *J. appl. Ecol.* **6**, 61–75.

Linacre, E. T., Hicks, B. B., Sainty, G. R. and Grauze, G. (1970). *Agric. Met.* **7**, 375–386.

Lind, E. M. *Uganda J.* **7**, 375–386.

Lourence, F. J., Pruitt, W. O. and Servis, A. (1970). Energy balance and the crop water requirements of fescue and rice grown in California. Univ. of Calif., Davis, Water Science and Engineering Paper No. 9002.

Luther, H. and Rzoska, J. (1971). Project Aqua. Internat. Biological Program Handbook No. 21 (Blackwell).

McDonald, C. C. and Hughes, G. H. (1968). Studies of consumptive use of water by phreatophytes and hydrophytes near Yuma, Arizona. U.S. Geological Survey Prof. Paper 486–F:F15–F17.

McNaughton, S. J. and Fullem, L. W. (1969). *Pl. Physiol.* **45**, 703–7.

Muckel, D. and Blaney, H. F. (1945). Utilization of the waters of the lower San Luis Rey Valley, San Diego County, Calif., U.S. Dept. of Agric.

National Resources Committee. (1938). Regional Planning pt. VI–The Rio Grande Joint Investigation in the Upper Rio Grande Basin in Colorado, New Mexico, and Texas, 1936–1937, Supt. Documents, Washington pp. 373, 393.

National Resources Planning Board. (1942). The Pecos River Joint Investigation: reports of the participating agencies, Washington, p. 200.

Ondok, J. P. (1970). *Preslia (Praha)* **42**, 256–261.

Penman, H. L. (1963). Vegetation and Hydrology. Commonwealth Agric. Bureau Tech. Commun. 53, Farnham Royal, Bucks., England.

Phreatophyte Subcommittee of Pacific Southwest Inter-agency Committee (1957). Symposium on Phreatophytes, Pacific Southwest Regional Meeting, Amer. Geophysical Union 14–15 Feb. 1957, Sacramento, p. 28.

Priestley, C. H. B. and Taylor, R. J. (1972). *Mon. Wea. Rev.* **100**, 81–92.

Rijks, D. A. (1969). *Quart. J. R. met. Soc.* **95**, 643–9.

Robinson, T. W. (1949). Areas and use of water by phreatophytes in the Western United States, U.S. Dept. of Interior, Geological Survey, 11 pp., Carson City, Nev.

Robinson, T. W. (1952). *Trans. Am. geophys. Union* **33**, 57–66.

Romanov, V. V. (1956). *Pochvovedenie, (Moscow)* **8**, 49–56. (Translated by Israel Program for Scientific Translations, Office of Tech. Services, U.S. Dept. of Commerce).

Romanov, V. V. (1962). *Gidrometeorizdat (Leningrad)* (ISPT, Jerusalem 1968). Reference from Dooge (1972).

Rose, C. W. and Chapman, A. L. (1968). *Agric. Met.* **5**, 391–409.

Shjeflo, J. B. (1968). Evapotranspiration and the water budget of prairie potholes in North Dakota. U.S. Geological Survey, Prof. Paper 585-B.

Stearns, H. T. and Bryan, L. L. (1925). Preliminary report of the geology and water resources of the Mid Lake Basin. U.S. Geological Survey, Water-Supply Paper 560-D, 134 pp.

Stewart, R. E. and Kantrud, H. A. (1972). Vegetation of prairie potholes, North Dakota, in relation to quality of water and other environmental factors. U.S. Geological Survey, Prof. Paper 585-D.

Thompson, G. D. and Boyce, J. P. (1972). Proc. 14th Congr. Internat. Soc. Sugar Cane Technol.

Thompson, K. and Gaudet, J. J. *Archiv. Hydrobiol.*, (in press).

Vecchioli, J., Gill, H. E. and Lang, S. M. (1962). *J. Am. Water Works Assoc.* **54**, 695–701.

Yeo, R. R. (1964). *Weeds* **12**, 284–287.

Young, A. A. and Blaney, H. F. (1942). Use of water by native vegetation. State of Calif. Dept. of Public Works, Div. Water Res., Bull. No. 50:51–54, 60–63, 83–89.

Zonderwijk, P. (1962). Management of reedlands in the Netherlands. Project Mar: Conference on the Conservation and Management of Temperature Marshes, Bogs and other Wetlands, Les Saintes Maries de La Mer, 12–16 Nov. 1962. **1**, 279.

12. Grassland

E. A. RIPLEY and R. E. REDMANN

Department of Plant Ecology, University of Saskatchewan, Saskatoon, Saskatchewan, Canada

Canadian Committee for the International Biological Programme Contribution No. 199.

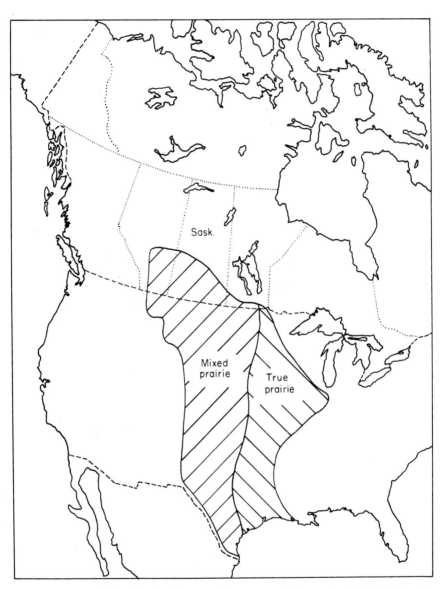

Fig. 1. Map of North America showing approximate extent of the grasslands in the eighteenth century.

I. INTRODUCTION

A. The North American Grasslands

1. *Description*

The great mid-continental grasslands of North America once stretched from the southern Canadian plains to Mexico (Fig. 1). The eastern portion, the true prairie, having a more favourable climate was dominated by tall grass species. To the west, the mixed prairie, characterized by progressively shorter grasses, dominated the landscape all the way to the Rocky Mountains. The shift to shorter grasses in the rainshadow of the Rocky Mountains was probably in response to the drier climate, but soil conditions and grazing by native ungulates, and later by cattle and sheep, also influenced the grassland composition. Most of the grassland on arable soil, in areas of adequate precipitation, has long since been plowed, but enough original grassland has remained to allow ecologists to reconstruct its former extent and composition (Shelford, 1963).

2. *Climate*

The climate of the North American grassland has been described in some detail by Borchert (1950). A more recent study by Bryson (1966) attempted to relate vegetation zones to frequencies of air-mass residence and modal positions of frontal zones. Except for the extreme southern part of the grasslands, the climate is distinctly continental. There are large annual and diurnal temperature ranges; precipitation is low and, during the summer, greatly exceeded by potential evaporation; solar radiation and wind speeds are relatively high.

Extending from latitude 25° N to 55° N the grassland spans a wide range of climatic conditions. Mean annual temperatures range from 0°C in the north to 20°C in the south. Annual precipitation varies from 250 mm in the foothills of the Rocky Mountains to 1000 mm just south of the Great Lakes. Mean annual snowfall is near zero in the south, increasing to 800 mm in the north and to considerably higher values near the mountains.

The northern tip of the North American grassland covers south-western Saskatchewan and south-eastern Alberta and falls within Köppen's 'middle latitude dry' climatic region. Carder (1970) discussed the climate of this area of Canada in relation to the natural vegetation.

In the mixed prairie region of south-western Saskatchewan (Fig. 2) the mean annual precipitation is 300 to 400 mm, of which approximately one-third falls as snow. Mean temperatures range from about $-15°C$ in January to $+20°C$ in July. The snowpack usually persists through the winter with

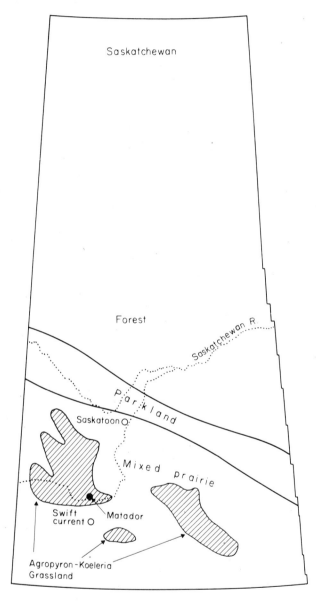

Fig. 2. Map of Saskatchewan showing major vegetation zones and the location of areas of Agropyron-Koeleria grassland.

little sublimation loss because of the low temperatures. A large fraction of the snowmelt enters the soil when the thaw arrives in late March (MacKay and Thompson, 1968). This is supplemented by the rains of June and July during which another third of the annual precipitation falls. The summer rainfall is almost equally split between convective and frontal storms. About one-third falls in rainstorms of less than 10 mm (Ripley, 1972a) so that canopy interception may cause an appreciable loss of 'potential' soil moisture, particularly during dry spells. The presence of snow cover for about four months during the winter (MacKay and Thompson, 1968) substantially reduces the annual potential evaporation loss and also insulates the soil from the extremely cold midwinter air masses. Although air temperatures drop as low as $-50°C$ during the winter, with a covering of 20 to 40 cm of snow the soil rarely cools below $-10°C$ (Ripley, 1972a).

B. Vegetation and Soil of the Matador Area

Most of the Canadian portion of the North American grassland belongs in the mixed prairie (Stipa-Bouteloua) association (Coupland, 1950). This type is sub-divided into faciations, defined mainly on the basis of climate (especially precipitation–evaporation ratio) and soil type (Coupland, 1961). In Saskatchewan this ratio increases from the drier south-west corner of the province to the north-east, and a gradient of vegetation and soil type also occurs (Richards and Fung, 1969). In the south-west, grass species such as *Bouteloua gracilis* and *Stipa comata* are dominant while in the areas to the north and east *Agropyron* spp. and *Stipa spartea* var. *curtiseta* are most important. Soil type, being related to climate and vegetation, also changes; brown soils in the south-west grade into black soils in the north. The Agro-pyron-Koeleria faciation once occupied areas of lacustrine clay soils (Fig. 2). Nearly all of this grassland type has been plowed, because the heavy clay soil has a high capacity for holding water and nutrients. One remaining area near Matador, Saskatchewan (Fig. 2) was selected in 1967 as the site of a grassland ecosystem analysis study (The Matador Project) under the International Biological Programme. *Agropyron dasystachyum* is the dominant species in the faciation; other important species are *A. smithii*, *Koeleria cristata*, *Carex eleocharis*, *Stipa viridula*, and *Artemisia frigida*.

A study by Coupland (1959) found considerable variation in species composition of the grassland of this area due to natural climatic variation. He compared observations in 1944 (following near normal precipitation during the period 1939 to 1943) with similar observations in 1955 (precipitation 20% above normal during 1950–54). The percentage of *Agropyron* spp. nearly doubled from the earlier to the later period.

Fig. 3. Photograph of the native vegetation at Matador, Sask. The grass has been clipped away near ground level leaving a slice 10 cm thick between the camera and the background. The camera was pointing towards the south.

A photograph of a section of the vegetation at Matador is shown in Fig. 3. The mean height is near 20 cm with a few stems and flowering stalks to 40 cm and occasionally higher. The dense mat of dead leaves below 10 cm is a distinctive feature of this grassland. Many dead leaves are caught in this mat and prevented from falling to the ground until assisted by high winds, heavy rain or hail, or the hooves of larger animals. The winter snowpack usually plays a major role in bringing 'standing dead' down to the soil surface. The dry prairie climate helps preserve the standing dead and decomposition is very slow until it reaches the somewhat moister soil surface.

The presence of this large 'standing dead' component (approximately twice the maximum green biomass) has a considerable effect on the microclimate near the surface and plays an important role in the surface–atmosphere energy exchange. Some of the effects of this natural mulch have been studied by Weaver and Rowland (1952) and include micro-climate modification, rainfall interception and infiltration effects, and influences on plant growth and species composition. Natural control of the dead vegetation was provided by occasional prairie fires (started by lightning strikes) in addition taken to be about 1·4, so rates of swamp evaporation faster than pan evaporation must exceed lake evaporation rates even more. Thus, early measurements on swamp and pan evaporimeters led to the conclusion that the vegetation substantially augmented the water loss (Gel'bukh, 1963).

This conclusion might be expected also from considerations of surface roughness (Linacre et al., 1970) and thermal response. The small weight of a canopy, compared with an equal volume of water, makes it heat more rapidly in the day, which presumably promotes evaporation. Indeed, summertime measurements by Young and Blaney (1942), at Prado in California, showed an earlier start of evaporation from tules each day, but a more rapid decline in the early afternoon presumably due to the relatively small thermal capacity of the canopy. In addition, the evaporation from the water surface within a swamp is expected to be faster because shallow water allows more rapid heating. Shjeflo (1968) observed an increased evaporation rate from shallow potholes in hot days. In an Australian swamp, the water surface during day-time was about 1°C *hotter* than the air above it in some summertime experiments, whereas a nearby lake surface was 3 to 5°C *cooler* (Linacre et al., 1970). The swamp water would probably have been even hotter in the absence of the reeds, because of its shallowness and turbidity (Rose and Chapman, 1968).

Gel'bukh (1964) used an empirical formula for lake evaporation and ignored the net-radiation flux. Also reed-field evaporation was calculated from an expression depending solely on the air dryness, irrespective of wind speed or net radiation. The ratio of the two expressions of unexplained origin, each containing empirical numerical constants, implies a swamp evaporation greatly in excess of lake evaporation. This deduction by Gel'bukh was justified by the notion that vegetation absorbs more solar energy than water, whereas in fact it will retain *less*, having a greater reflectivity.

More recently, Rijks (1969), working at the middle of a Ugandan swamp of papyrus (*Cyperus papyrus*) 500 m wide and 3·8 m high, obtained daily values of evaporation in the range 0·38 to 0·81 times the calculated lake evaporation rate. The ratio was about 0·48 when lake evaporation was between 2 and 4 mm/day, 0·61 for 4 to 6, and 0·66 for 6 to 7 mm/day, reflecting the dependence of $\lambda E/R_n$ on temperature mentioned earlier. Rijks concluded that the dead foliage in an old stand of papyrus acts as a major impedance to the evaporation process. He suggested that Penman's analysis of Migahid's observations (Penman, 1963), pointing to equality of water evaporation and swamp vegetation evaporation, was due to the plants in that experiment being luxuriant and freshly transplanted.

The evaporation of bogs is less than that of open water but approaches equality during wet summer months in Germany, according to Eggelsmann

(1963). In contrast, Bulavko (1971) reported that evaporation from a water-logged marsh is often 20 to 40 % more than from a lake. In neither case was the evidence given. Bay (1967) measured the monthly evaporation from two Minnesota bogs of about 9 hectares and found that the rates during May to October varied between 88 and 121 % of lake evaporation, estimated by an approximate formula. In short, rough equality with lakes is probably the most reasonable inference for bog evaporation.

A recent paper by Thompson and Boyce (1972) on the evaporation from South African sugar cane is relevant to the case of reed swamps, as the geometry is similar. With well-irrigated crops, the measured evaporation rate was about the same as that from a Class A pan evaporimeter, i.e. more than from a lake.

Hall et al. (1972) used the water-balance equation (1) to calculate the evaporation from 3 ha of mixed vegetation (not reeds) within a 4-ha swamp in New Hampshire. During a June—August period, precipitation was 20 cm, inflow 11 cm, and the fall in water level 35 cm. The submerged peat associated with the swamp was reckoned to contain 30 % by volume of water, and the evaporation rate from the 1 ha of clear water was calculated from the well known Penman equation as 20 cm. These figures imply an evaporation rate from the vegetation of 36 cm, i.e. appreciably more than from the open water.

The only comparisons of the *measured* evaporation of a swamp and the *measured* evaporation of open water in the same climate were by Eisenlohr (1966a) and Linacre et al. (1970). The former found the average evaporation loss for May–October was 60 cm for vegetated patches and 67 cm for pot-holes without vegetation. In other words, the implication is that swamp vegetation *reduced* evaporation. The measurements of Linacre et al. (1970) were in the middle of an Australian swamp of 3000 ha and, simultaneously, a 240 ha lake 16 km away. The ratio of evaporation rates in summer depended on whether the surrounding country was wet or dry. When the country was dry, the swamp evaporation over four day-time hours was only about a third that of the lake, whereas, after heavy rain had wetted the entire region, lake evaporation was reduced to about that of the swamp. This change emphasizes the importance of the oasis effect on the relatively small lake, and differences between lake and swamp vegetation are trivial in comparison.

In other words, it might be concluded that swamp vegetation *enhances* evaporation on the basis of Gelbukh's measurements, considerations of surface roughness and thermal response, analogous studies on sugar cane, and the measurements by Hall et al. But the contrary and more sound conclusion was reached in the later, more sophisticated field measurements by Shjeflo et al. In any case, the effects of advection can be overwhelming.

When advection is negligible, it is possible to use a simple formula to estimate swamp evaporation, derived from that of Penman. A formula for the evaporation rate for a lake given by Linacre (1969a) reduces to the following:

$$\lambda E_0 = \frac{10R_n + 100(T_{max} - T_{min})}{22 - 0.15(T_{max} + T_{min})} \quad (6)$$

assuming a wind speed of 3 m s^{-1}, where λ is the latent heat of evaporation, R_n is given by Eq. (4) and T_{max} and T_{min} are respectively the daily maximum and minimum temperatures at 2 m, say. If R_n is 145 W m^{-2}, T_{max} is $24^\circ C$ and T_{min} is $12^\circ C$, λE_0 is 100 W m^{-2}, i.e. E_0 is 11 cm/month. This resembles the 13 cm/month mentioned in Section III.F.

IV. THE RATE OF GROWTH OF SWAMP PLANTS

Thompson (1973, personal communication) calculated large apparent efficiencies of over 5% for the conversion of incident radiation in the photosynthesis of tropical papyrus. However, the fast growth rate is not necessarily a result of rapid photosynthesis, but may simply represent a remobilization of reserve material stored within the submerged parts. This is the case for reeds growing in Czechoslovakia (Dykyjova et al., 1970).

Considerable plant-physiological information is available about the photosynthesis of individual swamp plants (e.g. McNaughton and Fullem, 1969), as distinct from natural stands. The intake of carbon dioxide in photosynthesis by *Phragmites communis* in Austria has been examined by Burian (1971). The growth of swamp vegetation is favoured by abundant water and by vertical leaf orientation, which reduces intense heating by the sun and mutual shading of the leaves. Also swamps often occur in regions of much sunshine and warmth. As a consequence, the rate of net dry-weight production is as high as 2.7 to 3.3 kg m^{-2} per year for *Spartina alterniflora* in Georgia or *Phragmites communis* in Long Island, and 6.4 kg m^{-2} per year for cultivated *Scirpus americana* in Germany (Keefe, 1972). In energy terms, these rates may represent up to 1.5% of the incident radiation. The figure for *gross* photosynthesis, which makes no allowance for respiration, may be up to 6% in the case of *Spartina*. For *Typha* marsh, Bray (1962) calculated that the overall dry-matter production accounted for 1.1% of the incoming radiation.

V. CONCLUSIONS

A thorough search of the (English) literature has demonstrated the paucity of knowledge on the physics of swamps. Most work has been on evaporation rates, but the *process* of evaporation from a reed field, say, remains a matter of ignorance. It is clear that advection of energy from outside a small swamp greatly increases evaporation rates, as in the case of a small body of water. This complicates the comparison of swamp and lake evaporation. Compared with regional climate and local advection, the presence of swamp vegetation and its type are relatively minor influences on evaporation rates, at least whilst the vegetation is actively growing.

ACKNOWLEDGMENTS

Several dozen experts from many countries kindly supplied information for this survey. In particular, I thank Dr Julian Rzoska (International Biological Program, Central Office, London) for advice, and also Dr David Mitchell (University of Rhodesia) and Professor Samuel Snedaker (Center for Aquatic Sciences, Gainesville, Florida); Mr Harry Blaney for helpful discussions and guidance through the early literature held at the Water Resources Center at the University of California, Los Angeles; Commonwealth Bureau of Soils (Commonwealth Agricultural Bureaux, Harpenden, England) for compiling an annotated bibliography (number 1546, June 1972) on Soils, Vegetation and Reclamation of Swampy Areas (1967 to 1971): The Director of the Water Resources Scientific Information Service (Office of Water Resources Research, U.S. Dept. of the Interior) for a computer scan and printout of abstracts on all holdings concerning swamps; Dr Don Boelter (US Dept. of Agriculture, Grand Rapids, Minnesota), Dr Joseph Ondok (Czechoslovak Academy of Sciences), Dr S. S. Khodkin (International division, Hydrometeorological Service of USSR) for extensive source material; and Dr Keith Thompson of Makere University (Uganda) for expert comments on a draft of this paper.

REFERENCES

Anon. (1961). Instructions for Hydrometeorological Stations and Posts, No. 8: Manual of Hydrological Observations in Swamplands. Meteorological Translation No. 6, of Department of Transport, Canada, p. 25.

Anon. (1962). Project Mar: Conference on the Conservation and Management of Temperate Marshes, Bogs and other Wetlands, Les Saintes Maries de la Mer, 12–16 Nov. 1962, 2, 10.

Bavina, L. G. (1967). *Soviet Hydrology* 4, 348–370. (Am. Geophys. Union, 1968, translated from *Trudy GGI*: 69–96, 1967).

Bavina, L. G. (1971). Water Balance of Swamps and its Computation. Symp. on World Water Balance. Internat. Assoc. Sci. Hydrol. Unesco, Vol. 2, pp. 461–466.

12. Grassland

E. A. RIPLEY and R. E. REDMANN

Department of Plant Ecology, University of Saskatchewan, Saskatoon, Saskatchewan, Canada

Canadian Committee for the International Biological Programme Contribution No. 199.

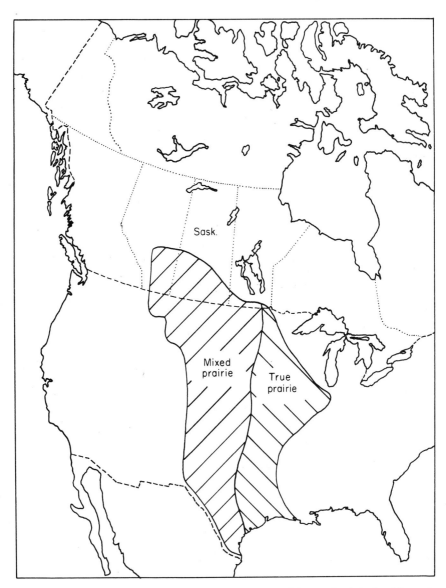

Fig. 1. Map of North America showing approximate extent of the grasslands in the eighteenth century.

to the factors mentioned above. A complete review of fire as a natural feature of grasslands was provided by Daubenmire (1968).

Because of the very different properties and roles of the green and dead fractions of the vegetation at Matador, they were treated separately for canopy structure analysis (Coupland *et al.*, 1973). The results of one sampling are presented in Fig. 4a. The cumulative leaf (foliage) area indexes at that

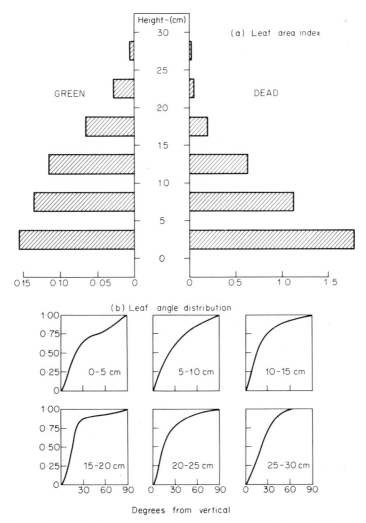

Fig. 4. Structure of the Matador vegetation on 17 August, 1971 in 5-cm layers. (a) Green and dead leaf (foliage) area indexes (note different scales). (b) Leaf angle distribution (green plus dead).

time were 0·5 for the green leaves and 3·8 for the dead leaves. The leaf angle distribution (Fig. 4b) was measured for green plus dead leaves. The erectophile nature of the vegetation may be clearly seen, the predominant leaf angle being 10 to 20° from the vertical. The relatively high frequency of greater leaf angles, particularly in the 0 to 5-cm layer, reflects the presence of the nearly horizontal mat of dead vegetation.

The difficulty of measuring leaf area in such a complex canopy prompted the search for a relation between the routinely observed green and dead plant biomass and the total green and dead leaf areas. A preliminary study suggested a value of about 90 cm^2 g^{-1} for green leaves and 80 cm^2 g^{-1} for the dead leaves. The lower value for the dead may be a result of leaf rolling. Applying this conversion to the biomass measurements during the 1970 and 1971 growing seasons yields the curves shown in Fig. 5 for the seasonal variation in total green and total dead leaf area indexes.

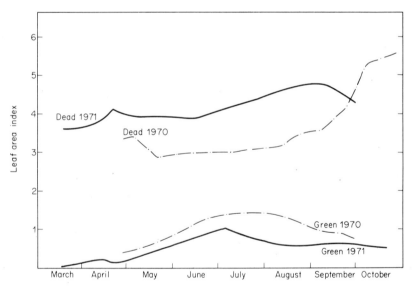

Fig. 5. Seasonal variation of green and dead leaf area indexes during the 1970 and 1971 growing seasons. Most of the data were estimated from biomass measurements (see Robins and Ripley, 1973).

Rainfall interception by the vegetation canopy plays an important role in the water balance of a prairie ecosystem. Although many papers have been written on this subject (e.g. Clark, 1940), the development of a suitable model for grassland canopy interception still remains to be done. McMillan and Burgy (1960) discuss canopy interception and, in particular, assess the loss of water intercepted by dead vegetation. They conclude that net interception loss may be appreciable when the soil is dry, wind speed is high and much dead vegetation is present. All of these conditions usually obtain in the northern Great Plains grasslands during mid- and late-summer. It is estimated that about one-fifth of the rainfall during this period does not reach the soil because of canopy interception loss, with a further fraction not available to the plant due to interception and subsequent evaporation from the litter and surface soil layers (Couturier and Ripley, 1973).

The soil structure plays an important role in water exchange with the atmosphere. The heavy clay soils of the Matador area, with their high montmorillonitic component, undergo large volume changes with moisture content. This affects the distribution of pores in the soil and hence infiltration and evaporation. A less obvious effect is the change in soil-surface elevation which is reported by White (1962) to be 4 cm and 8 cm for clay soils in Texas and Australia respectively. The interrelations between the soil structural properties and root morphology has been investigated by White and Lewis (1969). They contend that the formation of large vertical cracks as the clay soil dries out causes greater damage to lateral roots and thus gives an advantage to plants with predominantly vertical root systems (such as the *Agropyon* spp., Coupland and Johnson, 1965). They also mention the radial compression of the roots by the soil as it dries out which may constrict and damage roots. This damage would be minimal for roots with a suberized cortex as has been observed for *Agropyron*. The presence of the soil cracks (about 1 cm width and unknown depth) undoubtedly affects rainfall infiltration and may affect the exchange of carbon dioxide and water vapour.

On average the snowpack lasts about four months in south-western Saskatchewan, from late November to late March (MacKay and Thompson, 1968) and has a depth of 10 to 30 cm. With its low thermal conductivity, the snowcover protects the vegetation and soil from the extremely cold mid-winter weather. Murcray and Echols (1960), during a period of four days in Alaska when the air temperature did not rise above $-40°C$, measured temperatures in a 30-cm snowcover of $-30°C$ at 15 cm and $-8°C$ at the soil surface. Observations at Matador have shown that under a 20 to 30 cm snowcover temperatures rarely drop lower than $-10°C$ in the soil (and bottom 5 cm of the canopy) in spite of monthly mean air temperatures as low as $-25°C$ (Ripley, 1973).

The prairie snowcover is usually dry and powdery with a loose surface, and the high winds produce considerable motion of the surface snow. As a result the surface properties remain relatively constant throughout the winter. At the same time, however, the subsurface part of the snowpack is undergoing an ageing process and gradually increasing in density. Measurements at Matador and elsewhere (MacKay and Thompson, 1968; Gray *et al.*, 1969) show a transition from about 0·15 in December to 0·25 in February and 0·30 in late March. This increase in density reflects the agglomeration of the original snow flakes into larger and larger crystal aggregates.

> 'It proceeds in a variety of ways—migration of water vapour under the influence of surface free energy or a temperature gradient, cycles of melting and refreezing, compactures from the weight of overlying snow, etc.' Schlatter (1972).

Schlatter also shows that the penetration of solar radiation causes earliest snowmelt at a depth of 10 cm rather than at the surface.

A final feature of prairie snow to be examined is the oft-noted rapid thaw of soil covered by a snowpack (Ripley, 1971). This is discussed by Longley (1967) who postulates heat transfer by latent heat exchange to account for the rapid warming-up observed. Since the soil is usually cracked and dry in the late autumn (and hence through the winter) the spring snowmelt is able to penetrate to depths of a metre or more very quickly. Thus the spring thaw may be due largely to mass flow of water rather than to conducted heat.

C. Review of Grassland Micro-climate Studies

It should be emphasized that the aerial and edaphic micro-climates in an ecosystem are important not only as they affect growth and development of the primary producers (green vegetation) but also as they affect the activity of consumer and decomposer elements of the ecosystem. These, in turn, themselves affect the vegetation and micro-climate. For example, leaf consumers change the vegetation structure; decomposing organisms convert dead vegetation to litter and thence to humus and inorganic material; soil invertebrates and microorganisms change the structure of the soil, affecting its thermal and mass exchange properties; while the metabolism of soil organisms generates a considerable amount of carbon dioxide which diffuses upward to be largely assimilated by the green vegetation and synthesized into plant material.

Therefore, micro-climate descriptions are particularly appropriate in ecosystem studies if for no other reason than to inform biologists studying particular organisms of the actual field environment in which the organisms exist. This hopefully will lead to more studies of the influences of variation in micro-environment on organisms.

Very few detailed micro-climatic studies have been carried out on natural grassland. Perhaps the most intensive meteorological study ever undertaken on native prairie was the Great Plains Turbulence Field Program of 1953 in Nebraska (Lettau and Davidson, 1957). However, this study was micrometeorological in nature and measurements were confined to only seven periods of an average of 25 hours each during August and September, 1953.

In 1955, Waterhouse reported a study of micro-climate of a grass surface in Scotland. His original aim in the study was to assess 'the conditions under which grass-feeding insects live'. His measurements included canopy profiles of temperature, humidity and wind speed at various times of the year in several types of grassy vegetation. He noted the presence of a matted layer of dead grass to be important in maintaining a cool, humid environment, favourable to many insects, near the ground. Whitman (1969) and Whitman and Wolters (1967) discussed the results of a number of years' micro-climate measurements in the mixed grass prairie at Dickinson, North Dakota. Their measurements included temperatures at a number of levels above and below ground; humidity and wind speed at several levels; and profiles of evaporation as measured by black- and white-bulb Livingston atmometers. Whitman (1969) discussed the importance of the micro-climate in grassland ecosystems and expressed the need for more precise measurements of plant response to environmental factors.

D. The International Biological Programme and the Matador Project

As part of the International Biological Programme (IBP) a total ecosystem study of native prairie vegetation was established at Matador, Saskatchewan, Canada (Figs 1 and 2) in 1967. The main themes of this study (The Matador Project) were assessment of the productivity, at each trophic level, of the prairie and detailed studies of the photosynthesis and nitrogen fixation processes in the ecosystem. Field measurements ended in 1972 and the following two years were devoted to analysis, synthesis and mathematical modelling of the vegetation growth and its interaction with other ecosystem components. The overall aims of the photosynthesis studies under the IBP were defined by Specht (1967).

The authors of this chapter were involved in the meteorological (E.A.R.) and physiological (R.E.R.) studies of the project. As part of their work, field studies of micro-climate, evapotranspiration, and carbon dioxide assimilation were carried out over three growing seasons. Some of the results of these, and other, measurements are presented in the following pages.

The investigations carried out at the Matador Site are described in the annual reports of the Project (1967 to 1972) and in a series of final technical reports. The micrometeorological and climatological studies were described by Ripley and Saugier (1972) and Ripley (1972b) respectively. A brief summary is given here as a background for the results presented in the following sections of this chapter. A year-round 'climatological' station was put into operation in the summer of 1968. Measurements included precipitation; temperature profiles above and below ground; net, solar, and reflected radiation; sunshine; wind speed at 0·5, 2, and 10 metres; wind direction; screen temperature and humidity; and pan evaporation. A separate trailer-based micrometeorological facility has been in operation during the summers of 1970 to 1972. It was equipped to measure atmospheric gradients of temperature, humidity, wind speed and carbon dioxide concentration; soil temperature profiles and heat flux; leaf temperatures; global, reflected, net, and total incoming radiation above the canopy; global and net radiation profiles within the canopy; wind direction; precipitation; and atmospheric pressure. All micrometeorological measurements were logged automatically on magnetic tape by a digital data-acquisition system.

Carbon dioxide exchange of the sward was measured under hemispherical butyrate plastic chambers, attached to 60-cm diameter cylindrical base rings driven 10 cm into the soil. Carbon dioxide concentration in the closed system was monitored with an infra-red gas analyser. The system was automatically purged whenever inside CO_2 concentrations departed from ambient by more than 15 $\mu l/l$. Environment within the chamber was kept as close to ambient as possible. More detailed descriptions of the chamber design and performance appear elsewhere (Redmann, 1973). In the laboratory, an open gas analysis system was used, with three grass leaves per chamber. Differences of CO_2 and dew-point between ambient and chamber air were measured with an infra-red gas analyser and dew-point hygrometer. Osmotic potentials were measured with a cryoscopic osmometer; total water potential with a Spanner-type thermocouple psychrometer.

The topography of the Matador Field Station is discussed in detail by Ripley (1972b). The site has a minimum fetch of almost 2 km to the nearest coulées (ravines) and wheat field. As the highest micrometeorological measurement level was 6·4 m, this represented a minimum fetch–height ratio of over 300:1.

II. RADIATION

A. Fluxes Above the Canopy

Southern Saskatchewan has the greatest annual total of sunshine hours in Canada (in excess of 2200 hours or 50% of the total possible). The highest frequency of sunny days is during the months of July and August when anti-cyclonic conditions prevail and resident air masses are predominantly very dry.

The Matador area has an annual mean income of total short-wave radiation of about 14 MJ, ranging from 3·6 in December to 25 in July. In the native grassland, 15 to 20% of the incident short-wave radiation is reflected from the surface back to the atmosphere during the summer; the amount rises to 75% when the surface is covered with snow.

The combination of clear skies and low atmospheric humidity during the summer results in a relatively high net outgoing long-wave radiation (Sellers, 1965, p. 60). An example of the diurnal variation of the radiation balance components for a clear midsummer day is shown in Fig. 6a. The day (30 July 1971) was typical for the time of year: wind westerly 1 to 4 m s⁻¹ at a height of 1 m above the ground; vapour pressure about 7 mb; screen minimum temperature 9·6°C, maximum 31·6°C. The 24-hour totals of the various fluxes were: incoming short-wave radiation 27·3 MJ; reflected short-wave 4·5 MJ; incoming long-wave 27·5 MJ; outgoing long-wave 36·8 MJ. The net short-wave radiation for the day was 22·7 MJ (downwards); the net long-wave 9·3 MJ (upwards); and the net all-wave radiation 13·4 MJ (directed downwards).

A radiative surface temperature (T_s) has been calculated from the upward long-wave flux for the same day assuming a surface emissivity of 0·975 (Fuchs and Tanner, 1966). This is plotted in Fig. 6b along with measured green and dead leaf temperatures for comparison. Although the absolute value of T_s may be in error because of a poor choice of emissivity, this would have little effect on the amplitude of the diurnal curve which is considerably less than either of the leaf temperature traces. The cooler temperature of the green leaves during daylight is presumably a result of the lower atmospheric sensible heat flux (because of transpiration) in the vicinity of these leaves. The upward flux of long-wave radiation (at the measurement height of 2 metres) is composed of contributions from the green and dead leaves, the soil surface, and the air layer beneath the sensor. The soil-surface temperature has been found to exceed leaf temperature under the conditions of 30 July 1971 so that, other than measurement error, the only explanation of the difference shown in Fig. 6b is radiative flux divergence in the layer from the surface to 2 m.

Hourly totals of the net all-wave radiation have been plotted against hourly totals of global radiation in Fig. 7. The resultant is a very nearly straight line of slope 0·72 and intercept −91 which compares very closely with the values obtained by Nkemdirim (1972) for a patch of prairie on 3 July 1970. The absence of a noticeable morning/afternoon hysteresis, as found by Monteith and Szeicz (1961), is perhaps partly due to the higher wind speed at Matador and also to the higher downward flux of long-wave during the afternoon (Fig. 6a).

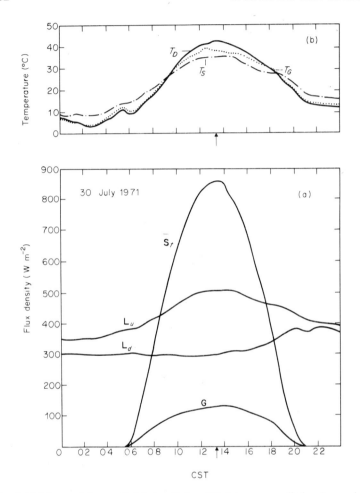

Fig. 6. (a) Measured fluxes of global radiation (S_t), long-wave radiation (upwards L_u, down-wards L_d) and soil heat for Matador on 30 July, 1971. (b) Diurnal variation of surface radiative temperature (T_S) and measured green (T_G) and dead (T_D) leaf temperatures for 30 July, 1971. The arrows represent the time of solar noon. The time used throughout this study is Central Standard Time (CST) = GMT + 6 hours.

The reflection coefficient of a natural surface for short-wave radiation is dependent on both wavelength and angle of incidence as well as on properties of the surface. Although reflection coefficients for many surfaces have been reported (Monteith, 1959; Bowers and Hanks, 1965; Geiger, 1965; Sellers, 1965, p. 21; Munn, 1966; Coulson and Reynolds, 1971), few measurements are available for natural prairie. Lettau and Davidson (1957) reported a mean

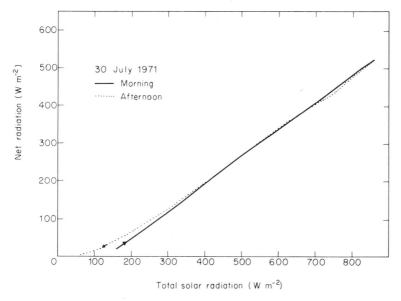

Fig. 7. Comparison of hourly integrations of net and total solar radiation for a sunny, midsummer day at Matador.

value of 0·27 for a mowed short-grass prairie site in Nebraska during early September, 1953. At the time of their measurements a large part of the grass had turned yellow. Nkemdirim (1972) studied radiation over a small patch of what was originally prairie in southern Alberta. Cut to a height of 5 cm and irrigated, this 'lawn' was found to have a mean daily reflection coefficient of about 0·23 with hourly values dropping to 0·18 at solar noon. Observations over natural (uncut, unirrigated) prairie at Matador (Ripley, 1973) showed a mean value of 0·17 for the reflection coefficient with a slight variation through the season. Cloudy days showed higher than average values (near 0·19) and sunny days lower than average values (0·16).

The diurnal variation in mean monthly reflection coefficient is shown in Fig. 8 for the months of July and August, 1969. The curves, with minima of 0·15, have the typical 'dish-shape' also reported by Rijks (1967), Davies and Buttimor (1969) and Nkemdirim (1972).

Mean daily values of the reflection coefficient were plotted against maximum solar altitude for twenty-four days during the 1971 growing season. A straight line was fitted to the points and used to 'correct' the reflection coefficient values for the seasonal variation in solar altitude. With this variation removed the 'corrected' values were plotted against the date (Fig. 9). A gradual decrease in the coefficient over the season is clearly evident.

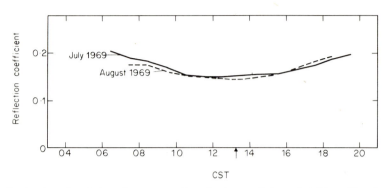

Fig. 8. Diurnal variation of mean monthly reflection coefficient of prairie at Matador for the months of July and August, 1969. The arrow shows the approximate time of solar noon.

No significant correlation could be found with different relative proportions of green and dead leaves. This is perhaps not surprising since the green leaf area never exceeded 20 % (averaging about 12 % over the season) of the total. It seems likely that the seasonal variation in reflection coefficient is due to a change in reflectivity and/or geometry of the dead vegetation. The dead component would have a higher proportion of older leaves in the spring with the mean age dropping as new leaves grow, mature and die. The winter

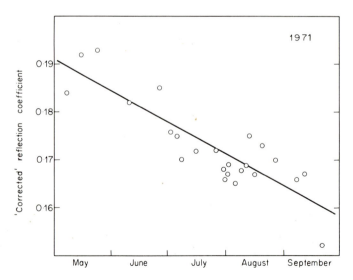

Fig. 9. Seasonal variation of 'corrected' reflection coefficient (see text) at Matador during the period May to September, 1971.

snowcover tends to compress the mat of dead vegetation. It is relatively flat in the spring becoming gradually deeper and more erect during the season. Since the reflection coefficient appears to increase as the dead leaves decompose and as the dead mat is flattened out, both of these effects could explain the observed trend shown in Fig. 9.

The half-hourly reflection coefficients for a number of days during 1971 were analysed as a function of solar altitude. For most of the days, considerably higher values of the reflection coefficient were found during the afternoons than during the mornings. The values for three days are presented in Fig. 10, the arrows indicating increasing time. The diurnal patterns for 20

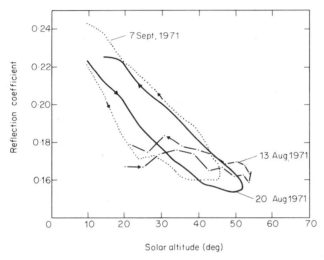

Fig. 10. Variation of surface reflection coefficient with solar altitude for three days in 1971. The arrows indicate increasing time.

August and 7 September (both clear days) are very similar, while that for 13 August (a cloudy day) shows considerably less variation with solar altitude and less morning/afternoon differential. The prevailing wind directions at Matador are west-north-west (WNW) and east-south-east (ESE). The former wind direction is associated more with higher wind speeds, sunny skies, and low humidities. Thus it is likely that the mat of standing dead vegetation would be aligned pointing towards the ESE most of the time. This has been observed although no observations have been made of any changes in this alignment in response to other wind directions. Reference may be made to the 10-cm 'slice' of the vegetation shown in Fig. 3 in which the left-hand edge of the photograph is towards the east. The eastward alignment of the vegetation may be clearly seen. When dry, the dead vegetation in this mat is sufficiently stiff to resist being moved by even moderate to strong winds. It is proposed that at least part of the observed variation in reflection coefficient with solar azimuth (at constant altitude) is due to vegetation alignment. If a large fraction of the leaves were pointing towards the ESE it would be expected that more of the radiation incident from this direction would be absorbed (and less reflected). Hence, this would be consistent with a lower reflection coefficient in the morning than in the afternoon. Other effects such as plant-water content, leaf rolling, and atmospheric turbidity may also be involved.

With the arrival of the first winter snow the surface reflection coefficient at Matador increases from about 0·17 to 0·78. Long-period measurements at two sites in Antarctica gave values varying from 0·81 to 0·97 for a complete snowcover (Schlatter, 1972), while Geiger (1965) noted a range of 0·75 to 0·95 for a fresh snowcover. Although the absorbed radiation becomes quite small during the winter, because of low solar angle and high reflection, a considerable fraction of it penetrates deep into the snowpack, particularly in the case of dry snow of low density. Bergen (1971), using a pair of photo-cells in the mountains of Colorado, calculated extinction coefficients varying from 0·078 cm^{-1} in mid-February (snow density 0·23) to 0·125 cm^{-1} in mid-March (density 0·31).

B. Fluxes within the Canopy

A number of profiles of total short-wave radiation, measured in native prairie at Matador on 28 June 1972, are shown in Fig. 11. These were

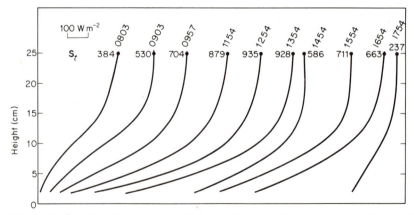

Fig. 11. Profiles of total short-wave radiation inside the vegetation canopy at Matador for a number of half-hour periods (end time shown) on 28 June, 1972. The curves are displaced from each other for clarity with the values of the flux density at 25 cm indicated on each curve.

measured with tubular solarimeters (Szeicz *et al.*, 1964; Szeicz, 1965) at heights of 2, 5, 7, 15, and 25 cm in addition to a conventional Kipp solarimeter at a height of 2 m. The total (green plus dead) leaf area index at the time was 4·1.

Expressing radiation extinction in the vegetation as a function of leaf area index [as Eq. (6) in Monteith, 1969]:

$$\mathbf{S}(z)/\mathbf{S}(0) = \exp\left(-\mathscr{K}'L(z)\right)$$

where $\mathbf{S}(z)$ is the downward flux of solar radiation at height z in the canopy; $\mathbf{S}(0)$ is the radiation flux above the canopy; $L(z)$ is the cumulative leaf area index from the top of the vegetation to the level z; and $\mathscr{K}'$ is the extinction coefficient. Values of $\mathscr{K}'$ were computed by least squares and are presented for 28 June 1972 in Fig. 12. The diurnal course of solar radiation above the canopy is shown in Fig. 13. The curve in Fig. 12 shows the 'dish' shape also noted by Anderson (1966) for an erectophile type of canopy. The dead vegetation mat, composed to a large extent of nearly horizontal leaves concentrated in the 5 to 15-cm layer (Fig. 4) leads one to expect variation of the extinction coefficient with height. However, this has not yet been investigated.

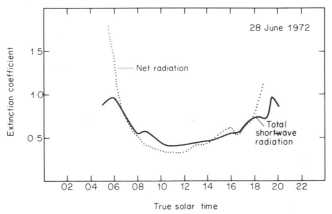

Fig. 12. Diurnal variation of solar and net radiation extinction coefficients for the vegetation at Matador on 28 June, 1972.

Following Monteith (1969) a midday (0900 to 1700 CST) value of $\mathscr{K}'$ for 28 June 1972 is about 0·47. The coefficient $\mathscr{K}$ used by Monteith depends only on vegetation geometry, and is defined as $\mathscr{K}'/(1 - \tau)$, where τ is leaf transmissivity. Using τ equal to 0·25 yields a value of $\mathscr{K} = 0·63$. This is much greater than the values quoted by Monteith (1969, Table 5-2) for rye grass,

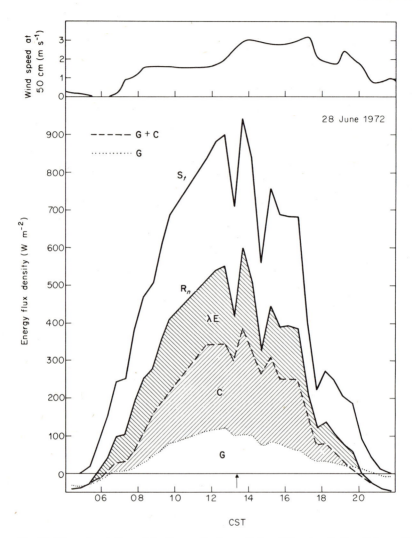

Fig. 13. Diurnal variation of the fluxes of solar radiation (S_t), net radiation (R_n), sensible (C) and latent (λE) atmospheric heat, and soil heat (G), and of wind speed at a height of 50 cm above the ground at Matador on 28 June, 1972. The arrow indicates solar noon.

but is similar to the values for rice and beans, which were presumably less erect. This relatively high value of $\mathscr{K}$ must be attributed to the dense mat of dead vegetation.

A similar treatment has been accorded net (all-wave) radiation (R_n) for which profiles are shown in Fig. 14. Using tubular net radiometers (Denmead,

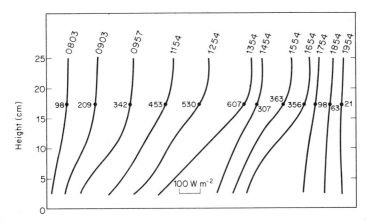

Fig. 14. Profiles of net radiation inside the vegetation at Matador. Absolute values are indicated for the 17·5-cm level. For further explanation see the caption to Fig. 11.

1967), 1 m in length, aligned in an east-west direction at heights of 5, 10 and 15 cm above the ground and a conventional net radiometer (Funk, 1959) at a height of 2 m, extinction coefficients were calculated by the same method as used for solar radiation above. The diurnal variation of the extinction coefficient for 28 June 1972 is shown in Fig. 12. The curve is very similar to that of solar radiation except for lower values during the morning. It shows fairly close agreement with those presented by Impens and Lemeur (1969) for corn and oats.

The change in spectral composition of solar radiation as it penetrates a vegetation canopy has been discussed by many authors (Federer and Tanner, 1966; Robertson, 1966; Daynard, 1969; Monteith, 1969). Due to the greater absorptance of visible than near infra-red radiation by vegetation there is much greater penetration of the near infra-red compared with the visible. The variation of radiation spectrum with depth in the canopy is important because of the differing action spectra of the many photomorphogenetic processes, and of photosynthesis. Some of these spectra were discussed by Salisbury and Ross (1969, p. 512) and by McCree (1971–72). While the vegetation at Matador is not sufficiently dense to affect the quality of the visible component (400 to 700 nm), the ratio of visible to far red (700 to 800 nm) will be changed considerably, particularly near the ground. This could be of importance in the photomorphogenesis of the shorter grasses and forbs of the community.

The soil-plant-atmosphere model of Lemon et al. (1971) has been adapted for use with the Matador vegetation. The first section of this model calculates the vertical distribution of solar and net radiation inside the vegetation.

Measured canopy radiation was found to be higher than predicted by the model, which assumed randomly distributed leaf area. To compensate for this, an empirical coefficient was added to the exponential term in the basic radiation penetration equation. It was also necessary to divide the vegetation into green and dead components, since the dead leaves are not involved in photosynthesis or transpiration and have different spectral properties than the green. Figure 15 compares the profiles of global and net radiation calculated using the model with measured values (Stewart, 1973). The agreement is good considering the likely magnitude of measurement and sampling errors.

C. Single Leaf Radiative Properties

Optical properties for individual leaves of grasses such as *Agropyron* spp. or *Koeleria cristata* have not been reported in the literature. However, certain generalizations can be made based on data for other species. As water is lost from fully turgid leaves of corn or sorghum, reflectance in the visible and infra-red increases (Carlson *et al.*, 1971; Sinclair *et al.*, 1971; Woolley, 1971). The increased reflectance is most prominent in the infra-red wavelengths, and is a result of absolute decreases in water content as well as changes in internal structure (intercellular space size and cell orientations). Sinclair *et al.* (1971) point out that increased reflectivity during maturation and senescence is a result of decreased absorption of infra-red owing to water loss accompanying senescence, and decreased absorption in the visible resulting from loss of chlorophyll.

The leaf surface reflectivity for the xeromorphic grasses at Matador is probably higher than that of these cultivated grasses. Reflectivity of crop species has been shown to be lower than xerophytes, which often have light-coloured waxy or hairy surfaces (Monteith, 1959; Pearman, 1966). *Agropyron smithii* in particular has an especially glaucous leaf surface. Another important xeromorphic characteristic of the leaves of many grasses occurring at Matador is their ability to roll, a response to turgor changes in special cells at the bottoms of longitudinal furrows along the upper leaf surface. Rolling and unrolling is a dynamic process which occurs in response to changes in plant-water status (Fig. 16). Although often interpreted as a means of reducing transpiration (Oppenheimer, 1960), the effect on radiation absorption seems much more significant. Surface area exposed to radiation is much reduced, and reflection from the rolled leaves should be considerably greater than from unrolled. The changes in leaf shape over the growing season at Matador, as well as declining leaf-water content (Fig. 17) and increased senescence due to water stress, can explain part of the increase in reflection coefficient during the growing season discussed earlier.

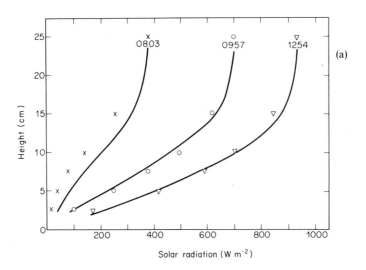

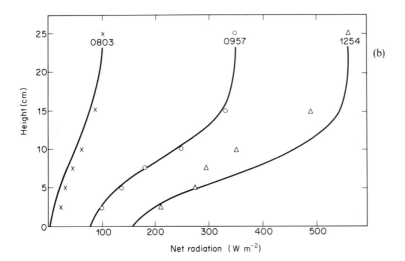

Fig. 15. Comparison of measured in-canopy radiation profiles with those computed by soil–plant–atmosphere model (Stewart, 1973): (a) solar radiation; (b) net radiation.

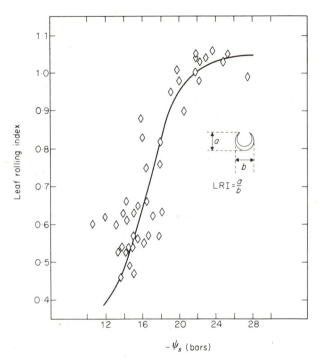

Fig. 16. Variation of leaf rolling index (defined as the ratio of apparent thickness to leaf width) with leaf osmotic potential (ψ_s) for *Agropyron dasystachyum* leaves at 1400 CST during the 1971 season.

III. MOMENTUM

A. Exchange of the Whole Stand

The great expanse of the treeless prairie presents little obstacle to the flow of air and, accordingly, winds near the surface layer are relatively greater than in other inland areas of the middle latitudes. In south-western Saskatchewan, where the mean annual wind speed ranges from 4 to 5 m s^{-1}, both very high winds and calm periods are relatively infrequent. Three years of wind speed data at a height of 10 m at Matador show that only 5% of the mean hourly winds exceeded 10 m s^{-1} and 7% were lower than 1 m s^{-1}. Thus the vegetation is exposed to moderate to strong winds, mainly from the west-north-west and east-south-east directions.

In the middle latitudes of the northern hemisphere, easterly winds accompany the approach of depressions, while westerly winds follow the passage

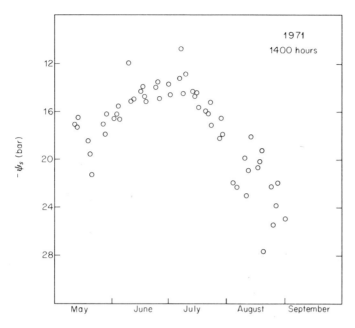

Fig. 17. Leaf osmotic potential (measured at 1400 CST) variation during the 1971 growing season.

of the depressions and the arrival of clearing, anticyclonic weather. One might surmise that the standing dead vegetation would be made flexible by the rains accompanying a depression and then would be fixed in position by the strong, dry, westerly winds following its passage. Although no detailed observations have been made, there is (as mentioned earlier) some evidence of a preferred alignment towards the east-south-east.

The presence of the standing dead component of the canopy plays an important role in determining the height distribution of the momentum flux near the surface. This, along with other factors, determines the micro-climate in the surface layer. Two major effects are the reduced wind erosion of the soil in summer and of the snowpack in winter.

Drag of the vegetation on the wind may be attributed separately to the dead and green vegetation components. Although the dead component has a greater biomass, its density is less (due to the absence of water); it is more streamlined (because of rolling); and it is more concentrated in the lower part of the canopy than the green component. A study of the variation of the aerodynamic roughness length (z_0) over a growing season shows greater correlation with green than dead leaf biomass.

Wind profile measurements were made above the canopy at Matador during the 1970 to 1972 field seasons. Analysis of these data, corrected for stability, gave mean values of zero-plane displacement (d) ranging from 15 cm in 1970 to 13 cm in 1971, and 10 cm in 1972. Part of this variation may have been due to actual changes in the vegetation structure from year to year but equally likely are errors in determining the height of the soil surface (because of soil microtopography). The variation in the height of the soil surface with soil-moisture content has already been mentioned. In the aerodynamic computations, d was computed by 'best-fit' of the theoretical profile to the observed data for a number of days selected for near-neutral stability, and this value was used throughout. The roughness length was computed independently for each summary period and showed considerable variation with date, atmospheric stability, and wind speed and direction. In general, values of z_0 ranged between 1 and 4 cm with a mean near 2·5 cm.

Examination of computed values of the friction velocity (u_*) during neutral stability conditions in late May, 1971, and the wind speed at a height of 2 m yields a value of the drag coefficient

$$C_{a,M}(2\text{m}) = (u_*/u_{2\text{m}})^2 = 0.0070;$$

with values ranging from 0·0093 at $u = 2 \text{ m s}^{-1}$ to 0·0046 at 15 m s^{-1}. This may be compared with tabulated values for various types of surfaces quoted by Priestley (1959, p. 21) and Webb (1965).

B. In-canopy Momentum Transfer

Eddy diffusivities were computed for the interior of the Matador vegetation by both the energy-balance and momentum-balance methods (Vol. 1, p. 100). For the momentum-balance method, profiles of wind speed and leaf area inside the canopy are required in order to estimate the momentum diffusivity, K_M. This may then be assumed equal to the heat, water vapour, and CO_2 diffusivities, K_H, K_V and K_C, and combined with the appropriate measured gradients to yield the fluxes. In the energy-balance method a weighted mean of the heat, water vapour, and CO_2 diffusivities is computed from the profile of net radiation inside the canopy and the soil heat flux. This, again, may be combined with the approximate gradients to yield fluxes.

In-canopy wind speed profiles (measured with hot-wire anemometers) were available for only a few periods during midsummer, 1972. A typical profile is shown in Fig. 18. The dashed line represents the extrapolation of the logarithmic profile, above the vegetation, to the zero plane, $z_0 + d$. The height of the vegetation canopy (h) has been taken to be 17·5 cm, in agreement with the expression $z_0 = 0.36\,(h-d)$ suggested by Thom (1971)

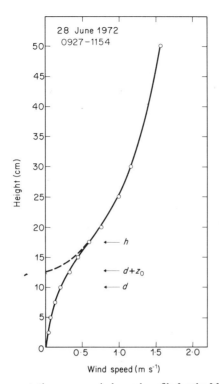

Fig. 18. Representative canopy wind-speed profile for the Matador vegetation.

with $d = 10$ cm and $z_0 = 2.7$ cm. The canopy wind profile was fitted to the expression

$$u(z) = u(h) \exp \left\{ - [0.385(z/h) + 0.72] L(z) \right\} \qquad z \leqslant h$$

where $L(z)$ is the total leaf area index above the level z. Above the canopy the usual logarithmic profile was assumed

$$u(z) = u(h) \{ \ln \left[(z - d)/z_0 \right] / \ln \left[(h - d)/z_0 \right] \} \qquad z \geqslant h.$$

Assuming these relationships remained valid over a period of several weeks the wind speed at one height above the canopy (usually at $z = 50$ cm) was measured and the entire profile computed from this. Measurements of leaf area v height were than combined to yield $K_M(z)$. The results for two half-hour periods on 26 June 1972, compared with simultaneous determinations of the diffusivity by the energy balance method, are shown in Fig. 19. The wind speeds were virtually identical for the periods so they have been combined.

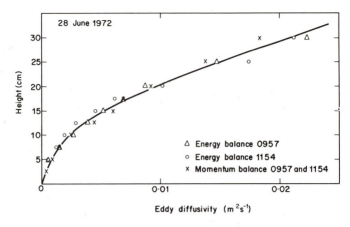

Fig. 19. In-canopy profiles of eddy diffusivity computed by the energy-balance and momentum-balance methods.

The agreement between the two estimates of K inside the canopy is surprisingly good. Just above the canopy, however, the energy-balance values increase more rapidly than those calculated by the momentum-balance. A similar increase was noted by Wright and Brown (1967) to be related to wind speed and is discussed by them. Other periods of the day did not show such good agreement and spatial sampling errors may have been responsible for much of this disagreement. The small scale of the mixed-prairie vegetation makes instrumentation difficult. Inherent variability in the grassland canopy structure (although homogeneous on a large scale) means that much greater spatial averaging is required to produce adequate in-canopy data than was possible in this study.

IV. HEAT AND WATER

A. Fluxes above the Canopy

The fluxes of heat, water vapour and carbon dioxide above the prairie canopy were computed by both the aerodynamic and energy-balance techniques. For the aerodynamic method the Businger–Dyer stability correction (Paulson, 1970) was applied during unstable conditions and a correction by Webb (1965) during stable conditions. A comparison of the fluxes of heat and water vapour computed by the two methods for 10 June 1971 is presented in Fig. 20. At that time the leaf area index of the green vegetation was about 0.7 and of the dead about 3.9 (Fig. 5). Soil moisture

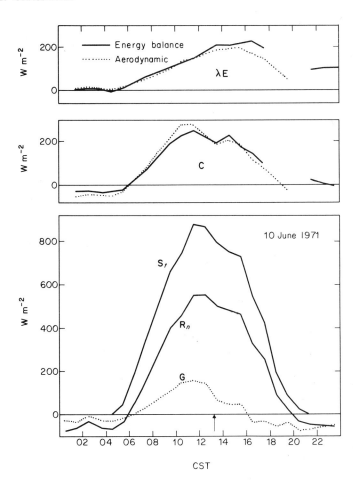

Fig. 20. Above-canopy measured fluxes of solar and net radiation and soil heat, and computed fluxes of sensible and latent heat by the energy-balance and aerodynamic methods for 10 June, 1971 at Matador. The arrow represents the time of solar noon.

conditions were favourable, 31 mm of rain having fallen during the previous five days. The last rain fell during the early morning of 9 June. The data used to plot Fig. 20 were in the form of hourly averages.

It can be seen from the radiation traces that there were numerous cloudy spells during the late morning and afternoon. The day was windy, with the 6 m wind varying from $4 \, \mathrm{m \, s^{-1}}$ in the early part of the day to 6–$8 \, \mathrm{m \, s^{-1}}$ during the afternoon, dropping to 1–$2 \, \mathrm{m \, s^{-1}}$ in the evening. The energy-balance and aerodynamic flux estimates show reasonably good agreement over most of the day. Gaps occur in the evening when the Bowen ratio is near -1 in the energy balance case, and because of extreme stability (Richardson number greater than 0·2) in the aerodynamic case.

Such analyses have been carried out for all of the data collected during the summer of 1970. The fluxes of solar radiation, net radiation, sensible and latent heat (energy-balance method), and soil heat were averaged for 10-day periods, and plotted (Fig. 21a). The variation of green leaf area during the summer is shown for comparison in Fig. 21b. The total soil-water content of the 0- to 135-cm soil layer is in Fig. 21c. The circles represent neutron probe measurements (MacDonald et al., 1973) while the line shows the change in soil moisture computed from precipitation and the daily (energy-balance) fluxes of latent heat. Through drainage was assumed to be negligible. The agreement is quite satisfactory and systematic errors, if present, appear to be quite small. Finally in Fig. 21d the daily precipitation is shown for the period.

Figure 21 shows that there was a gradual drying out of the soil during the period with brief interruptions due to a light rain in early August and a heavier rain in early to mid-September. The green leaf area index decreased gradually after the summer peak in late July, then more rapidly during late August, and more gradually again in September. Examination of the flux curves shows a few interesting features. While solar radiation (S_t) increased from the first to the second period, net radiation (R_n) remained nearly constant. This was likely due to an increase in the upward flux of long-wave radiation (i.e. radiative surface temperature). The soil heat flux (G) was relatively constant at just over 10 W m^{-2} until late August when it started to decrease, becoming negative in early September. The evaporative flux (λE) decreased from near 70% of the net radiation in late July to 50% by late August. There was approximate equality of the sensible and latent heat fluxes throughout September.

A large part of the decrease in λE over the season was due to decreasing transpiration and transpiring leaf area (mainly the result of water stress (see Fig. 17) and high leaf temperatures). The remainder can be attributed to the decrease in evaporation from the soil due to the drying of the upper layers. By late August, soil-water potentials had reached -110 bars in the 0 to 15-cm layer and -40 bars in the 15 to 30-cm layer (MacDonald et al., 1973). Approximately 65% of the total root biomass has been found above 30 cm (Coupland, 1972). However, the steadying of the Bowen ratio near unity in late August, in spite of continued drought indicates that the vegetation was able to use soil water from lower depths. Another explanation would be that stomatal closure and perhaps internal plant and soil resistances would limit evaporation during much of the day during the high evaporative

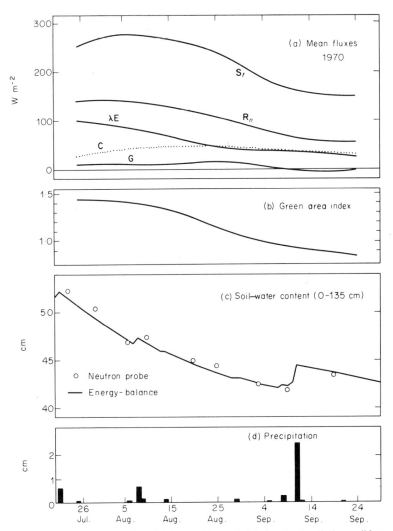

Fig. 21. 1970 seasonal variation of: (a) mean fluxes of global and net radiation, soil heat, sensible and latent heat; (b) green leaf (foliage) area index; (c) total (to a depth of 135 ċm) soil-water content; and (d) precipitation.

demand period of late July and early August. These effects would be reduced in the cooler month of September and evaporation might continue through the midday period.

Canopy resistances (Monteith, 1965) were calculated from the saturation deficits at the 'zero-plane' and the evaporative fluxes computed by the energy-balance method. The 'zero-plane' was defined as the level $(d + z_T)$ where z_T is the roughness parameter for heat exchange given by

$$z_T = z_0 \exp(-kB^{-1}) \qquad \text{(Thom, 1972)}.$$

The reciprocal Stanton number (B^{-1}) was taken equal to 5 (Chamberlain, 1966) and $k = 0.41$ is von Karman's constant. The use of z_T instead of z_0 allows for differences in the source-sink distributions of mass and heat, and momentum in the canopy and is justified on the basis that bluff-body forces, which are involved in momentum transfer, have no counterpart in heat and mass transfer.

Diurnal plots of canopy resistance are shown in Fig. 22 for a number of days in 1970. The low values in the morning for a number of cases is likely

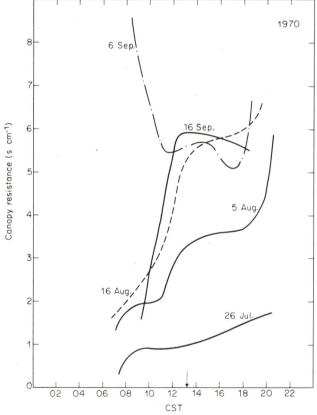

Fig. 22. Diurnal variation in canopy resistance for several days during the 1970 growing season at Matador.

caused by the re-evaporation of dew. The afternoon values increase from 1 s cm^{-1} in late July to 3·5 s cm^{-1} in early August, and to 5–6 s cm^{-1} during the remainder of the period.

B. Lysimetry

Weighing lysimeters have been used by many workers to measure directly the water losses from a piece of land. For a summary of earlier types of weighing lysimeters see Gangopadhyaya (1966). More recently it has been recognized that unless conditions inside the lysimeter are carefully matched to those outside, the results obtained may not be representative. A recently constructed lysimeter, designed to match temperature and moisture conditions inside the container with those outside, is described by Mukammal *et al.* (1971).

In order to use a lysimeter for studying natural grassland ecosystems it is essential that it be installed with as little disturbance as possible to the soil and vegetation (especially rooting characteristics) both inside and outside the container. This requirement considerably increases the difficulty and cost of construction.

The only known lysimeter, satisfying the above requirements, operating in natural grassland with a sensitivity sufficient to measure hourly evaporation has recently been constructed in Colorado. A brief description of this lysimeter is contained in a recent report (Armijo *et al.*, 1971).

C. Fluxes within the Canopy

In-canopy profiles of solar and net radiation, wind speed, and diffusivity for 28 June 1972 were discussed previously. Above-canopy fluxes for that day are presented in Fig. 13 together with soil heat flux and wind speed at 50 cm. The sensible and latent heat fluxes were calculated by the energy-balance method. For the daylight period as a whole the net radiation was partitioned into sensible heat (43 %), latent heat (37 %), and soil heat (20 %).

The continuation of a positive soil heat flux into the evening on this day differs from previous data (see Fig. 20, for example) and is somewhat questionable. In 1972, G was measured with flux plates located just under the surface whereas in previous years these were positioned at a depth of 5 cm and supplemented with thermocouples in the 0- to 5-cm layer. The difference may also be the result of a somewhat denser dead vegetation mat in 1972.

Several temperature profiles, both above and below ground, are presented for the day in Fig. 23a. All profiles are half-hour means ending at the time indicated. The diurnal variation at a soil depth of 20 cm was about 1·5°C

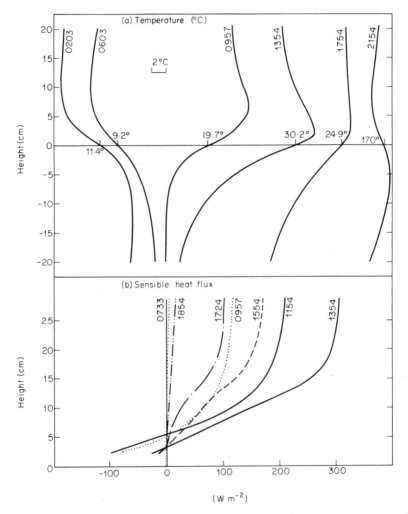

Fig. 23. (a) Temperature profiles in the canopy and upper layer of soil for a number of half-hour periods (end times shown) on 28 June, 1972. The actual temperatures at the surface are indicated on each curve. (b) Canopy profiles of sensible heat flux for the same day.

while that at mid-canopy height exceeded 25°C. Comparison of the 0957 and 1354 traces shows that the warmest canopy level was at a height of about 7 cm at the earlier time and moved downward to about 2 cm by 1354. This was due mainly to deeper penetration of solar radiation at the higher solar elevation. An inversion had developed in the upper part of the canopy by 2154. This profile shape is similar to those during the early morning.

Sensible heat flux profiles for a number of periods on 28 June are shown in Fig. 23b. The shallow depth of the vegetation permitted only a few sampling levels, resulting in lack of detail. However, it may be seen that the greatest sensible heat flux was near the 10-cm level. Also, during the late morning period the flux was negative below 5 cm and became less so as the inversion was destroyed during the afternoon. One might expect that an inversion in the lower part of the canopy would tend to produce a build-up in water-vapour and CO_2 concentrations in this area during the mornings (as a result of soil evaporation and respiration). This has not yet been investigated in any detail.

A number of vapour pressure profiles are shown in Fig. 24a. The strong gradient at 0803 was likely due to the re-evaporation of dew or to soil evaporation (16 mm of rain had fallen on 24 June). The decrease in vapour pressure in the early afternoon appears to be the result of more vigorous turbulent mixing bringing drier air down from higher levels (wind speed is shown in Fig. 13). The highest vapour pressure was found at the surface throughout the day and reached as much as 9 mbar above the value at 30 cm. The computed latent heat fluxes are shown in Fig. 24b. Scarcely any reliable data were obtained for the lower levels, so that little can be said about soil evaporation. Again the 10-cm level appears to be the major source region although there is evidence that this moved somewhat higher during the afternoon. This shift may also be seen on the heat flux curves (Fig. 23b). It was likely due to instrument errors but may have been a result of neglecting buoyancy effects inside the canopy.

D. Individual Leaf Exchanges

The aerodynamic exchange characteristics of narrow-leaved grasses (such as those found at Matador) differ considerably from broader-leaved plants. Leaves of *Agropyron smithii* are about 5 mm wide; *A. dasystachyum* leaves are approximately 3 mm wide; and both are 15 to 20 cm long. As a result, the boundary layers surrounding individual leaves are shallow, especially under the windy conditions which usually prevail at Matador. Sinclair (1970) studied convective heat transfer from narrow leaves, and determined formulae for calculating the heat transfer coefficient (h_c) in still air that differ from the free convection formulae published previously (Gates, 1962). The h_c calculated for a 5-mm wide leaf, for example, is four times higher when Sinclair's equation is used. However, formulae for forced convection are similar. Sinclair re-emphasizes the point that narrow leaves are efficient heat exchangers, and that this feature has ecological importance to plants growing under conditions of high radiation and temperature.

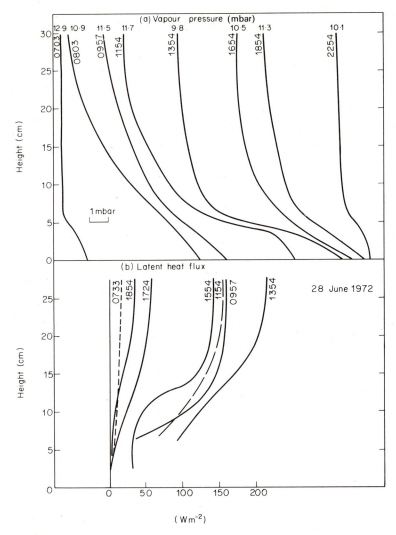

Fig. 24. (a) Canopy vapour pressure profiles for 28 June, 1972. The actual vapour pressure at 30 cm is indicated at the top of each curve. (b) Latent heat flux profiles for the same day.

As in the case of all plant species occupying dry habitats, latent heat transfer is not a reliable means of energy dissipation for the grasses at Matador. The grasses exhibit a number of xeromorphic characteristics such as having the stomata on the upper surface sunken into longitudinal furrows. Stomata on the lower surface are more exposed. Another feature is the ability of the leaves to roll when under moisture stress, which was mentioned earlier in connection with radiation absorption. Since rolling changes the effective dimension and shape of the leaf, sensible heat transfer is also affected.

Leaf rolling influences transpiration rate through its effect on leaf resistance to water-vapour diffusion. Leaf resistances of *Elymus cinereus* were measured in the laboratory. This species has larger leaves than *A. dasystachyum*, but also exhibits prominent rolling. Transpiration of individual leaves with and without water stress was measured with potometers, and boundary-layer resistance (r_V) was evaluated using filter paper models. Total resistance during the steady-state period after cutting was 5.8 s cm^{-1} under low stress (leaf base in distilled water) and 124 s cm^{-1} under high stress (leaf base in -18 bar sucrose solution). Values of r_V ranged from 0.26 s cm^{-1} for a flat leaf model (2×15 cm), to 0.84 s cm^{-1} for a model of the same dimension rolled to near complete closure (ends open). Even for the rolled model, r_V is small compared to stomatal resistance, and the significance of rolling for retarding water loss is negligible. This argument becomes more convincing when one observes that rolling occurs subsequent to stomatal closure and transpiration reduction.

The leaf rolling phenomenon is significant when transpiration is slow due to moisture stress, sensible heat transfer is reduced due to still air conditions, and air temperature and radiation are high. Conditions such as these occur in late summer at Matador. For example, if the air temperature at a height of 10 cm was 34°C, the green leaf temperature 37°C and the wind was calm, a rolled leaf (diameter 2 mm) would lose sensible heat at a rate of 87 W m^{-2}; an unrolled leaf (width 5 mm) would have to be at 39·8°C in order to transfer the same amount of heat. Since the radiation load of the rolled leaf is less, its temperature is less likely to rise to a lethal level, and excessive leaf respiration is also reduced.

The exact lethal temperature of *A. dasystachyum* was not determined but, as in other plants, it is certainly a duration-dependent characteristic which also is influenced by pre-conditions (Lange, 1967). Wright (1970) measured heat-induced mortality in bunch grasses, and found for example that growing points of *Stipa comata* could survive heating to 49°C for about 100 minutes. The shoot bases of *A. dasystachyum*, being located near the soil surface, are often in a high temperature regime. Air temperature at 2 cm approached 50°C on several occasions in early August, 1971.

It is unlikely that roots of the grasses growing at Matador are under high temperature stress, except perhaps very near the surface. The monthly mean soil temperature at a depth of 5 cm for July, 1971 was 17·7°C and for August 20·5°C. Other factors such as soil physical properties (water content, shrink-swell of clay, aeration, etc.) are more important than high soil-root temperature in controlling root function. On the other hand, low soil temperatures during the spring probably limit root growth, despite the adaptation to low soil temperatures reported for several native grasses (Smoliak and Johnson, 1968). Both soil temperature and moisture conditions are most favourable during the period from late May to late July, and shoot growth, root growth and organic matter decomposition tend to be concentrated in this period. Soil temperature also plays an important role in controlling the dynamics of carbohydrate transfer within plants, periods of root dormancy, root growth and tillering (Troughton, 1957; Ueno and Yoshihara, 1967; Cooper and Tainton, 1968; Davidson, 1969).

V. CARBON DIOXIDE

A. Flux above the Canopy and Soil Respiration

Three basic approaches may be made to the determination of short period ($\sim$ hourly) photosynthesis of vegetation canopies:

   (1) the response of plant parts may be measured under various environmental conditions in the laboratory and used, with a model of the canopy structure and micro-climate, to compute the photosynthesis of the whole canopy;
   (2) an area of intact vegetation in the field may be enclosed in a transparent chamber and the change in CO_2 concentration of air circulating through the chamber used to estimate photosynthesis;
   (3) micrometeorological techniques may be used to measure fluxes of CO_2 both above and within a vegetation canopy.

All three methods have been used in the studies at Matador. The limitations of the various approaches have been discussed by many authors (see, for example, Šetlík, 1970 and Šesták et al., 1971).

The diurnal variation of net atmospheric carbon dioxide flux above the prairie vegetation for four days in 1971, as measured by the aerodynamic method, is presented in Fig. 25. The curves are drawn through points representing 30-minute averages. Gaps longer than one hour are shown as dashed lines. For comparison, hourly averages obtained by the chamber method are plotted on the same graphs. Comparison of the midday estimates by the two methods shows that the aerodynamic value is about 30% greater than the chamber value in each case. Some noteworthy features are the much greater aerodynamic flux (respiration) at night and the tendency for the aerodynamic estimate to become positive later in the morning (and negative earlier in the evening) than the chamber. For one of the few comparisons of the chamber and micrometeorological estimates of net CO_2 assimila-

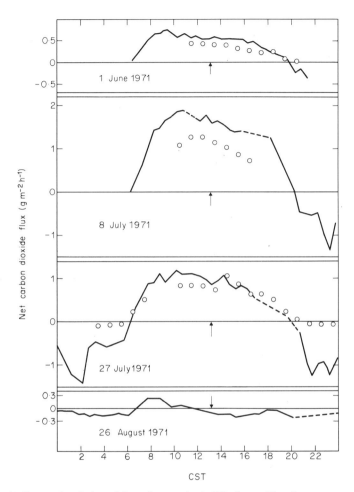

Fig. 25. Seasonal variation of diurnal atmospheric CO_2 flux at Matador as measured by micro-meteorological methods (solid lines) with chamber measurements (circles) indicated for comparison. The 25-cm midday wind speeds, for the three days with comparative data, were 1·4, 0·8, and 2·5 m s^{-1} respectively; while green leaf area indexes were 0·6, 1·0, 0·7. The arrows represent solar noon.

tion, one may refer to Lemon (1965) who reports on data collected in 1961 in a corn field. Although no night-time chamber values are shown, the other features of the comparison are remarkably similar to the Matador results. Whether the differences are due to errors in one or both methods is yet to be determined. It may be noted that although hourly assimilation values estimated by the two methods differ greatly, integrations over a day or more show better agreement, because the day-time and night-time differences compensate for each other.

All of the days shown in Fig. 25 were mainly sunny. Both soil-moisture stress and small green leaf area contributed to the slow rates of assimilation on 1 June. By 8 July, growth conditions were near optimum, with the micro meteorological estimate of net (atmospheric) CO_2 assimilation reaching a peak of near $2\,\mathrm{g\,m^{-2}\,h^{-1}}$. As the water stress increased, and green leaf area diminished during July and August, the net assimilation decreased, until by the 26 August there was a net downward flux from the atmosphere only during the morning hours. Part of the reason for this was undoubtedly the rapid plant respiration during hot midday periods in August. Day-time leaf temperatures were usually greater than 30°C and frequently higher than 40°C. Another factor that should be kept in mind, is the important contribution of soil (root and microorganism) respiration to the overall plant intake of CO_2.

Soil respiration, measured at Matador by the method of CO_2 concentration profiles in the soil (de Jong and Schappert, 1972) gave values of 0·07, 0·29, 0·31, and $0·14\,\mathrm{g\,m^{-2}\,h^{-1}}$ for the same four days shown in Fig. 25 (MacDonald et al., 1973). These were estimated from late afternoon measurements at two-week intervals. These values fall between those of the chamber and aero-dynamic methods. Errors in all methods need to be examined more closely and more comparative data analysed before an accurate assessment can be made.

B. Flux Within the Canopy

Ambient air CO_2 concentrations (at a height of 6 m) were rarely less than $300\,\mu l/l$ or greater than $350\,\mu l/l$. Concentrations at mid-canopy (10 cm) ranged from about $10\,\mu l/l$ below ambient to more than $150\,\mu l/l$ above. A series of CO_2 profiles for 28 June 1972 is shown in Fig. 26a, together with the mean wind speeds just above the canopy. Absolute concentrations were not measured during this period. The region of lowest concentration is seen to lie between the 5- and 10-cm levels, indicating greatest plant photosynthetic activity in this region. The concentration at this height was about $8\,\mu l/l$ below ambient at 0903, dropping gradually during the day as the wind speed increased to about $3\,\mathrm{m\,s^{-1}}$. The concentration just above the soil surface remained between 10 and $15\,\mu l/l$ above the ambient atmospheric values throughout the day.

The CO_2 flux distribution inside the canopy (calculated using the energy-balance estimate of diffusivities) is shown in Fig. 26b. The lack of humidity and net radiation profiles near the soil surface prevented calculation of the

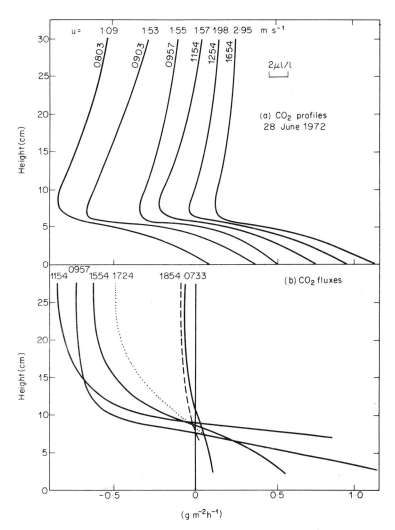

Fig. 26. (a) Canopy CO_2 profiles for 28 June, 1972. Absolute concentration is not available for this period but may be estimated from previous data to be near $320 \, \mu l/l$. Wind speeds at a height of 30 cm are shown at the top of each curve. (b) CO_2 fluxes for the same day computed by the energy-balance method.

fluxes in the lowest layers (and hence an estimate of soil respiration). However, a few values are shown for the 2·5 cm level, indicating fluxes from the soil, litter, and lower canopy as great as the flux from the atmosphere.

C. Photosynthesis of Single Leaves and of the Sward

The photosynthetic response of leaves of *Agropyron dasystachyum* was measured in the laboratory using leaf cuvettes. Leaves became light saturated between 100 and 200 W m^{-2} (PAR), and the maximum photosynthetic rate averaged 2.2 g CO$_2$ m^{-2} h^{-1} (Fig. 27). Cooper and Tainton (1968) in their review of light and temperature requirements of grasses reported that many temperate species become light saturated at about 250 W m^{-2}, and show maximum photosynthetic rates between 2.0 and 3.0 g m^{-2} h^{-1}; tropical grasses fail to saturate, and exhibit higher photosynthetic rates. North American grasses are frequently classified as cool season or warm season types, based on the prevailing temperatures during the period of optimum growth. Warm season grasses are mainly of tropical origin, and most probably retain those physiological characteristics peculiar to the group, namely high potential photosynthetic rates, failure to light saturate, high optimum temperatures for growth, and low CO$_2$ compensation points. Most grasses

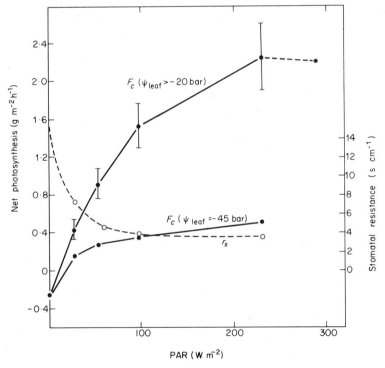

Fig. 27. Dependence of stomatal resistance and net CO$_2$ assimilation of single leaves of *Agropyron dasystachyum* on photosynthetically active radiation (PAR) intensity.

of the northern Great Plains fit into the cool-season category. The light response curve of *A. dasystachyum*, a cool season grass, is similar to those of typical temperate grasses (see Cooper and Tainton, 1968). Light saturated net photosynthesis of this group is lower than the warm season and tropical grasses. The difference is generally attributed, at least in part, to the existence of high photorespiration in genera such as *Agropyron* (Hesketh and Baker, 1967; Downton and Tregunna, 1968). Steward *et al.* (1971) emphasize that caution should be used in interpreting experiments on apparent photo-respiration, because different pre-treatments can produce entirely different CO_2 release patterns in subsequent light periods.

Another important characteristic affecting the photosynthetic rate at saturation is the pathway resistance to CO_2 diffusion. Resistances of *A. dasystachyum* were determined from laboratory data on net photosynthesis and transpiration. Stomatal resistance assumes a minimum value of about 3.5 s cm^{-1} in bright light (Fig. 27), a value similar to a number of xerophytes studied by Whiteman and Koller (1967), but Saugier (personal communication) has recently found smaller minimum values of about 2 s cm^{-1} for this species. Boundary-layer resistance is quite small (about 0.25 s cm^{-1}), owing to the narrow leaf dimensions, as discussed earlier. Mesophyll resistance (r_m) at light saturation was calculated to be 4.2 s cm^{-1} (using CO_2 concentration at the chloroplasts equal to zero), or 3.2 s cm^{-1} (using an intercellular CO_2 concentration equal to $30 \mu l/l$. These values are lower than those reported for many crop species (El-Sharkawy and Hesketh, 1965). The relatively low value of r_m may be the result of the smaller cell size characteristic of xerophytes. A negative relationship between cell size and photosynthesis has been reported by Wilson and Cooper (1969) and Wilson (1970). A value of $r_s/(r_s' + r_m)$ calculated at 0.35 is high relative to other species (such as oats, calculated at 0.24 using data in El-Sharkawy and Hesketh, 1965), and emphasizes the xerophytic nature of *A. dasystachyum*.

The increase in stomatal resistance with declining water status discussed earlier also results in lowered photosynthetic rates. Some photosynthesis continues even under severe water stress (Fig. 27), although it is unlikely that this would be sustained over longer periods, especially under laboratory conditions where rooting is confined. Redmann (1971) reported some photosynthetic activity in clones of two native Great Plains grasses (*Bouteloua gracilis* and *Stipa viridula*) at water potentials as low as -30 bars.

Light response curves for intact swards measured in the field (Fig. 28) differ from the curve of individual leaves, in that peak photosynthetic rates are less than half those of individual leaves, and the curve shapes are affected by uncontrolled variables. An apparent feature of the net photosynthesis response to global radiation shown in Fig. 28 is a hysteresis effect which

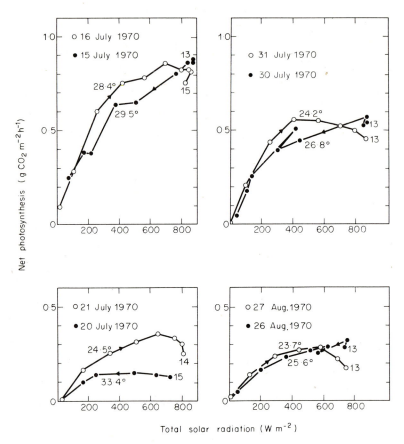

Fig. 28. Light response curves of the intact sward at Matador for a number of days in 1970. The data points are at hourly intervals and the time of one point on each curve is indicated (e.g. 13 means the period 1230 to 1330 CST). The arrows show the direction of increasing time. Mean canopy air temperatures (solar irradiance exceeding 200 W m^{-2}) are shown on each curve.

produces different responses to radiation changes before and after solar noon. A large part of the difference must result from temperature effects, a fact that is especially apparent on 20–21 August, 1970. Photosynthetic rates for radiation levels of 500 to 700 W m^{-2} were over twice as high on the morning of 21 August as on the afternoon of 20 August. Average air temperatures (at mid-canopy) for the two periods were 24·5 and 33·4°C respectively. The declines in net photosynthesis on the morning curves of other days are also associated with increasing air temperatures. In addition to the high air temperature causing increased respiration, and thereby decreasing net photosynthesis, diurnal changes in leaf-water status are also involved. Leaf-water potentials were frequently several bars higher before than after solar noon. Other explanations could also be proposed, including differences in light interception due to foliage configuration and inhibition of photosynthesis due to carbohydrate accumulation. It seems likely, however, that differences in temperature and water status can explain most of the variation.

The seasonal pattern of net carbon dioxide exchange of the intact sward, measured in the field (Fig. 29), shows a steady increase in carbon dioxide

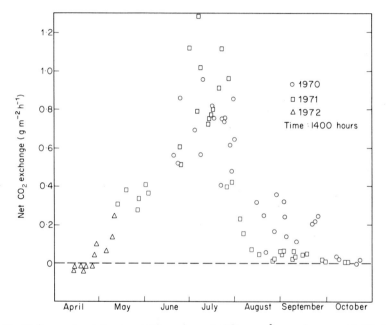

Fig. 29. Seasonal variation in net CO_2 exchange (g CO_2 per m^2 ground area per h) of the Matador vegetation as measured by the chamber method over the three years 1970 to 1972. The values plotted are means for the 4-hour period centred at 1400 CST.

assimilation up to a peak of about 1 g CO_2 m^{-2} h^{-1} in mid July of both 1970 and 1971. The subsequent decline was a result of a decrease in soil-water status which is normal during that part of the growing season (Fig. 17). The late summer drought was more severe in 1971 (Ripley, 1972c) and net CO_2 uptake was also lower in that year. High air temperatures, associated with the dry periods, served to accentuate the reduction in net photosynthesis, as discussed previously. Some degree of recovery can occur later in the season, depending on the extent of late season precipitation.

The relationships between net photosynthesis and plant respiration, and meteorological and soil moisture variables have been established, so that a complete seasonal record of CO_2 assimilation can be obtained using a static semi-mechanistic mathematical model. In this way CO_2 exchange data are used to estimate primary production of the grassland system, and results may be compared with harvesting techniques. The relationships used in a preliminary model are summarized below.

Input to the model included daily soil moisture as volume % in the 0 to 15 cm layer (M) and green shoot weight (W) in g m^{-2}; income of solar radiation S_t (W m^{-2}) and canopy air temperature at a height of 10 cm (T) in °C. Calculations were made on an hourly basis of osmotic potential (ψ_s), pressure potential (ψ_p) and leaf-water potential (ψ_1); photosynthesis in relation to solar radiation (P_S), temperature (P_T) and leaf-water potential (P_ψ); respiration in response to temperature and leaf-water potential (R_T).

The relationships used in the model are:

$$\psi_s = 6\cdot1 + (288/M)$$

$$\psi_p = 0\cdot36\,M - 0\cdot0103\,S_t - 0\cdot9$$

$$\psi_1 = \psi_s - \psi_p$$

$$P_S = 12\{1 - \exp[0\cdot0034(S_t - 35)]\}$$

$$P_T = [(45-T)/20]^{0\cdot5} \exp\{0\cdot25[1 - ([45-T]/20)^4]\}$$

$$P_\psi = 0\cdot5 + 0\cdot32\arctan[-0\cdot57(\psi_1 - 17\cdot5)]$$

combining the three photosynthesis functions

$$P_{ST\psi} = P_S \times P_T \times P_\psi$$

and computing respiration

$$R_T = 0\cdot06 \times T \times P_\psi.$$

Carbohydrate equivalent was then calculated on a daily basis.

$$C = 6\cdot8 \times 10^{-4}\,W(\Sigma P_{ST\psi} - \Sigma R_T).$$

The output of the model (Fig. 30) shows accumulated carbohydrate gradually increasing until early August, when the rate of increase drops off. Fluctuations in daily photosynthesis result from changes in global radiation and temperature, but there is an average increase in photosynthesis until late July, when in response to increased moisture stress, it drops precipitously. Short recovery periods in late season are simulated, but by October 15, photosynthesis is zero.

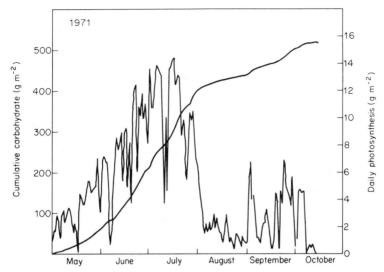

Fig. 30. Model simulation of daily photosynthesis and seasonal carbohydrate accumulation for the 1971 season at Matador.

During the period 26 May to 28 September, 1971, production estimated by CO_2 exchange methods (PP) was about twice that of shoot production estimated from harvesting data (PH) (Coupland, 1972). The difference, material translocated below ground, agrees with estimates of translocation at Matador measured by Warembourg and Paul (1974), using ^{14}C tracer techniques. During earlier and later portions of the growing season, PH is greater than PP. It seems reasonable that much of early shoot growth is at the expense of stored reserves, and that the excess carbohydrate is being translocated from below ground. It is somewhat more surprising that late season growth is being similarly subsidized.

Each of these two approaches (PP and PH) gives different kinds of information. Harvesting tells much about shoot dynamics over the season; CO_2 exchange gives an estimate of total primary productivity, and explains the effects on productivity of both short and long-term environmental changes. Together, the methods give an estimate of underground production, which is difficult to obtain by harvesting techniques alone in natural communities.

ACKNOWLEDGMENTS

The authors are grateful for the funding provided for this study by the National Research Council of Canada through its Canadian Committee for the International Biological Programme as part of the Matador Project. They would particularly like to thank B. Saugier and D. L. Spittlehouse for their assistance in the collection and analysis of field data.

REFERENCES

Anderson, M. C. (1966). *J. appl. Ecol.* **3**, 41–54.
Armijo, J. D., Twitchell, G. A., Burman, R. D. and Nunn, J. R. (1971). *Am. Soc. Engrs*, Paper No. 71-582, 16 pp.
Bergen, J. D. (1971). *J. geophys. Res.* **76**, 7385–7388.
Borchert, J. R. (1950). *Ann. Ass. Amer. Geogr.* **40**, 1–39.
Bowers, S. A. and Hanks, R. J. (1965). *Soil. Sci.* **100**, 130–138.
Bryson, R. A. (1966). *Geogr. Bull.* **8**, 228–269.
Carder, A. A. (1970). *J. Range Man.* **23**, 263–267.
Carlson, R. E., Yarger, D. N. and Shaw, R. H. (1971). *Agron. J.* **63**, 486–489.
Chamberlain, A. C. (1966). *Proc. R. Soc.* A **290**, 236–265.
Clark, O. R. (1940). *Ecol. Monogr.* **10**, 243–277.
Cooper, J. P. and Tainton, N. M. (1968). *Herb. Abstr.* **38**, 167–176.
Coulson, K. L. and Reynolds, D. W. (1971). *J. appl. Meteorol.* **10**, 1285–1295.
Coupland, R. T. (1950). *Ecol. Monogr.* **20**, 271–315.
Coupland, R. T. (1959). *In* 'Grasslands', pp. 291–306, Publ. 53, Amer. Assoc. Advance-ment Sci., Washington, D.C.
Coupland, R. T. (1961). *J. Ecol.* **49**, 135–167.
Coupland, R. T. (1972). Matador Project, Fifth Annual Report, pp. 21–26.
Coupland, R. T. and Johnson, R. E. (1965). *J. Ecol.* **53**, 475–507.
Coupland, R. T., Ripley, E. A. and Robins, P. C. (1973). Matador Project, Technical Report No. 11.
Couturier, D. A. and Ripley, E. A. (1973). *Can. J. Pl. Sci.* **53**, 659–663.
Daubenmire, R. (1968). *Adv. Ecol. Res.* **5**, 209–266.
Davidson, R. L. (1969). *Ann. Bot.* **33**, 561–571.
Davies, J. A. and Buttimor, P. H. (1969). *Agric. Met.* **6**, 373–386.
Daynard, T. B. (1969). *Can. J. Bot.* **47**, 1989–1994.
Denmead, O. T. (1967). *Aust. J. Instrum. Control.* **23**, 61.
Downton, W. J. S. and Tregunna, E. B. (1968). *Can. J. Bot.* **46**, 207, 215.
El-Sharkawy, M. and Hesketh, J. (1965). *Crop Sci.* **5**, 517–521.
Federer, C. A. and Tanner, C. B. (1966). *Ecology* **47**, 555–560.
Fuchs, M. and Tanner, C. B. (1966). *Agron. J.* **58**, 597–601.
Funk, J. P. (1959). *J. scient. Instrum.* **36**, 267–270.
Gangopadhyaya, M., Harbeck, G. E., Nordenson, T. J., Omar, M. H. and Uryvaev, V. A. (1966). W.M.O. Tech. Note. No. 83, Geneva.
Gates, D. M. (1962). 'Energy Exchange in the Biosphere'. Harper and Row, New York.
Geiger, R. (1965). 'The Climate near the Ground'. Harvard University Press, Cambridge.
Gray, D. M., Norum, E. I. and Dyck, G. E. (1969). 'Snow Measurement in the Prairie Environment', 18 pp., Presented at Can. Soc. Agric. Engr. Conference, Saskatoon.
Hesketh, J. and Baker, D. (1967). *Crop. Sci.* **7**, 285–293.
Impens, I. and Lemeuer, R. (1969). *Arch. Met. Geophyst. Bioklimt.* **17**, 403–412.

Jong de, E. (1972). Matador Project, Fifth Annual Report, pp. 68–72.

Jong de, E. and Schappert, H. J. V. (1972). *Soil Sci.* **113**, 328–333.

Lange, O. L. (1967). *In* 'The Cell and Environmental Temperature' (A. S. Troshin, ed.), pp. 131–141, Pergamon Press, London.

Lemon, E. R. (1965). *In* 'Plant Physiology' (F. C. Steward, ed.), Vol. IVA, pp. 203–227, Academic Press, New York.

Lemon, E. R. (1968). Proc. 'Copenhagen Symposium, pp. 381–389, UNESCO, Copenhagen.

Lemon, E. R., Stewart, D. W. and Shawcroft, R. W. (1971). *Science* **174**, 371–378.

Lettau, H. H. and Davidson, B. (1957). 'Exploring the Amosphere's First Mile'. Pergamon Press, London.

Longley, R. W. (1967). *Soil Sci.* **104**, 379–382.

McCree, K. J. (1971/72). *Agric. Met.* **9**, 191–216.

MacKay, G. A. and Thompson, H. A. (1968). *Trans. Am. Soc. Agr. Engrs.* **11**, 812–815.

McMillan, W. D. and Burgy, R. H. (1960). *J. geophys. Res.* **65**, 2389–2394.

Monteith, J. L. (1959). *Quart. J. R. met. Soc.* **85**, 386–392.

Monteith, J. L. (1965). *Sympt Soc. Exp. Bot.* **19**, 205.

Monteith, J. L. (1969). *In* 'Physiological Aspects of Crop Yields' (J. D. Eastin *et al.*, eds), Amer. Soc. Agronomy, Madison, Wis.

Monteith, J. L. and Szeicz, G. (1961). *Quart. J. R. met. Soc.* **87**, 159–170.

Mukammal, E. I., MacKay, G. A. and Turner, V. R. (1971). *Boundary-Layer Meteorology* **2**, 207–217.

Munn, R. E. (1966). 'Descriptive Micrometeorology'. Academic Press, New York.

Murcray, W. B. and Echols, C. (1960). *J. Meteorol.* **17**, 563–566.

Nkemdirim, L. C. (1972). *Arch. Met. Geophysk. Bioklimt.* **20**, 23–40.

Oppenheimer, H. R. (1960). *In* 'Plant-Water Relationships in Arid and Semi-Arid Conditions—Reviews of Research', pp. 105–138, UNESCO, Paris.

Paulson, C. A. (1970). *J. appl. Meteorol.* **9**, 857–861.

Pearman, G. I. (1966). *Aust. J. Biol. Sci.* **19**, 97–103.

Priestley, C. H. B. (1959). 'Turbulent Transfer in the Lower Atmosphere', Univ. of Chicago Press, Chicago.

Redmann, R. E. (1970). Matador Project, Third Annual Report, pp. 43–45.

Redmann, R.E. (1971). *Can. J. Bot.* **49**, 1341–1345.

Richards, J. H. and Fung, K. I. (1969). 'Atlas of Saskatchewan', University of Saskatchewan, Saskatoon.

Rijks, D. A. (1967). *J. appl. Ecol.* **4**, 561–568.

Ripley, E. A. (1971). Matador Project, Fourth Annual Report, pp. 228–230.

Ripley, E. A. (1972a). *Atmosphere* **10**, 113–127.

Ripley, E. A. (1972b). Matador Project, Tech., Rep. No. 2, 51 pp.

Ripley, E. A. (1972c). Matador Project, Fifth Annual Report, pp. 65–67.

Ripley, E. A. and Saugier, B. (1972). Matador Project, Technical Report No. 4, 68 pp.

Robertson, G. W. (1966). *Ecology* **47**, 640–643.

Salisbury, F. B. and Ross, C. (1969). 'Plant Physiology'. Wadsworth Publishing Co., Belmont, Cal.

Schlatter, T. W. (1972). *J. appl. Meteorol.* **11**, 1048–1062.

Sellers, W. D. (1965). 'Physical Climatology'. University of Chicago Press, Chicago.

Šesták, Z., Čatský, J. and Jarvis, P. G. (1971). 'Plant Photosynthetic Production— Manual of Methods'. W. Junk N. V., The Hague.

Šetlík, I. (1970). 'Prediction and Measurement of Photosynthetic Productivity'.

Shelford, V. E. (1963). 'The Ecology of North America'. University of Illinois Press, Urbana.

Sinclaiz, R. (1970). *Aust. J. biol. Sci.* **23**, 309–321.

Sinclair, T. R., Hoffer, R. M. and Schreiber, M. M. (1971). *Agron. J.* **63**, 864–868.

Smoliak, S. and Johnston, A. (1968). *Can. J. Pl. Sci.*, **48**, 119–127.

Specht, R. L. (1967). *Photosynthetica* **1**, 132–134.

Steward, F. C., Craven, G. H., Weerasinghe, S. P. R. and Bidwell, R. G. S. (1971). *Can. J. Bot.* **49**, 1999–2007.

Szeicz, G. (1965). *J. appl. Ecol.* **2**, 145–147.

Szeicz, G., Monteith, J. L. and dos Santos, J. M. (1964). *J. appl. Ecol.* **1**, 169–174.

Thom, A. S. (1971). *Quart. J. R. met. Soc.* **97**, 414–428.

Thom, A. S. (1972). *Quart. J. R. met. Soc.* **98**, 124–134.

Troughton, A. (1957). Comm. Bur. Pastures and Field Crops, Bull. **44**, 163 pp.

Ueno, M. and Yoshihara, K. (1967). *J. Brit. Grassl. Soc.* **22**, 148–152.

Warembourg, F. (1972). Matador Project, Fifth Annual Report, pp. 48–49.

Waterhouse, F. L. (1955). *Quart. J. R. met. Soc.* **81**, 63–71.

Weaver, J. E. and Rowland, N. W. (1952). *Bot. Gaz.* **114**, 1–19.

Webb, E. K. (1965). *Met. Monogr.* **6**, 27–58.

White, E. M. (1962). *Soil Sci.* **94**, 168–172.

White, E. M. and Lewis, J. K. (1969). *J. Range Man.* **22**, 401–404.

Whiteman, P. C. and Koller, D. (1967). *J. appl. Ecol.* **4**, 363–377.

Whitman, W. C. (1969). *In* 'The Grassland Ecosystem—A Preliminary Synthesis' (R. L. Dix and R. G. Beidleman, eds), pp. 40–64, Range Sci. Ser. No. 2, Fort Collins, Colorado.

Whitman, W. C. and Wolters, G. (1967). *In* 'Ground Level Climatology' (R. H. Shaw ed.), pp. 165–185, Amer. Assoc. Advancement Sci., Washington, D. C.

Wilson, D. (1970). *Planta* **91**, 274–279.

Wilson, D. and Cooper, J. P. (1969). *New Phytol.* **68**, 645–657.

Woolley, J. T. (1971). *Pl. Physiol.* **47**, 656–662.

Wright, H. A. (1970). *Ecology* **51**, 582–587.

Wright, J. L. and Brown, K. W. (1967). *Agron. J.* **59**, 427–432.

13. Tundra

M. C. LEWIS and T. V. CALLAGHAN

Institute for Plant Ecology, University of Copenhagen, Denmark, and The Nature Conservancy, Merlewood Research Station, Grange-over Sands, England

I. INTRODUCTION

Tundra is a term originally used for treeless plateau areas in northern Finland but its meaning has become enlarged to describe the whole of the vegetation zone lying between the northern limit of the boreal forest and the permanent ice caps (Pruitt, 1970); it includes a land area of 6 million km^2 mainly lying above 65° N latitude (Bliss, 1971). Fragmentary areas of tundra are also found on the coast and adjacent islands of Antarctica as well as at high altitudes in all latitudes (Greene and Longton, 1970).

Climatically, tundra is characterized by short, cool summers, mean temperature rising above zero for only 2 to 3 months, and long cold winters with extensive snowcover. The arctic 'treeline' coincides reasonably with several climatological and geophysical factors, e.g. mean July isotherm of 10°C, the southern limit of permafrost, the southern limit of auroral activity in the ionosphere (Sater *et al.*, 1971), and the isoline for areas with air temperatures above zero for less than 140 days (Hare and Ritchie, 1972). The southern limit of tundra also corresponds closely with a marked decrease in annual absorbed solar radiation (less than 2100 MJ m^{-2}) and annual net radiation (less than 700 MJ m^{-2}) resulting primarily from the high albedo of the more persistent snow cover in spring (Hare, 1970; Hare and Ritchie, 1972). For North America, 'treeline' also coincides with a decline in the frequency of warmer, summer air masses (Larsen, 1971; Hare, 1968) and the southern surface limit of frontal activity around the cold polar vortex (Bryson, 1966).

Within these broad limits, tundra climate is very variable ranging from oceanic to continential and from low to high arctic (60° to 84° N) (Sater *et al.*, 1971; Orvig, 1970; Vowinckel and Orvig, 1964). Vegetational gradients following these major climatic trends are much modified by local topographical and microclimatic variation, particularly snowcover, wind exposure and soil moisture (Britton, 1957; Böcher, 1963; Wielgolaski, 1972b; Aleksandrova, 1970; Wielgolaski and Rosswall, 1972; Billings and Mooney, 1968; Bliss, 1971). Communities range from dense willow scrub up to 2 m height in the most favourable situations, through wet meadow and dwarf scrub heath to sparse mats and cushions of the drier, exposed sites. The most extreme types may consist more or less exclusively of lichens and mosses.

Tundra, although treeless, is thus almost as varied as temperate zone vegetation and defies generalizations. This chapter will be restricted mainly to a general comparison of two contrasting sites situated at around 70° N and with a climate intermediate between oceanic and continental.

Tundra meadow. Winter snow up to 0·5 m depth, persisting from September to June; moist soil during summer; the summer vegetation cover continuous, dominated by deciduous grasses and sedges with a minor dicotyledonous component and an understorey of bryophytes.

Fellfield. Wind exposed; sparse, intermittent winter snowcover; soils well drained and drying out in summer; vegetation cover less than 30% and discontinuous, occurring in discrete patches; low mats and cushions of non-deciduous dwarf shrubs, herbs and graminoids dominant, with deciduous shrubs and herbs as a minor component; lichens and mosses also important.

The choice of these two specific sites was also based on the availability of micrometeorological records; tundra meadow and fellfield have been intensively investigated at Point Barrow, Alaska (Brown and Bowen, 1972), Devon Island, Canada (Bliss, 1972) and Disko, West Greenland (Lewis and Callaghan, 1971), particularly in recent years as part of the International Biological Programme.

II. RADIATION

A. Radiation Balance

The main limitations to plant growth are low temperatures and shortness of the growing season (Sørensen, 1941). Both these aspects are primarily determined by the extreme radiation regime, a generalized version of which is illustrated in Fig. 1 (Weaver, 1969; Courtin, 1972; Kelley and Weaver, 1969;

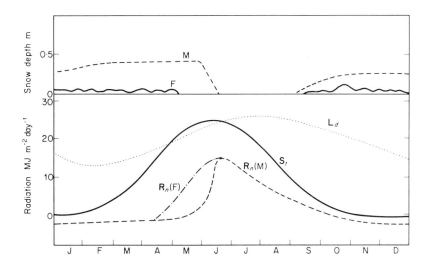

Fig. 1. The generalized annual course of solar irradiance (S_t) and downward long-wave radiation (L_d) for a typical tundra station together with the patterns of snowcover and net radiation (R_n) at the meadow (M) and fellfield (F) sites (Kelley and Weaver, 1969; Weaver, 1969; Courtin, 1972; unpublished Disko data).

Vowinckel and Orvig, 1970; Fletcher *et al.*, 1966; Budyko, 1958). The most distinctive feature is the single annual cycle of solar irradiance with a great amplitude between the long period of winter darkness and the continuous daylight over 2 to 3 months about the summer solstice. Sixty percent of the annual solar receipt (total averaging 3200 MJ m^{-2}, range 2300 to 3800 MJ m^{-2}: Marshunova, 1966; Hare and Ritchie, 1972) is concentrated in these three midsummer months, when daily values may equal or even exceed those for lower latitudes. The lower solar elevations (maximum for summer solstice at 69° N is 44° compared with 72° for 41° N) and associated smaller flux densities are compensated by the greater daylength (a maximum of 24 hours at 69° N compared with 15 hours at 41° N). Daily summer solar irradiance is much modified by cloudiness, the June mean ranging between 19 for oceanic areas and 24 MJ m^{-2} day^{-1} for continental areas (Vowinckel and Orvig, 1964).

The variation in net radiation is determined by the relationship between the annual patterns of solar irradiance and snowcover. The annual balance for tundra is positive, ranging from 0 to 700 MJ m^{-2} (Hare and Ritchie, 1972). At the sheltered *meadow* site, the long period of snowcover results in high albedo (80 to 85 %) and negative net radiation from September to May (Kelley and Weaver, 1969; Weaver, 1969; Courtin, 1972). Even during spring when downward short-wave fluxes are large, the high albedo results in negative net radiation. During late May and June, net radiation values become positive and the snow begins to melt. At this time, wide fluctuations in albedo are associated with the extent of pools of water on the snow surface (between 15% and 90% in area but with an average of 80%) (Weller *et al.*, 1972). This period marks the beginning of the growing season which is restricted to a short period of declining solar irradiance after the summer solstice. Summertime albedo increases to well over 20% as the surface becomes drier and vegetation cover increases. On an annual basis, net radiation for this site is around 360 MJ m^{-2}, with approximately one half of the solar irradiance being absorbed (Weaver, 1969).

The sparsity of snowcover at the exposed *fellfield* site is responsible for a lower albedo and a net radiation balance which is positive up to two months earlier than at the meadow. Thus in comparison with the meadow, a much greater proportion of the annual insolation is absorbed and the potential growing season is longer (Courtin, 1972). After mid-June, the radiation balances of the contrasting sites are similar. The effects of cloudiness on the daily pattern in early July are shown in Fig. 2. Due to the greater pathlength of solar radiation at higher latitudes, the ratio of diffuse to direct irradiance on clear days is greater, with a diurnal sinusoidal cycle ranging from 0·36 at noon to 0·77 at midnight (Miller and Tieszen, 1972). This is associated with a shift towards longer wavelengths at night. Through the growing season, day length falls from 24 hours lasting until late July, to 16 hours by the end of August.

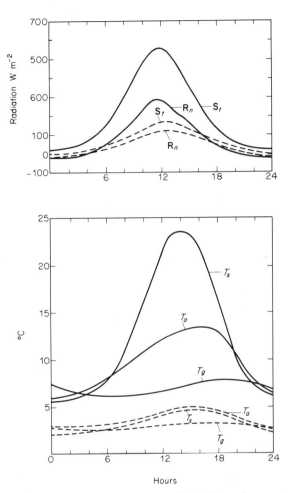

Fig. 2. Daily pattern of solar irradiance (S_t), net radiation (R_n), and temperature for typical clear and cloudy days during midsummer. T_a air temperature at 2 m; T_s surface temperature; T_g soil temperature at 0·2 m depth; (——— clear, ----- cloudy).

B. Canopy Development and Interception

The efficiency of light interception by tundra vegetation is severely limited by size and functional duration of the canopy, both of which are much lower in magnitude than for temperate counterparts (Miller and Tieszen, 1972; Sørensen, 1941). Seasonal changes in canopy characteristics for the two contrasting sites are outlined in Fig. 3.

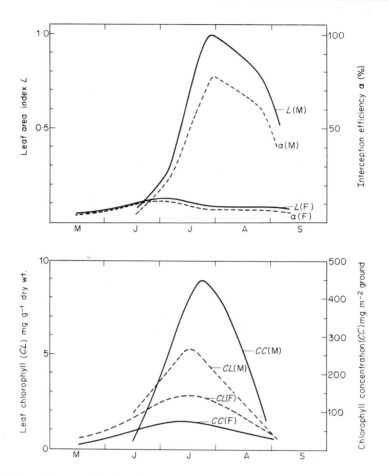

Fig. 3. Seasonal changes in leaf area index, interception efficiency, leaf chlorophyll levels (mg g^{-1} dry weight) and chlorophyll concentration (mg m^{-2} ground) at the meadow (M) and fellfield (F) sites (Tieszen, 1972b; Svoboda, 1972). ———— leaf area index and chlorophyll concentration: – – – – – interception fraction and leaf chlorophyll.

The deciduous grasses of the meadow develop a full canopy within a few weeks of snowmelt. This rapid development is facilitated by the existence of partially developed leaves at snowmelt, and in the initial stages appears to depend on translocation from below-ground reserves (Tieszen, 1972a, b; Muc, 1972).

Peak leaf area index (L) for this type of community approaches unity, within a maximum height of 20 cm; chlorophyllous non-blade structures contribute up to 20 % of the total. The vertical distribution of total live leaf area is as follows: 0 to 5 cm 55 %, 5 to 10 cm 30 %, 10 cm 15 %. The canopy is predominantly erectophile, leaf angles (θ) ranging between 60 and 80° (Caldwell et al., 1974; Tieszen, 1972a, b). As leaf turnover rates are rapid, senescent and dead leaves become an increasingly important component of the canopy after mid-season, particularly at lower levels (Tieszen, 1972b; Miller and Tieszen, 1972).

The optical properties of the leaves are similar to those of temperate grasses although there is some evidence of higher chlorophyll levels (Tieszen, 1970; Tieszen and Johnson, 1968). Leaf chlorophyll levels rise steeply after snowmelt, maximum levels being reached in mid-July, 2 to 3 weeks earlier than maximum leaf area (Tieszen, 1972b; Muc, 1972). The maximum chlorophyll concentration on a ground area basis falls between these two peaks. During August, leaf senescence and declining leaf chlorophyll levels result in a rapid reduction in chlorophyll concentration.

Figure 3 shows seasonal changes in light interception. Early in the season, efficiency is greatly limited by the size of the canopy. However, interception during mid-season may exceed 80 % and calculations indicate that a small maximum leaf area combined with steeply inclined leaves tends to optimize interception over periods of 24 hours in regions where low solar elevations are predominant. This is supported by field measurements which indicate that as little as 11 % of the daily solar irradiance and 2 to 6 % of photosynthetically active wavelengths penetrate the fully developed canopy. Extinction, following the normal laws, is greatest below 5 cm height where the greatest proportion of the total leaf area occurs (Miller and Tieszen, 1972).

Vertical and seasonal differentiation of canopy structure is generally poorly developed in tundra. However, interception by mosses and the more horizontal leaves of dicotyledons, minor components in the lower strata of the meadow canopy, appears to be more efficient early in the season and at noon relative to midnight (Tieszen, 1972b); the mosses may also utilize late-season irradiance preferentially. Differences in the phenology of leaf development and thus interception may exist between the vegetative and reproductive phases of the same species (Callaghan, 1972).

At fellfield sites, the maximum leaf area index for higher plants is an order of magnitude less than in the meadow (approximately 0·1, see Fig. 3). The canopy is very discontinuous, large areas of bare ground (e.g. 70 %) separating dense mats and cushions. The non-deciduous canopies show very little seasonal variation in size. Leaf turnover rates are slow, leaves remaining active, although at declining levels for several years; dead leaves may remain attached for even longer periods and may form an important component of the total interception system (Svoboda, 1972).

The very shallow, more or less erectophile canopy ($h < 3$ cm) is displayed either parallel to the ground or in domes. In these plants, seasonal chlorophyll production replaces the annual total leaf turnover of the deciduous species. Individual leaf chlorophyll levels rise in spring to values between 2 and 4 mg g^{-1} dry weight (less than half the grass values); levels decline from mid-season throughout the winter, often in association with the development of red pigments (Mayo *et al.*, 1973), the cycle being repeated from the following spring. Chlorophyll concentration on a ground area basis follows the same pattern with a very approximate mid-season peak of 50 mg m^{-2} (Svoboda, 1972; Bliss, 1966). Seasonal variation in interception by living and dead components, like canopy structure, is small. Average interception by live leaves probably rarely exceeds 15% on a ground area basis, but within the vegetated patches, values may be high over a vertical distance of 1 to 3 cm. Thus the plants have a greater potential for utilizing the more favourable radiation conditions early in the season.

The minor deciduous component (e.g. *Salix* spp.) may slightly increase interception from early June until late August. The physiological features of lichens and mosses, which are important in some fellfield communities, suggest more efficient utilization earlier and later in the growing season (see Section V.B), but this does not involve a change in interception levels.

III. MOMENTUM

In tundra regions, average wind velocities some distance above the surface may not be significantly greater than in other climatic regions (Sater *et al.*, 1971) but long fetches associated with low vegetation often result in relatively strong winds and large shear stresses close to the surface. Small differences in the degree of wind exposure due to microtopography may have an over-riding effect on micro-climate (Corbet, 1972; Holmen, 1957; Warren Wilson, 1959).

During winter, periods of prolonged calm associated with strong thermal inversions, alternate with persistent strong winds (mean 14 m s^{-1} or more and 800 to 2000 km run of wind per day). Turbulence near the surface is greatly reduced by the smoothing effect of snow and ice on the topography (Savile, 1972).

Momentum gradients and the annual pattern of transfer are very dissimilar at the two sites. During winter, the plants of the meadow are completely protected by snowcover whilst those at the fellfield site are more or less exposed to the full force and effect of wind. In summer, the difference in wind exposure is less but still important (Courtin, 1972).

Weller *et al.* (1972, 1974) and Weller and Cubley (1972) measured momentum gradients for tundra meadow communities. The roughness parameter undergoes appreciable changes during the season from 0·01 cm for a smooth snow surface, increasing to 0·3 with relief irregularities caused by melting, 0·4 for the water-covered tundra immediately after melt, and up to 2·4 for the fully developed leaf canopy. Mid-season values for zero-plane displacement are around 5 cm in dense meadow communities and 1·2 cm for sparser types. The bare and vegetated areas of exposed sites have values of z_0 between 0·2 and 0·7 cm, with only a slight increase from winter to summer; corrections for zero-plane displacement do not apply (unpublished data).

The ecological implications of wind exposure are manifold and variable with season:

(a) *Redistribution of snow*: the depth and duration of snowcover is the main determinant of micro-climate through its effect on radiation balance, heat and water balance, length of growing season, etc. (Corbet, 1972; Savile, 1972).

(b) *Soil and litter erosion*: transport of soil and litter from exposed to sheltered sites may significantly modify nutrient cycling and soil properties; deposition on the snow surface affects albedo and heat conductivity and thus melt rate (Warren Wilson, 1958); extreme erosion may result in uprooting of whole plants.

(c) *Dispersal*: wind may be a very potent agent particularly in long distance dispersal of propagules (Savile, 1972).

(d) *Mechanical effects on individual plants*: modification of growth and injury may result from ice-crystal abrasion, flailing and wind-pruning of live and dead parts (Bliss, 1962; Savile, 1972); changes in the geometrical display of leaves due to wind distortion may significantly alter radiation interception (Caldwell, 1970; Bliss, 1971).

(e) *Effect on temperatures*: increased convection coefficients reduce temperatures and the advantages of radiant heating on which tundra plants rely so heavily (Warren Wilson, 1959; Kevan, 1970; Corbet, 1972).

(f) *Evaporation*: increased evaporation may result from larger diffusion coefficients, although this effect may be counteracted by the reduction in water-vapour gradients associated with lower surface temperatures.

(g) *Gaseous exchange of leaves*: wind may directly initiate stomatal closing, with a reduction in transpiration and photosynthesis (Tranquillini, 1969; Caldwell, 1970).

The direct and indirect effects of wind produce great differences in growth conditions at the two sites. The plants themselves, through growth form, shoot density, and the accumulation of standing dead material may very significantly modify momentum transfer and shearing stresses and thus their own micro-climate (see Section VI).

Föhn Winds:

Föhn winds are a feature of some arctic areas; they mainly occur in winter, and are characterized by the sudden appearance of strong winds (up to 30 m sec^{-1}), low humidities, and high temperatures. Within a few hours, winter air temperatures may rise by as much as 20°C to levels above freezing leading to substantial snowmelt (Orvig, 1970). The steep vapour-pressure gradient results in high evaporation and sublimation rates for water, snow and ice. A föhn may persist for several days and drastically alter the environment for many plants particularly as regards depth of snow, and water potentials. More normal conditions may not be reinstated that season and the after-effects may be catastrophic for many plant communities. Erosion and abrasion may be particularly active during such periods. The effects of a summer föhn were studied by Courtin (1972); he reports that, during a short-lived föhn, leaf-water potentials fluctuated as much in six hours as over the entire growing season.

IV. HEAT AND WATER

The general energy balance of tundra is dominated by the transition from a snow-covered to a snow-free surface and vice-versa. The snow not only determines the radiation balance, but also the major annual features of the sensible and latent heat fluxes. The difference in the energy balances and micro-climatic conditions of the contrasting sites reflects to a large extent dissimilarities in snow cover. (Kelley and Weaver, 1969; Weller and Cubley, 1972; Vowinckel and Orvig, 1970; Weller *et al.*, 1972, 1974; Budyko, 1958; Fletcher *et al.*, 1966; Hare, 1968).

A. Temperature

The great annual amplitude in mean air temperature between the coldest month (normally February) and the warmest (normally July) is the most striking feature; this varies from 50°C for high arctic, and more continental regions, to 15°C for more sub-arctic, oceanic parts. Mean air temperatures are above zero for only three months of the year with continental summers characterized by a more pronounced summer maximum (July mean 10 to 14°C) than more oceanic areas (July mean generally below 10°C).

'Freeze up' in September is marked by a sudden drop in air temperature following the first snowfall and the attendant increase in albedo. Air temperatures fall gradually through the winter to a mean minimum in February or

March ranging between $-10°C$ for oceanic areas and $-30°C$ for continental areas. (Sater et al., 1971; Sørensen, 1941; Orvig, 1970).

Generally, tundra soils are characterized by an upper organic layer of varying thickness, over a mineral horizon. These two horizons comprise the active layer which undergoes a very distinct annual freeze-thaw cycle: from a very low minimum coinciding with the spring equinox, its temperature rises above 0°C for a short period in midsummer (Corbet, 1972; Sørensen, 1941; Nakano and Brown, 1972; Kelley and Weaver, 1969; Matveeva, 1972). Below, a permanently frozen layer extends to a considerable depth [e.g. 300 m (Brewer, 1958)]. This permafrost zone exhibits annual temperature fluctuations of decreasing amplitude with depth, but never rises above 0°C. The large variation in net radiation through the year results in a greater annual soil temperature fluctuation and a deeper level of constant temperature (e.g. 20 m) than for lower latitudes (Chang, 1957). Exposed sites have greater seasonal ranges of soil temperatures resulting in lower annual means (Sørensen, 1941; Corbet, 1972).

Commencement, rate and depth of thaw are very variable between different sites. In general, soils with a thinner organic layer and lower moisture contents (lower thermal capacity), associated with the more exposed habitats and shallower snow cover, thaw earlier and to a greater depth but freeze earlier and faster in the autumn.

B. Water

Precipitation in tundra regions varies considerably between oceanic and continental areas (ranging from 100 to over 1000 mm year^{-1}) (Sater et al., 1971; Orvig, 1970; Hare and Hay, 1971) but is generally less than 200 mm per year. Much falls as snow in the autumn and oceanic areas may also experience a spring peak (Sater et al., 1971). During the summer, rain is usually slight and sporadic due to the low water-holding capacity of cold air masses. Thus, solid precipitation is stored as snow and ice through the winter months and is not released until the thaw. Great differences in the water availability of the tundra surface result from both the unevenness of snow-cover and the extensive redistribution of liquid water after melt; run-off from elevated areas is greatly facilitated by the shallowness of the active layer and by the impermeability of the permafrost. Low-lying areas normally become waterlogged at this time. This spatial variability is further accentuated by the lowering of the permafrost boundary as summer progresses, resulting in drainage from the surface layers (Corbet, 1972).

Thus, during the growing season at low-lying meadow sites, water levels are permanently high whilst the elevated, more exposed sites are susceptible to prolonged summer droughts particularly in continental areas and where the active layer is deep (Bliss, 1956; Weller et al., 1972; Addison, 1972).

The relative humidity of the air during the growing season is normally high (except in continental areas) a reflection of the abundant open water in the form of permanent lakes, saturated meadows, pools etc. However, the actual pressure of the water vapour is generally low, rising to 6 to 8 mbar in July and August (Courtin, 1972; unpublished data).

Winter air vapour pressures are very low (less than 0·3 mbar) reflecting the extremely low dew-point of the snow or ground surface where sublimation occurs and the absence of radiation to promote evaporation. During calms, the air close to the surface may be supersaturated whilst that above the inversion may have a relative humidity of 40 % (Vowinckel and Orvig, 1970). Frequent sea mists occur in many oceanic tundra areas. The incidence and amount of dew formation becomes less at higher latitudes (Corbet, 1972).

C. Energy Balances of Whole Stands

The following description of energy balances is based on the scheme of Weller and Cubley (1972) and Kelley and Weaver (1969) in which the tundra year is divided into six characteristic phases. The generalized balances for the contrasting sites, presented in Fig. 4, are based mainly on data from Kelley and Weaver (1969), Weller et al. (1972, 1974), Weller and Cubley (1972), Courtin (1972), Addison (1972), Ahrnsbrak (1968), Mather and Thornthwaite (1958) and results for Disko (Lewis, unpublished). The net radiation component has been discussed earlier (see Section II.A).

1. Phase I. Midwinter

The negative net radiation balance results in steep and persistent thermal inversions between bulk air and the surface of snow or bare ground, in association with relatively strong atmospheric stability (Richardson numbers of 0·6 to 0·8) and weak or non-existent diurnal fluctuations (Weller and Cubley, 1972). Sensible heat flux towards the surface balances a large proportion of the net radiation loss. Heat transfer through the snow is by both molecular conduction and convection, the latter possibly predominant (Kelley and Weaver, 1969; Benson, 1969). The flux levels due to sublimation are corroborated by ice deposition rates on exposed pans, values of 0·05 mm ice per day being observed (Weller and Cubley, 1972; Weller et al., 1974).

Insulation due to snowcover during this period has an over-riding influence on the temperature regime of the ground surface. The degree of insulation depends on the depth and density of the snow, the conditions of bare areas contrasting most sharply with those under snow of greater than 50 cm depth, the so-called hiemal threshold (critical insulating snow thickness).

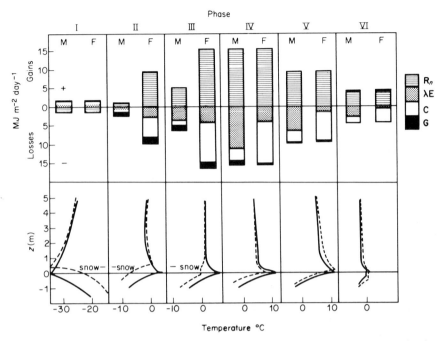

Fig. 4. Upper part: total energy balances of the meadow (M) and fellfield (F) surfaces for six characteristic phases of the year. Energy fluxes towards the surface are plotted positively and away from the surface negatively. R_n net radiation flux; C sensible heat flux in air; G sensible heat flux in soil or snow; λE latent heat flux in melting and evaporation. Lower part: Typical temperature profiles above and below the surfaces during the six characteristic phases based on daily mean temperatures (Weller and Cubley, 1972; Weller et al., 1972, 1974; Kelley and Weaver, 1969; Courtin, 1972; Addison, 1972; unpublished Disko data).

The low conductivity of the snow layer maintains much warmer conditions at the ground surface. Transfer of stored sensible heat from the underlying soil is the main source of this winter warmth. The extreme dryness of the air above the snow may be contrasted with the more or less saturated air below the snow cover. This protection from thermal extremes and desiccation is critical to the survival of many plants (Corbet, 1972; Sørensen, 1941).

2. Phase II. Premelting (Late April to Early June)

Net radiation becomes positive as solar irradiance rapidly increases; the atmosphere is less stable. At snowcovered sites, albedo is still high and temperature at the snow surface is around zero but decreases rapidly with depth in the snow and soil (Weaver, 1969). The proportionately large downward sensible heat flux associated with this steep gradient increases the temperature of the snow and soil. Some of the energy is used in melting and evaporation of the snow. The convective flux induces a slow rise in air temperature. At snow-free sites, net radiation as well as the downward and upward sensible heat fluxes are proportionately larger; the mean temperature of the soil surface rises above freezing at this time. This marks the beginning of the fellfield growing season, the sheltered site must wait for the snow to disappear, several weeks later (Sørensen, 1941).

3. Phase III. Snow Melting (Early to Mid-June)

Mean air temperature rises above zero. At the sheltered site, the snow is melting and the surface is covered with pools of water; the albedo changes give a marked increase in net radiation (Weller et al., 1972). A great proportion of radiant energy is dissipated as a latent heat flux in the melting and evaporation of snow and water. It appears that up to 90 % of this latent heat flux is used for melting rather than evaporation due to the large difference between the latent heats required for evaporation and melting. Assuming a snow depth of 33 cm and a mean density of 0.28 g cm^{-3} which melts in ten days, Weller et al. (1972) and Weller and Cubley (1972) calculate melting rates equivalent to 9 mm of water per day requiring 3 MJ m^{-2} day^{-1} whereas evaporation at 0.2 mm day^{-1} used only 0.5 MJ m^{-2} day^{-1}. Evaporation thus contributed very little to the total ablation. Snowmelt and soil-surface thaw are accelerated by the downward percolation of melt water; absorption of short-wave radiation penetrating the snow may also be important as up to 8 % of total solar irradiance may reach the soil surface below 40 cm snow ($\mathscr{K}$ approximately 0.02 cm^{-1}) (Kelley and Weaver, 1969). Only a small amount of energy is available for raising temperature levels.

At the fellfield site, the Bowen ratio is much larger, reflecting the contrast in water levels at the sites, and the downward soil-flux is a larger component (Addison, 1972). The surface temperature is considerably above air for most of the time, and the soil has already thawed to a depth of 30 to 50 cm (Sørensen, 1941; Courtin, 1972; see Section VI.D).

4. Phase IV. Post-melting (Mid-June)

The meadow site is covered with pools of water and net radiation has increased by an order of magnitude from Phase II (Kelley and Weaver, 1969; Weaver, 1969). The latent heat flux is still dominant, the evaporation of melt water reaching a maximum of 6 and mean values of 4.5 mm day^{-1}. This represents a 40-fold increase in the latent heat flux within 2 to 3 weeks over the melting phase. There is close agreement in the evaporation rates computed from energy balance formulae and actual evaporation measurements (Weller and Cubley, 1972; Weller et al., 1972; Mather and Thornthwaite, 1956).

More energy is used to heat the air (mean 2 to 3°C) and the soil, the temperature of which increases rapidly at this time. The balance at the dry site is similar to Phase III.

5. *Phase V. Summer* (*Late June to Late August*)

Radiation levels, solar angles and day-length decline through this period; air temperature falls gradually from a maximum in mid-July. The atmosphere is less stable, Richardson numbers ranging from 0·01 to −0·16 with a mean of −0·05 (Weller and Cubley, 1972). At the meadow site, evaporation rates are still rapid, approaching 3 mm day^{-1} for mid-July and in close agreement with pan estimates (Weller and Cubley, 1972). The upward sensible heat flux is proportionately greater than earlier in the year. The mean daily temperature of the soil surface is 5 to 7°C above the air; the average daily amplitude (July mean 8°C) may be 4 to 5 times greater than for air. This mainly reflects the disparity in maxima rather than minima at the two levels (Sørensen, 1941), thermal inversions being less developed in the longer days of higher latitudes (Corbet, 1972). The layer of maximum temperature and maximum fluctuation is displaced upwards in the more closed meadow communities to heights between 4 and 10 cm (Courtin, 1972). The effects of slope and aspect on these diurnal temperature fluctuations are similar but smaller than for temperate regions (Corbet, 1972; Romanova, 1972).

By late July, daily downward heat flux falls in association with a general steadying in thaw depth (25 to 30 cm maximum) and mean soil temperatures; although steep thermal gradients occur in the middle of the day (−20 to 25°C within 20 cm depth) transfer tends to be reversed at night (Kelley and Weaver, 1969; Courtin, 1972). The amplitude of diurnal fluctuations declines rapidly with soil depth as the permafrost is approached (Corbet, 1972).

At the fellfield site, the Bowen ratio β increases further, as summer drought develops (mean $\beta = 5·0$ compared with 0·5 for the meadow). Lysimeter determinations give a mean daily evaporation of less than 1 mm over the season with large fluctuations up to rates of 3 mm day^{-1} associated with precipitation (Addison, 1972). Vegetation cover, particularly lichens and mosses, dampens these fluctuations by retaining water close to the surface (Addison, 1972).

Increased eddy diffusion coefficients related to wind exposure compensate for the lower latent heat flux, so that temperatures, gradients and fluctuations are not greater than for the meadow. The zone of maximum fluctuation is more or less at the ground surface, although it may be slightly elevated in cushions. Maximum thaw-depths may exceed 50 cm.

6. *Phase VI. Freeze-up* (*Late August to Early September*)

Net radiation now averages around $4 \, \mathrm{MJ \, m^{-2} \, day^{-1}}$ and air temperature frequently falls below zero. At the wet site this energy is again primarily dissipated as latent heat. On a daily basis, there is a small net upward sensible heat flux towards the soil surface.

As the surface freezes, temperatures throughout the active layer rapidly drop to zero. As freezing progresses downwards and occasionally upwards, a steadily decreasing layer of soil remains isothermal around $0°C$ as the latent heat of fusion compensates for sensible heat loss. This phenomenon is called the 'zero curtain' and persists for two to four weeks depending particularly on soil-water content. Once the soil is totally frozen the cold wave can penetrate into the perennially frozen ground. Diurnal fluctuations which are non-existent during the zero-curtain phase, resume but may disappear again as snowcover approaches the hiemal threshold. (Kelley and Weaver, 1969; Nakano and Brown, 1972; Cook, 1955; Corbet, 1972).

In autumn, there is greater synchrony between the two sites, air and surface temperature becoming negative at approximately the same time. The zero curtain is less obvious at the drier site (Sørensen, 1941; Cook, 1955).

D. Soil Processes

Certain aspects of heat transfer are more important in arctic soils than those of other regions (Nakano and Brown, 1972; Kelley and Weaver, 1969; Mather and Thornthwaite, 1958). In soils undergoing a phase change (freeze-thaw), latent heat predominates, accounting for 95 to 98% of the total heat change; this means that variations in the thermal diffusivity of frozen soils have very little effect on thaw depth. The latent heat function is dependent on the unfrozen water content as a function of temperature. Experimental evidence indicates that more than 50% of the soil water in peat remains unfrozen down to $-10°C$. The thermal diffusivity of peat increases 2 to 3 times when changing from the frozen to the unfrozen state (Nakano and Brown, 1972). This great increase is undoubtedly associated with the intensive downward evaporative heat flux which occurs in the wet organic layer in summer to a greater extent than in other regions due to the much steeper thermal gradients. The rate and depth of thaw are also correlated with vegetation cover such that the greater insulation of the denser canopy reduces downward heat fluxes (Bliss, 1956; Tyrtikov, 1969). Moss layers are particularly effective insulators (Tikhomirov, 1952); the substantial amounts of standing dead and litter accumulation characteristic of some tundra communities will have a similar effect (Miller, 1972). Thus the establishment of closed vegetation will lead to a deterioration in soil conditions, which in turn may result in slow cyclic changes in the vegetation cover (Benninghoff, 1952). Removal of the above-ground vegetation may considerably increase thaw depth (Corbet, 1972). Microtopography, through its effect on intercepted radiation, affects soil temperatures so that thaw is faster and deeper under hummocks than hollows (Matveeva, 1972).

Intense freeze-thaw processes within the upper layers lead to general instability and soil movement (patterned ground etc.), physical soil disturbances which severely stress plants and affect their distribution patterns (Raup, 1971; Washburn, 1969; Rønning, 1969; Tedrow and Cantlon, 1958). Corbet (1972) states that these upper layers are ones 'in which disintegration is maximal and decomposition minimal and in which there is little physical stability'.

E. Transpiration and Water Potentials

As described earlier, great differences in water regimes exist between the sites. Plants of the meadow enjoy a plentiful supply of water during the growing period and spend the winter in the relatively moist mild conditions below the snow. They may be contrasted with the plants of dry, windswept fellfield which may experience severe drought both summer and winter (Sørensen, 1941; Billings and Mooney, 1968; Bliss, 1971). Winter drought stress due to deeply frozen soils and cold dry winds may be particularly severe in these exposed sites. The effects of long term exposure on plants with evergreen leaves when soil water is virtually unobtainable may be as demanding as extreme desert conditions. Evergreen plants of exposed sites, at least in alpine areas, may maintain leaf-water potentials above lethal levels (ψ_{leaf} averaging -30 bars) (Lindsay, 1971) partly by much reduced cuticular transpiration and by the development of roots which can absorb from a thin layer of melt water which may develop over frozen soils during sunny winter days, particularly under thin crusts of an ice 'greenhouse' (Larcher, 1957; Shamurin, 1966; Tranquillini, 1964). Plants covered by snow avoid such stress and maintain high water levels throughout winter (ψ_{leaf} of -20 bars) (Lindsay, 1971).

At exposed sites the droughts of later summer are an important ecological factor (ψ_{soil} falling to below -20 bars) (Bliss, 1971). Sørensen (1941) observed that growth may stop well before temperatures have dropped to unfavourable levels. Addison (1972) (see Fig. 5) clearly demonstrated that transpiration rates, and therefore presumably photosynthetic activity, are depressed as leaf-water potentials drop below -20 bars. There are high positive correlations between transpiration rate, leaf-water potential and precipitation. Plants growing on the wetter meadow show steady leaf-water potentials

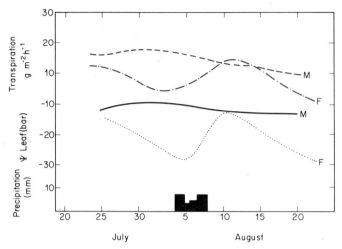

Fig. 5. Seasonal changes in transpiration rate and leaf-water potential in relation to precipitation for typical meadow (*Carex stans*) and fellfield (*Dryas integrifolia*) plants based on the lysimeter measurements of Addison (1972) from Devon Island, Canada.

normally above -10 bars and generally higher transpiration rates throughout the same period; transpiration and photosynthesis are curtailed when ψ_{soil} falls below -8 bars and ψ_{leaf} below -14 bars (Addison, 1972; Kuramoto and Bliss, 1970).

Again, there appears to be a clear distinction between the plants of wet and dry sites which is correlated with growth habit. The grasses, sedges and herbaceous plants of the wetter sites appear to have higher transpiration rates providing water is not limiting but are not able to control water loss to the same extent as the dwarf shrubs etc. of drier sites. This may well be correlated with differences in structure between the plants, (see Section VI). There also appears to be a difference in tolerance of lower water potentials, the dry site plants being capable of photosynthesis down to at least -20 bars and surviving much lower water potentials (Bliss, 1971; Billings and Mooney, 1968).

Evaporation rates from a water-saturated carpet of moss or fruiticose lichen are up to 20% higher than for a free water surface, probably because of greater roughness. At other times, rates fall to levels comparable with the general surface (Addison, 1972).

F. Plant Temperatures, Growth and Development

Tundra plants exist within a shallow layer above or below the soil surface, the warmest stratum of the environment. As Sørensen (1941) so clearly demonstrated, the mean temperature of the soil surface is a much better indicator of conditions for plant growth than air temperature. Nevertheless, mean temperatures during the growing season are low and close to the lower cardinal point for many metabolic processes. In this range, small changes in temperature will produce large plant responses (Warren Wilson, 1957).

The effects of temperature on plant growth and development are complex, partly due to the variation in temperature dependence of different physiological processes. Photosynthesis and transpiration rates of tundra plants are more closely coupled to daily patterns of radiation than air temperatures (see Sections IV.E and V.B). However, other processes associated with growth and development including differentiation, respiration, cell division and expansion, translocation and incorporation of assimilates into new or existing structures are more subject to temperature limitation (Monteith and Elston, 1971).

The longer term effects of tundra micro-climatic conditions on overall growth have been assessed by the growth analysis of native and introduced temperate species. Crop plants growing under non-limiting nutrient conditions accumulate dry matter more slowly in the tundra than under otherwise similar conditions in temperate regions; relative growth rates and leaf expansion are depressed to a much greater extent than assimilation rates (Warren Wilson, 1966; Lewis and Greene, 1970). Differences in performance between tundra sites, whilst depending on species, were clearly correlated with micro-climatic variation particularly the level and duration of maximum temperatures, and were most marked at the beginning and end of the season (Lewis and Callaghan, unpublished data). Results for tundra populations of *Phleum alpinum* showed similar trends, relative growth rate being correlated with the temperature regime as influenced by wind exposure at contrasting sites; the lower temperature dependence of net assimilation rate was not as clear (Callaghan and Lewis, 1971b). Warren Wilson (1966) proposed that the differential effect of temperature on growth and assimilation rates, results in the accumulation of soluble photosynthates to the high levels characteristic of tundra plants, leading to a reduction in assimilation rate.

Soil temperatures and depth of thaw radically affect the availability of water and nutrients as well as metabolic rates of underground organs (Billings and Mooney, 1968; Bliss, 1971). The temperatures of roots are closely coupled to those of the soil. It appears that the roots of tundra plants are concentrated within the warmer, uppermost 10 cm of soil, although many species also have the ability to produce roots which follow the retreating permafrost layer and maintain rapid growth at temperatures close to 0°C; these may be particularly important in tapping water from deeper, moister layers (Dadykin, 1954; Bliss, 1956; Dennis and Johnson, 1970; Chapin, 1974a, b).

Measurements on tundra plants growing naturally indicate that growth and developmental rates correlate better with soil temperature than with air temperature (Bliss, 1966; Dennis and Johnson, 1970). This agrees closely with the hypothesis advanced by Brouwer (1962) that the limiting effects of lower soil temperatures on root growth and uptake are more critical determinants of growth than those of air temperatures on above-ground parts. Numerous findings have demonstrated that the growth of arctic plants is limited by nutrient availability particularly of nitrogen and phosphorus (Bliss, 1966; Warren Wilson, 1966; Lewis and Greene, 1970; Callaghan and Lewis, 1971b; Larsen, 1964). This dependence probably results from the combined effect of the slow rates of decomposition and nutrient cycling and reduced root activity in cold soils (Chapin, 1972; Douglas and Tedrow, 1959). Limited microbial fixation of nitrogen may also partly account for the deficiency (Russell, 1940); recent studies indicate that blue-green algae and lichens are the main nitrogen fixation agents in tundra (Kallio and Käremlampi, 1972; Granhall and Selander, 1973).

The energy balance and thus the temperature of individual aerial plant parts may differ widely from those of the general stand in a manner which radically affects their physiological functioning. These differentials and their variation between different organs may radically affect plant performance, often in a way tending to harmonize overall development (Evans, 1963; Monteith and Elston, 1971). Radiant heating is particularly critical to the growth and even survival of tundra plants (Warren Wilson, 1957) which may show structural modifications aimed at optimizing temperature levels (see Section VI).

The temperature of leaves, flowers, etc., will be closely correlated with irradiance levels (see Fig. 6); substantial elevations occur particularly with strong direct irradiance, whilst during overcast periods plant temperatures are normally within a degree or two of ambient air temperature. The effects of orientation with respect to incident radiation are the same as for other latitudes, although the diurnal range of temperature tends to be smaller (Corbet, 1972).

Leaf temperatures of arctic plants have been particularly well studied. (Tikhomirov et al., 1960; Bliss, 1962; Kevan, 1970; Warren Wilson, 1957; Savile, 1972). During the middle of sunny days, excesses of 20°C above air temperature have been recorded in living leaves; the highest values have been observed for dense cushion plants and mats where structural features reduce diffusive fluxes (Longton, 1972; Biebl, 1968; Mayo et al., 1973; Courtin in Bliss, 1971) (see Section VI). The leaf temperatures of more open

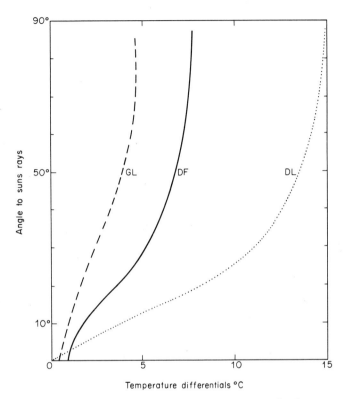

Fig. 6. Generalized temperature differentials on a clear summer day between ambient and a meadow grass leaf (GL), a *Dryas* leaf within the mat (DL) and an heliotropic *Dryas* flower (DF), 5 cm above the mat surface (based on Kevan, 1970; Biebl, 1968; Warren Wilson, 1957; Corbet, 1972); values are plotted in relation to the angle of incident solar irradiance.

growth forms (e.g. grasses) average 3 to 5°C above air temperature under similar conditions (Bliss, 1966). There is some evidence to suggest that the reddish coloration of certain leaves, particularly non-deciduous ones early in the season, may significantly increase absorptivity leading to more marked radiant heating (Shamurin, 1966; Mayo *et al.*, 1973). Pubescence may be an additional aid to heat conservation through the 'greenhouse' effect and reduction of convection (see below). The depressive effect of high temperatures on net photosynthesis is discussed in Section V.B; however, the deleterious effects are probably offset by the enhanced rates of translocation, leaf growth and development.

The establishment of warm teleo-climates for growing points, developing flowers and fruits may be particularly important; the essential activities of these structures are more dependent on temperature than those of mature leaves but must often attain a predetermined morphogenetic goal within a limited period of time (e.g. production of a full leaf complement, completion of sexual reproductive cycle) (Savile, 1972). Shoot meristems and flower buds may be well protected by surrounding leaves, often within cushions, or may occur near the ground surface (e.g. grasses) and may be several degrees above ambient air temperature for a considerable proportion of time (Warren Wilson, 1957). The flower buds of some species often open whilst sheltered within the warmer leaf canopy, scape elongation being delayed (Savile, 1972). It has been suggested that rapid stem elongation may be facilitated in some species by the development of hollow stems which enable temperatures of up to 20°C above ambient air temperature to be maintained within the stem (Billings and Godfrey, 1967).

The temperatures of flowers may be elevated by their shape and heliotropic properties (Tikhomirov et al., 1960; Hocking and Sharplin, 1965; Kevan, 1970; Bliss, 1962). The parabolic corollas of many flowers serve to focus incident radiation on the reproductive organs at the centre, the heating effect speeding up development processes as well as attracting pollinators. The heating effect is closely related to flower coloration, higher temperature differentials resulting from deeper pigmentation; temperatures within the corolla may be more than 6°C above ambient. Bell-shaped and tubular flowers may act as miniature greenhouses, sheltered from wind effects (Kevan, 1970; Corbet, 1972). The reproductive structures of a number of species are protected by a dense outer covering of more or less transparent hairs. As elegantly demonstrated by Krog (1955) for *Salix* catkins, these hairs transmit incoming radiation which is absorbed by the enclosed dark scales. The hairs are relatively opaque to longwave emission, thus providing a greenhouse effect as well as reducing convection coefficients and can give rise to temperature differentials of up to 25°C.

The stronger reliance of plants on radiant heating means that the cooling effect of wind on plant performance is particularly pronounced. The temperatures of leaves and flowers may be strikingly cooler on an exposed bluff compared with an adjoining hollow. Warren Wilson (1959) reports that leaf temperatures may differ by as much as 7°C between sheltered and exposed situations, resulting in significant reductions in net assimilation and relative growth rates (Warren Wilson, 1966).

V. CARBON DIOXIDE

A. Carbon Dioxide Fluxes for Whole Stands

In common with other regions, the atmospheric concentrations of carbon dioxide measured at a height of 16 m, follows a marked annual cycle with high winter and low summer levels about a mean of 316 μl/l (Kelley, 1968; Kelley *et al.*, 1968; Johnson and Kelley, 1970; Coyne and Kelley, 1971, 1972). The minimum concentration of 308 μl/l occurs in August and corresponds closely with the peak of photosynthetic activity of the tundra and adjacent ocean. With cessation of growth in the autumn, carbon dioxide concentration rises rapidly from August to December, levels off during winter and again increases rapidly to a maximum of over 320 μl/l in early June, coincident with snowmelt and surface thaw.

The shape and amplitude of this annual cycle can be explained by the influence of winter snowcover on the exchange of carbon dioxide between the tundra surface and the atmosphere. The results of Kelley *et al.* (1968) indicate that the average monthly carbon dioxide concentration at ground level has two distinct peaks, one in early winter and another in June, and is always above atmospheric concentration. Experimental observations (Coyne and Kelley, 1971) suggest that summer respiration of roots and micro-organisms within the active layer saturates the soil solution with carbon dioxide. Some of this CO_2 is released upon freezing and escapes at varying rates from the soil (leaky ice bubble effect), the highest rates being in the autumn and spring. Restricted exchange through the snow results in accumulation of carbon dioxide at the soil surface; exchange rates through the snowcover are positively correlated with wind velocity. The spring peak is probably due to the release of trapped carbon dioxide with the differential warming of the soil under the snow, but may also result from biological activity as temperatures approach or even exceed zero. From experimental results on the freezing of tundra soil cores, Coyne and Kelley (1971) estimate that as much as $18 \text{ g } CO_2 \text{ m}^{-2}$ could be released in this way, particularly from coarse-textured soils with high moisture content. The high levels of carbon dioxide under the snow during spring may enhance photosynthesis in those plants which appear to be active under the snow (Shamurin, 1966).

Simultaneous measurements of wind, temperature and carbon dioxide profiles above the meadow canopy throughout the growing season formed the basis of an aerodynamic approach to the measurement of vertical CO_2 fluxes and primary production (Coyne and Kelley, 1972; Tieszen, 1972a). Flux values were calculated according to the usual equation, with correction to neutral stability using a term based on Richardson's number. No comparable information is available for fellfield.

Figure 7 presents the seasonal pattern of net daily carbon dioxide flux at the meadow site. From slightly negative values (i.e. flux from surface to atmosphere) at the beginning of the growing season, fluxes steadily increase to a mid-season peak of $+8 \text{ g m}^{-2} \text{ day}^{-1}$ followed by a rapid decline during August. In general, carbon dioxide flux was correlated with solar radiation

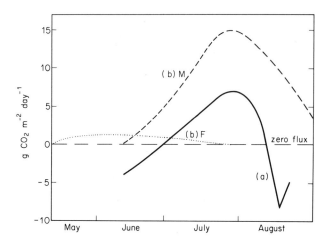

Fig. 7. Seasonal patterns on a ground area basis of (a) net daily atmospheric CO_2 flux based on the aerodynamic method, where flux towards the surface is plotted positively (Coyne and Kelley, 1972), (b) estimated daily net photosynthesis by all vascular plants at the meadow (M) (Tieszen, 1972a, c) and fellfield (F) [very rough estimate based on Mayo et al., (1973) and Svoboda (1972)]; neither net photosynthetic curves are corrected for canopy shading effects.

levels through the day; however, as mean daily temperatures increased this relationship became obscured by the effect of temperature on soil respiration rates (decomposer activity). At higher temperatures, positive fluxes decrease as proportionately more of the carbon dioxide fixed in photosynthesis comes from soil sources rather than from the atmosphere. This effect may be more pronounced in tundra areas where soil respiration is strongly temperature limited; Van Zinderen Bakker (1972) reports Q_{10} values approaching 5 for soil respiration over the temperature range 0 to 10°C falling to 2 between 10 and 20°C; decreases in soil-moisture content as the season progresses may also increase soil respiration (Van Zinderen Bakker, 1972).

Net fixation of CO_2 is the sum of fluxes from the atmosphere and soil microbial respiration. For the latter, an average daily value of $8 \text{ g } CO_2 \text{ m}^{-2}$ was estimated during periods of minimal photosynthesis after correction for temperature effects on decomposer activity (Coyne and Kelley, 1972); this figure agrees reasonably with direct measurements for wet arctic meadow made by Van Zinderen Bakker (1972). Integration of atmospheric and soil CO_2 fluxes over a 60-day growing season gives an approximate estimate for net CO_2 incorporation of $620 \text{ g } CO_2 \text{ m}^{-2}$ or 380 g dry matter (based on 60% conversion factor) for the Point Barrow meadow community, with peak season levels of 14 to $16 \text{ g } CO_2 \text{ m}^{-2} \text{ day}^{-1}$ (see Sections V.B and C).

B. Photosynthesis and Respiration

Tundra light regimes are not more limiting, at least for most of the growing season, than in temperate regions. The available information on *in situ* photosynthesis of leaves and individual plants indicates efficient carbon dioxide fixation at low temperatures (Billings and Mooney, 1968; Bliss, 1971; Tieszen, 1972a, 1973; Mayo *et al.*, 1973; Shvetsova and Voznesenskii, 1970; Zalenskij *et al.*, 1972).

The daily course of net photosynthesis in higher plants from contrasting sites (see Fig. 8) closely follows irradiance at least over the lower temperature range (see below) and daily carbon dioxide incorporation for meadow plants shows a strong linear correlation with total irradiance up to maximum values (see Fig. 9) (Tieszen, 1973). Photosynthetic activity is far less limited by low

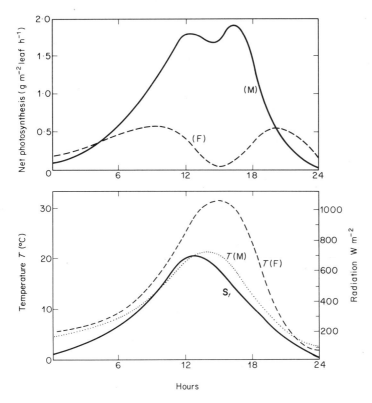

Fig. 8. Daily patterns of solar irradiance (S_t), leaf temperatures (T) and net leaf photosynthesis for a typical meadow grass (M) and fellfield *Dryas* (F); based on Tieszen (1972a, 1973) and Mayo *et al.* (1973).

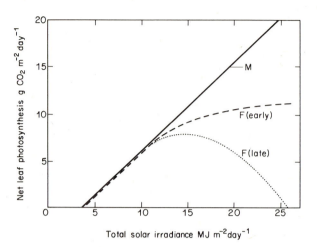

Fig. 9. Generalized relationship between net daily leaf photosynthesis and total daily solar irradiance for a typical meadow grass (M) and fellfield *Dryas* (F); for *Dryas*, the relationship differs between early and late season; based on Tieszen (1972a, 1973) and Mayo *et al.* (1973).

temperatures than in temperate species; net uptake of carbon dioxide continues down to $-4°C$ or less and rates approaching the maximum have been observed at ambient air temperatures as low as 3°C (Pisek *et al.*, 1967; Tieszen, 1973).

All tundra plants seem to have the C_3 photosynthetic system with photo-respiration accounting for a large proportion of fixed carbon dioxide (Tieszen, 1972a). Dark respiration rates are generally higher at all temperatures than in alpine or temperate counterparts. Values of Q_{10} range between 1·8 and 3·0 with some evidence that the higher values apply over the lower temperature ranges (Larsen, 1964; Billings and Mooney, 1968). Respiration rates are more restricted by low temperatures than photosynthesis and are low for most of the time in the field (Tieszen, 1972a, 1973). In consequence, light compensation points are between 10 and 20 W m^{-2} when estimated from the irradiance level at which the diurnal curve of net photosynthesis approaches zero (see Fig. 8), i.e. conditions of weak irradiance primarily at night when temperatures are also low. Compensation points on a daily basis as estimated from the intercept in Fig. 9 are considerably higher (around 45 W m^{-2}) due to increased respiration (including photorespiration) associated with the higher temperatures and irradiance around noon (Tieszen, 1972a, 1973; Mayo *et al.*, 1973). However, the net result is that net leaf photosynthesis remains positive throughout most of the 24-hour day, negative values being restricted to heavily overcast nights, and towards the end of season (Tieszen, 1972a,

1973; Mayo *et al.*, 1973; Shvetsova and Voznesenskii, 1970). Average daily totals of net carbon dioxide uptake may reach comparatively high levels in some species, (see below), reflecting the efficiency of net photosynthesis at low temperature and the integrated effect of the long light period.

Several investigations (Tieszen, 1972a, 1973; Mayo *et al.*, 1973; Svetsova and Voznesenskii, 1970) demonstrate that in the middle of clear warm days, net photosynthesis may be depressed. The cause is not definitely known but recent evidence (Shvetsova and Voznesenskii, 1971) points to enhanced respiration at higher leaf temperature rather than stomatal closure or inhibition due to assimilate accumulation (Warren Wilson, 1966).

There are considerable differences in the photosynthetic response to environmental factors between plants from the contrasting sites (Shvetsova and Voznesenskii, 1970; Zalenskij *et al.*, 1972; Bliss, 1971; Moser, 1973). Net photosynthesis in the deciduous leaves of meadow grasses and sedges increases steeply after the initial growth phase up to maximum rates approaching those of temperate grasses (Tieszen, 1972a, 1973; Mayo *et al.*, 1973; Hadley and Bliss, 1964; Callaghan and Lewis, 1972a, b; Callaghan, 1972). For at least part of the season, they may fix more carbon dioxide over the 24-hour photoperiod than temperate counterparts; photosynthetic efficiency as estimated from the slope in Fig. 9 approaches 2 % of irradiance between 400 and 700 nm [based on 1 g CO_2 equivalent to 0·6 g dry weight of calorific value 19 kJ g^{-1} dry weight, and with 45 % of total irradiance between 400 and 700 nm (Tieszen, 1972b)].

Seasonal carbon dioxide uptake on a community basis (g CO_2 m^{-2} ground) can be estimated by integrating daily leaf totals and leaf area index (see Figs. 3 and 9) over a 60- to 70-day growing season. Estimates between 500 and 700 g CO_2 m^{-2} (or 300 to 400 g dry weight) after correction for canopy shading effects and photosynthesis of the moss understorey agree reasonably with values from aerodynamic investigations (see Section V.A and Fig. 7) (Tieszen, 1972a, 1973; Miller and Tieszen, 1972).

In the non-deciduous plants of fellfield, optimal net photosynthetic rates are associated with lower irradiance and temperature levels; light saturation points, optimum temperature range and lower leaf-water potential limit are generally lower than for meadow plants. (Mayo *et al.*, 1973; Shvetsova and Voznesenskii, 1970; Zalenskij *et al.*, 1972; Bliss, 1971). In fact, during June and July, night-time photosynthetic rates often exceed those during bright sunny periods, when leaf temperatures within the mat may rise to 30°C and photosynthetic depression is very marked (see Fig. 8). Maximal rates of net leaf photosynthesis are lower than for the meadow plants and occur early in the season; they are comparable with rates for semi-arid sclerophyllous shrubs (Mayo *et al.*, 1973).

The fellfield plants may begin photosynthesizing in early May when surface temperatures rise above zero for at least part of the day (Sørensen, 1941). 'Ice-greenhouses' may be important in this context (Shamurin, 1966). They appear to be more efficient earlier in the season, when strong sunshine is associated with low air temperatures, than later when excessive leaf temperatures, decreasing leaf water potentials, and lengthening dark periods restrict photosynthesis. The average efficiency of net leaf photosynthesis over cloudy days early in the season may approach 2% but falls to well below 1% in brighter light and under warmer, drier conditions later in the season. Based on average daily net photosynthesis and leaf area index values, a very approximate seasonal total for higher plants is 70 g CO_2 m^{-2}. The seasonal pattern contrasts strongly with that for the meadow where peak incorporation occurs after mid-July.

The photosynthetic behaviour of lichens and mosses is poorly understood. Maximum net photosynthetic rates, light saturation and compensation points, lower temperature limits and optimal temperature range seem to be generally below those of higher plants. The photosynthetic activity is very dependent on surface water, dew and high relative humidities for the maintenance of high thallus water contents (Mayo et al., 1973; Bliss, 1971; Richardson and Finegan, 1972; Kallio and Karemlampi, 1972; Ahmadjian, 1970; Barashkova, 1971; Clarke et al., 1971; Rastorfer, 1970). They are particularly abundant in more oceanic areas with frequent sea mists and low clouds. Growth may occur under shallow snow and ice and intermittently during winter when the sun may cause local melting of the surface and rock faces.

C. Primary Production

The limitations of temperature and the short growing season are reflected in the small annual values of total net primary production which range (very approximately) between 4 and 400 g m^{-2} year^{-1}, representing efficiencies of 0·055 to 0·5% based on the total annual irradiance between 400 to 700 nm (see Section II.A) (Bliss, 1970; Aleksandrova, 1970; Rodin and Basilevich, 1967; Wielgolaski, 1972a, b; Billings and Mooney, 1968; Clarke et al., 1971). Assuming a three-month growing season, average efficiency over this period rises to 0·01 to 1·0%, with peak daily values approaching 2% (Jones and Gore, 1972).

Net primary production at the meadow site may amount to 300 g m^{-2} (including mosses) a large proportion of which is incorporated into below-ground structures after the main period of above-ground production; this downward translocation is very important for the maintenance of the below-ground reserves which were depleted in rapid growth immediately after snowmelt, as well as growth of roots and rhizomes (Johnson and Kelley, 1970; Dennis and Johnson, 1970; Dennis and Tieszen, 1972; Tieszen, 1972a, b; Muc, 1972). This estimate is rather higher than most published values and is based on the assumption that below-ground production for tundra vegetation, particularly turnover rates for roots and rhizomes, have been generally underestimated; however, it agrees reasonably with independent estimates based on carbon dioxide flux (see Section V.A) and direct photosynthesis measurements (see Section V.B). It represents an average efficiency of approximately 1 % over the growing season, and is compounded of a photosynthetic efficiency which remains fairly constant at close to 2 % until mid-August and the seasonal pattern of interception efficiency (see Fig. 3) which is the main limitation until mid July (see Section II.B) (Miller and Tieszen, 1972).

The very much smaller fellfield production (very approximately 50 g m^{-2} year^{-1}) equivalent to an average seasonal efficiency of 0.1 % is strongly limited both by interception (less than 10 %, see Fig. 3) and by photosynthetic efficiency which falls rapidly from June onwards (seasonal average 1 %) (Syoboda, 1972; Chepurko, 1972). Rough estimates of production based on photosynthetic measurements are in reasonable agreement (assuming 70 g CO_2 m^{-2} equivalent to 42 g dry weight m^{-2}, see Section V.B).

VI. THE TUNDRA COMPLEX AND ADAPTATION

'The key to successful adaptation to tundra or alpine environments is the development and operation of a metabolic system which can capture, store and utilize energy at low temperatures and in a short period of time.' (Billings and Mooney, 1968).

The micrometeorological distinctions between the two sites contrasted in this chapter are paralleled by clear differences in the overall strategy of the associated plants (Billings and Mooney, 1968; Bliss, 1971).

The meadow grasses and sedges, in their structure and general growth habit, show close similarities with their temperate counterparts. They have responded to the rigorous demands of the tundra environmental complex mainly by acclimation of biochemical processes. The lower temperature limits and optimum ranges of several metabolic processes including photosynthesis, respiration, uptake and translocation, and extension growth are clearly adjusted to the severe conditions. This is demonstrated by the ability of plants to photosynthesize at temperatures below zero, to produce a full leaf complement within one month of snowmelt, to remobilize and translocate from root and rhizomes whilst soils are still frozen and to absorb water and nutrients from soils close to 0°C.

Table I

Summary of field data on net photosynthetic and dark respiration rates for leaves of typical meadow and fellfield plants, and their relationship to irradiance, leaf temperature, leaf water potentials and diffusive resistances for whole plants (Billings and Mooney, 1968; Shvetsova and Voznesenskii, 1970; Bliss, 1971;* Courtin in Bliss, 1971; Tieszen, 1972a, 1973; Mayo et al., 1973; Zalenskij et al., 1972; Addison, 1972)

Plant type	Maximum Net Photosynthesis $g\,m^{-2}\,leaf\,h^{-1}\;g\,m^{-2}\,leaf\,day^{-1}$	Light Saturation Point at Lower Temperatures Wm^{-2}	Light Compensation Point at Lower Temperatures Wm^{-2}	Lower Temperature Limit °C	Optimum Temperature Range °C	Lower Limit of ψ_{leaf} (bar)	Diffusive* Resistances of Whole Plants $s\,cm^{-1}$	Leaf Dark Respiration Rate at 5°C $g\,CO_2\,m^{-2}\,h^{-1}$
Deciduous, meadow graminoids	1·0–2·0 10–30	300–400	10–20	−5	15–25	−12 to −15	$r_H = 0.06$ $r_V + r_s = 0.69$ Carex bigelowii	0·10–0·15
Non-deciduous, fellfield plants (mainly dwarf shrubs)	0·3–1·0 3–15	100–300	10–20	−5	8–10	−20 to −25	$r_H = 0.99$ $r_V + r_s = 3.46$ Diapensia lapponica	0·15–0·25

In the fellfield plants, biochemical specialization is associated with a distinctive growth habit, characterized by non-deciduous leaves within low, compact mats and cushions. The low stature, compactness and more aerodynamic shape may be adaptations to the severe stresses of the exposed site (Savile, 1972; Svoboda, 1972; Courtin and Bliss, 1971). The mechanical effects of wind including ice abrasion, as well as erosion of litter and soil particles will be reduced. The small amounts of snow which become trapped between the branches may be important for summer soil-water availability to the plant.

The reduction in evaporation and sensible heat transfer due to the greater diffusive resistance of the compact growth form (see Table I; Courtin in Bliss, 1971) will be advantageous both summer and winter. Thus, this growth form determines its own micro-climate so as to maintain moister, calmer, warmer conditions within the canopy (Bliss, 1971). Svoboda (1972) suggests that by tenacious standing dead material and retention of litter these plants have developed a 'semi-closed system with respect to recycling of minerals released after decomposition'.

This boundary-layer feature combined with greater leaf diffusive resistances (see Table I) although reducing CO_2 exchange may be important in the maintenance of a favourable water balance (Bliss, 1971; Courtin in Bliss, 1971). In fact, their structure and physiological behaviour may be compared with the sclerophyllous xerophytes of lower latitudes in which slow growth rates, facilitated by persistent leaves are combined with greater water efficiency (Billings and Mooney, 1968; Lewis, 1972); approximate values for the ratio of g CO_2 fixed per g water loss, based on the resistances given in Table I, are twice as high for the cushion as the sedge (Courtin in Bliss, 1971).

The reproductive cycle of all tundra plants is particularly stressed by the unfavourable conditions since it involves a complex series of morphogenetic changes in a definite temporal sequence. The final goal of producing viable seed may be reached only in favourable years (Callaghan, 1972). The risk has led to the development of vegetative propagation as an effective alternative to the sexual cycle and the virtual exclusion of the annual habit. (Sørensen, 1941; Billings and Mooney, 1968; Bliss, 1971; Savile, 1972). Other mechanisms whereby the reproductive cycle adapts to the rigorous conditions of a short, cold growing season are varied. The most important are the ability of reproductive structures to overwinter at various stages of development, even advanced ones (Hodgson, 1966), as well as modification of individual developmental phases (e.g. reduction in culm length, and smaller inflorescences with fewer flowers) so as to facilitate completion of the normal cycle within a shorter period (Callaghan and Lewis, 1971a).

ACKNOWLEDGMENTS

We wish particularly to express gratitude to those scientists who have given us permission to cite preliminary results, particularly for the IBP Projects at Devon Island and Point Barrow (see Brown and Bowen, 1972; Bliss, 1972), and to Ann Lewis and Kirsten Bach Jensen (University of Copenhagen) for help in preparing the manuscript. The first author is most grateful to Dr Ole Mattson for valuable discussions and critical reading of the manuscript, whilst the second author is appreciative of the administrative help and advice given by Dr O. W. Heal. We are indebted to the University of York, particularly Professor M. H. Williamson, Professor D. H. Valentine of the University of Manchester and the University of Copenhagen (Arctic Station Committee) for the many facilities made available to us.

We gratefully acknowledge the financial support provided for the IBP Bipolar Botanical Project by the Royal Society, London, the University of Birmingham and the British Antarctic Survey; this project formed part of the British contribution to the International Biological Programme. We also wish to acknowledge the financial support given by the Royal Society, London, for an IBP (Tundra Biome) visiting fellowship awarded to the junior author.

REFERENCES

Addison, P. A. (1972). *In* Bliss ed. (1972). pp. 73–88.
Ahrnsbrak, W. F. (1968). Technical Report No. 37, 1–50. Univ. of Wisconsin, Dept. of Meteorology, Madison, Wisconsin.
Ahmadjian, V. (1970). *In* 'Antarctic Ecology' (M. W. Holdgate, ed.), Vol. 2, pp. 801–811. Academic Press, London.
Aleksandrova, V. D. (1970). *In* 'Productivity and Conservation in Northern Circumpolar Lands' (W. A. Fuller and P. G. Kevan, eds), pp. 93–114. I.U.C.N., Morges, Switzerland.
Barashkova, E. A. (1971). IBP Tundra Biome Translation No. 4, 1–10. Univ. of Alaska, College, Alaska.
Benninghoff, W. S. (1952). *Arctic* 5, 34–44.
Benson, C. (1969). Research paper No. 51, 1–80. Arctic Institute of North America, Montreal.
Biebl, V. R. (1968). *Flora* 157, 327–354.
Billings, W. D. and Godfrey, P. J. (1967). *Science N.Y.* 158, 121–123.
Billings, W. D. and Mooney, H. A. (1968). *Biol. Rev.* 43, 481–529.
Bliss, L. C. (1956). *Ecol. Monogr.* 26, 303–337.
Bliss, L. C. (1962). *Arctic* 15, 117–144.
Bliss, L. C. (1966). *Ecol. Monogr.* 36, 125–155.
Bliss, L. C. (1970). *In* 'Productivity and Conservation in Northern Lands' (W. A. Fuller and P. G. Kevan, eds), pp. 75–85. I.U.C.N., Morges, Switzerland.

Bliss, L. C. (1971). *Ann. Rev. Ecol. Systematics* **2**, 405–438.
Bliss, L. C. (ed.) (1972). Devon Island IBP Project, High Arctic Ecosystem 413 pp. Dept. of Botany, Univ. of Alberta.
Böcher, T. W. (1963). *Meddr. Grønland* **148**, (3), 1–289.
Brewer, M. (1958). *Trans. Am. geophys. Union* **39**, 19–26.
Britton, M. E. (1957). *In* 'Arctic Biology' 18th Biology Colloquium, pp. 26–61. Oregon State Univ., Corvallis.
Brown, J. and Bowen, S. (eds). (1972). Proceedings 1972 Tundra Biome Symposium, Univ. of Washington, 211 pp. U.S. Tundra Biome, Hanover, N.H.
Brouwer, R. (1962). *Inst. Biol. Sch. Ond. van Landbouw. Wag. Med.* 1–203.
Bryson, R. A. (1966). *Geog. Bull. Ottawa*, **8**, 228–269.
Budyko, M. I. (1958). 'The Heat Balance of the Earth's Surface' (N. A. Stepanova, transl.). U.S. Dept of Commerce, Washington.
Caldwell, M. M. (1970). *Pl. Physiol.* **46**, 535–537.
Caldwell, M. M., Tieszen, L. L. and Fareed, M. (1974). Arctic Alpine Res. **6**, 151–159.
Callaghan, T. V. (1972). Unpublished Ph.D. Thesis, Univ. of Birmingham, U.K., 239 pp.
Callaghan, T. V. and Lews, M. C. (1971a). *Br. Antarct. Surv. Bull.* **26**, 59–75.
Callaghan, T. V. and Lewis, M. C. (1971b). *New Phytol.* **70**, 1143–1154.
Chang, J. (1957). *Trans. Am. geophys. Union* **38**, 718–723.
Chapin, F. S. (1974a). *Science N.Y.* **183**, 521–523.
Chapin, F. S. (1974b). *Ecology* **55**, 1180–1198.
Chepurko, N. L. (1972). *In* Wielgolaski and Rosswall, eds (1972). pp. 236–247.
Clarke, G. C. S., Greene, S. W. and Greene, D. M. (1971). *Ann. Bot.* **35**, 99–108.
Cook, F. A. (1955). *Arctic* **8**, 237–249.
Corbet, P. S. (1972). *Acta arct.* **18**, 1–43.
Courtin, G. M. (1972). *In* Bliss, L. C., ed. (1972) pp. 46–72.
Courtin, G. M. and Bliss, L. C. (1971). *Arctic Alpine Res.* **3**, 81–89.
Coyne, P. I. and Kelley, J. J. (1971). *Nature, Lond.* **234**, 407–408.
Coyne, P. I. and Kelley, J. J. (1972). *In* Brown and Bowen, eds (1972). pp. 36–39.
Coyne, P. I. and Kelley, J. J. (1975). *J. appl. Ecol.* **12**, 587–612.
Dadykin, V. P. (1954). *Voprosy botaniki* **2**, 455–489.
Dennis, J. G. and Johnson, P. L. (1970). *Arctic Alpine Res.* **2**, 253–266.
Dennis, J. G. and Tieszen, L. L. (1972). *In* Brown and Bowen, eds (1972). pp. 16–21.
Douglas, L. A. and Tedrow, J. C. F. (1959). *Soil Sci.* **88**, 305–312.
Evans, L. T. (1963). *In* 'Environmental Control of Plant Growth' (L. T. Evans, ed.), pp. 421–435. Academic Press, London.
Fletcher, J. O., Keller, B. and Olenicoff, S. M., (eds) (1966). 'Soviet Data on the Arctic Heat Budget and its Climatic Influence'. Memorandum RM-5003-PR, 205 pp. The Rand Corporation, Santa Monica.
Granhall, U. and Selander, H. (1973). *Oikos* **24**, 8–15.
Greene, S. W. and Longton, R. E. (1970). *In* 'Antarctic Ecology' (M. W. Holdgate, ed.), Vol. 2, pp. 786–800. Academic Press, London.
Hadley, E. B. and Bliss, L. C. (1964). *Ecol. Monogr.* **34**, 331–357.
Hare, F. K. (1968). *Quart. R. met. Soc.* **94**, 439–459.
Hare, F. K. (1970a). *Trans. R. Soc. Can.* **7**, 32–38.
Hare, F. K. (1970b). *Trans. R. Soc. Can.* **8**, 393–400.
Hare, F. K. and Hay, J. E. (1971). *Can. Geogr.* **15**, 79–94.
Hare, F. K. and Ritchie, J. C. (1972). *Geog. Rev.* **62**, 333–365.

Hocking, B. and Sharplin, C. D. (1965). *Nature, Lond.* **206**, 215.

Hodgson, H. J. (1966). *Bot. Gaz.* **127**, 64–70.

Holmen, K. (1957). *Meddr. Grønland* **124** (a), 1–149.

Johnson, P. L. and Kelley, J. J. (1970). *Ecology* **51**, 73–80.

Jones, H. E. and Gore, A. J. P. (1972). *In* Wielgolaski and Rosswall, eds (1972). pp. 35–47.

Kallio, P. and Kärenlampi, L. (1972). *In* Wielgolaski and Rosswall, eds (1972). pp. 276–280.

Kelley, J. J. (1968). *In* 'Arctic Drifting Stations' (J. E. Sater, ed.), pp. 155–166. The Arctic Institute of North America, Montreal.

Kelley, J. J. and Weaver, D. F. (1969). *Arctic* **22**, 425–437.

Kelley, J. J., Weaver, D. F. and Smith, B. P. (1968). *Ecology* **49**, 358–361.

Kevan, P. G. (1970). Unpublished Ph.D. Thesis. Univ of Alberta, Canada. 399 pp.

Krog, J. (1955). *Physiologia Pl.* **8**, 836–839.

Kuramoto, T. T. and Bliss, L. C. (1970). *Ecol. Monogr.* **40**, 317–347.

Larcher, W. (1957). *Veröff. Mus. Ferdinandeum Innsbr.* **37**, 49–81.

Larsen, J. A. (1964). Technical Report No. 16, 1–70. Univ. of Wisconsin, Dept. of Meteorology, Madison, Wisconsin.

Larsen, J. A. (1971). *Arctic* **24**, 177–194.

Lewis, M. C. (1972). *Sci. Prog. Oxf.* **60**, 25–51.

Lewis, M. C. and Callaghan, T. V. (1971). *In* 'Proceedings of IBP Tundra Biome Workshop Meeting on Analysis of Ecosystems, Kevo, Finland' (O. W. Heal, ed.), pp. 34–50, Tundra Biome Steering Committee.

Lewis, M. C. and Greene, S. W. (1970). *In* 'Antarctic Ecology' (M. W. Holdgate, ed.), Vol. 2, pp. 838–850. Academic Press, London.

Lindsay, J. H. (1971). *Arctic Alpine Res.* **3**, 131–138.

Longton, R. E. (1972). *Br. Antarct. Surv. Bull.* **27**, 51–96.

Marshunova, M. S. (1966). *In* Fletcher *et al.*, eds (1966). pp. 51–150.

Mather, J. R. and Thornthwaite, C. W. (1956). *Publs. Clim. Drexel Inst. Technol.* **9** (1), 1–51.

Mather, J. R. and Thornthwaite, C. W. (1958). *Publs. Clim. Drexel Inst. Technol.* **11** (2), 1–239.

Matveeva, N. V. (1972). IBP Tundra Biome Translation No. 6, 1–10. Univ. of Alaska, College, Alaska.

Mayo, J. M., Despain, D. G. and van Zinderen Bakker, E. M. (1973). *Can. J. Bot.* **51**, 581–588.

Miller, P. C. (1972). *In* Brown and Bowen, eds (1972). pp. 13–15.

Miller, P. C. and Tieszen, L. (1972). *Arctic Alpine Res.* **4**, 1–18.

Montieth, J. L. and Elston, J. F. (1971). *In* 'Potential Crop Production' (P. F. Wareing and J. P. Cooper, ed), pp. 23–42. Heinemann, London.

Moser, W. (1973). *In* 'Ökosystemforschung' (H. Ellenberg, ed.), pp. 203–223. Springer-Verlag, Berlin.

Muc, M. (1972). *In* Bliss, L. C., ed. (1972). pp. 113–145.

Nakano, Y. and Brown, J. (1972). *Arctic Alpine Res.* **4**, 19–38.

Orvig, S. (ed.). (1970). 'World Survey of Climatology' Vol. 14, 'Climates of the Polar Regions'. Elsevier, Amsterdam.

Pisek, A., Larcher, W. and Unterholzner, R. (1967). *Flora* **157**, 239–264.

Pruitt, W. O. (1970). *Trans. R. Soc. Can.* **7**, 13–25.

Rastorfer, J. R. (1970). *Bryologist* **73**, 544–556.

Raup, H. M. (1971). *Meddr. Grønland* **194** (1), 1–92.

Richardson, D. H. S. and Finegan, E. J. (1972). *In* Bliss, ed. (1972). pp. 197–213.
Rodin, L. E. and Bazilevich, N. I. (1967). 'Production and Mineral Cycling in Terrestrial Vegetation' (English translation edited by G. E. Fogg). Oliver and Boyd, Edinburgh.
Romanova, E. N. (1972). IBP Tundra Biome Translation No. 7, 1–10. Univ. of Alaska, College, Alaska.
Russell, R. S. (1940). *J. Ecol.* **28**, 269–288.
Rönning, O. I. (1969). *Arctic Alpine Res.* **1**, 29–44.
Sater, J. E., Ronhovde, A. G. and Van Allen, L. C. (1971). 'Arctic Environment and Resources' The Arctic Institute of North America, Washington D.C.
Savile, D. B. O. (1972). Monograph No. 6 Canada Dept. of Agriculture, Ottawa, 1–81.
Shamurin, V. F. (1966). *In* 'Adaptations of the Plants of the Arctic to the Environment' (B. A. Tikhomirov, ed.), pp. 5–125. Nauka, Moscow.
Shvetsova, V. M. and Voznesenskii, V. L. (1970). *Bot. Zh. SSSR* **55**, 66–76.
Svoboda, J. (1972). *In* Bliss, ed. (1972). pp. 146–184.
Sörensen, T. (1941). *Meddr. Grønland* **125** (a), 1–305.
Tedrow, J. C. F. and Cantlon, J. E. (1958). *Arctic* **11**, 166–179.
Tieszen, L. L. (1970). *Am. Midl. Nat.* **83**, 238–253.
Tieszen, L. L. (1972a). *In* Wielgolaski and Rosswall, eds (1972). pp. 52–62.
Tieszen, L. L. (1972b). *Arctic Alpine Res.* **4**, 307–324.
Tieszen, L. L. (1973). *Arctic Alpine Res.* **5**, 239–251.
Tieszen, L. L. and Johnson, P. L. (1968). *Ecology* **49**, 370–373.
Tikhomirov, B. A. (1952). *Bot. Zh. SSSR* **37**, 629–638.
Tikhomirov, B. A. ed. (1971). 'Biogeocenoses of Taimyr Tundra and Productivity'. Nauka, Leningrad.
Tikhomirov, B. A., Shamurin, V. F. and Shtepa, V. S. (1960). *Izv. Akad. Nauk. SSSR Biol.* Ser. **3**, 429–442.
Tranquillini, W. (1964). *A. Rev. Pl. Physiol.* **15**, 345–362.
Tranquillini, W. (1969). *Zentbl. ges. Forstw.* **86**, 35–48.
Tyrtikov, A. P. (1969). *cited in* Matveeva (1972).
Van Zinderen Bakker, E. M. (1972). *In* Bliss, ed. (1972). pp. 89–104.
Vowinckel, E. and Orvig, S. (1964). *Arch. Met. Geophysk. Bioklimt.* **13**, 352–377.
Vowinckel, E. and Orvig, S. (1970). *In* Orvig, ed. (1970).
Warren Wilson, J. (1957). *J. Ecol.* **45**, 499–531.
Warren Wilson, J. (1958). *J. Ecol.* **46**, 191–198.
Warren Wilson, J. (1959). *J. Ecol.* **47**, 415–427.
Warren Wilson, J. (1966). *Ann. Bot.* **30**, 383–402.
Washburn, A. L. (1969). *Meddr. Grønland* **176** (4), 1–300.
Weaver, D. F. (1969). 'Radiation Regime over Arctic Tundra, 1965'. Scientific Report. Dept. of Atmospheric Science, Univ. of Washington, Seattle.
Weller, G. and Cubley, S. (1972). *In* Brown and Bowen, eds (1972). pp. 5–12.
Weller, G., Cubley, S., Parker, S., Trabant, D. and Benson, C. (1972). *Arctic* **25**, 291–300.
Weller, G., Benson, C. and Holmgren, B. (1974). *J. appl. Meteorol.* **13**, 854–862.
Wielgolaski, F. E. (1972a). *In* Wielgolaski and Rosswall, eds (1972). pp. 9–34.
Wielgolaski, F. E. (1972b). *Arctic Alpine Res.* **4**, 291–305.
Wielgolaski, F. E. and Rosswall, T. (1972). 'Proceedings IV International Meeting on the Biological Productivity of Tundra'. IBP Tundra Biome Steering Committee, Stockholm.
Zalenskij, O. V., Shvetsova, V. M. and Voznessenskij, V. L. (1972). *In* Wielgolaski and Rosswall, eds (1972). pp. 182–186.

Index

Because each chapter in this book deals with the same set of topics more or less in the same sequence, it was convenient to replace the conventional alphabetical index with a pair of tables. Each row in the tables refers to a specific type of vegetation and corresponds to the division by chapters. Each column refers to a specific physical state, process, or parameter. The entries in the table are page numbers.

Reading along a row shows what topics are covered in a particular chapter and reading down a column shows what chapters contain a particular topic. Additional references to tabular or graphical material for other types of vegetation are identified by italic letters attached to column headings and the corresponding references are listed below.

INDEX REFERENCES

[a] Ross, Yu. K. and Nilson, T. (1967). *In* 'Photosynthesis of Productive Systems' (A. A. Nichiporovich, ed.), pp. 75–99. Israel Program of Scientific Translation, Jerusalem.

[b] Evans, L. T. (ed.) (1975). 'Crop Physiology'. Cambridge University Press, Cambridge.

[c] Milthorpe, F. L. and Moorby, J. (1974). 'An Introduction to Crop Physiology', p. 82. Cambridge University Press, Cambridge.

[d] Szeicz, G. (1974). *J. appl. Ecol.* **11**, 1117–1156.

[e] Monteith, J. L. (1975). 'Principles of Environmental Physics' pp. 57 and 66. Edward Arnold, London.

[f] Geiger, R. (1965). 'The Climate Near the Ground'. Harvard University Press, Cambridge, Mass.

[g] Stanhill, G. (1969). *J. appl. Meteorol.* **8**, 509–514.

[h] Tanner, C. B. and Pelton, W. L. (1960). *J. geophys. Res.* **65**, 3391–3400.

[i] Van Eimern, J. (1964). 'Untersuchungen über das Klima in Pflanzenbeständen'. Ber. dt. Wetterd, Offenbach, No. 96.

[j] Szeicz, G., Endrodi, G. and Tachman, S. (1969). *Water Resources Res.* **5**, 380–394.

[k] Rutter, A. J. (1975) *In* 'Vegetation and the Atmosphere', Volume I p. 137. Academic Press, London.

[l] Cowan, I. R. and Milthorpe, F. L. (1966). *In* 'Environmental Biology' (P. L. Altman and D. S. Dittmer, eds), pp. 464–465. Fed. Amer. Soc. Exp. Biol.

[m] Cooper, J. P. (ed.) (1975). 'Photosynthesis and Productivity in Different Environments'. Cambridge University Press.